Advanced Mechanical Models of DNA Elasticity

Advanced Mechanical Models of DNA Elasticity

Yakov M. Tseytlin
Doctor of Technical Sciences
Independent Researcher
Member of International Society of Automation (ISA)
Durham, CN, USA

ELSEVIER

AMSTERDAM • BOSTON • HEIDELBERG • LONDON
NEW YORK • OXFORD • PARIS • SAN DIEGO
SAN FRANCISCO • SINGAPORE • SYDNEY • TOKYO
Academic Press is an imprint of Elsevier

Academic Press is an imprint of Elsevier
125 London Wall, London EC2Y 5AS, UK
525 B Street, Suite 1800, San Diego, CA 92101-4495, USA
50 Hampshire Street, 5th Floor, Cambridge, MA 02139, USA
The Boulevard, Langford Lane, Kidlington, Oxford OX5 1GB, UK

British Library Cataloguing-in-Publication Data
A catalogue record for this book is available from the British Library

Library of Congress Cataloging-in-Publication Data
A catalog record for this book is available from the Library of Congress

ISBN: 978-0-12-801999-3

Publisher: Shirley Decker-Lucke
Acquisition Editor: Jill Leonard
Editorial Project Manager: Fenton Coulthurst
Production Project Manager: Lucía Pérez
Designer: Maria Inês Cruz

Typeset by Thomson Digital

Dedication

This book is dedicated with love to Mark, Arthur, Franklin, Zoya, Lev, and to bright memory of Adel, Ivan, Michel, and Meyna.

Contents

Biography

YAKOV M. TSEYTLIN is a mechanical engineer, educator, and research scientist. Born in Leningrad, Russia, he arrived in the USA in 1992, becoming naturalized in 1997. His qualifications and professional positions include: MS in Mechanical Engineering, Leningrad Polytechnic Institute, PhD, 1965; Doctor of Technical Sciences, All-Russian Research Institute of Metrology, Russia, 1991; Chief of research laboratory, senior designer, senior and lead researcher, Leningrad Instrumental Plant, All-Russian Research Institute of Metrology, Russia; Senior designer, senior researcher, Federal Products Co. USA, 1992; Project engineer, Automatic Machinery Co. US, 1999; Assistant to Associate professor at the Leningrad Polytechnic Institute; Visiting professor at the Leningrad Institute of Precise Mechanics and Optics; All-Russian Institute of Advanced Education in Standardization and Metrology for managers and engineers; Adviser and Opponent of post-Graduate Dissertations at the Central Research Institute for Fuel Apparatus, All-Russian Research Institute of Metrology, Leningrad Polytechnic Institute.

He has contributed numerous articles to professional journals, and authored four monographs. He is also a member of the International Society of Automation (with recognition awards, ISA). His achievements include the development of methods and concepts in micro elasticity, DNA elasticity modeling, atomic force microscopy, and information criterion of the measurement uncertainty negligibility.

Preface

Early on, the mechanical properties of DNA (deoxyribonucleic acid) attracted the interest of both physicists and biologists, because of their importance to numerous biological processes, such as DNA transcription, gene expression and regulation, and DNA replication. DNA is a very complicated biological object, with complex structure and functions that contain and translate genetic information to the RNA (ribonucleic acid) and proteins. The DNA molecule is a long polymer made from repeating units called nucleotides.

Despite the fact that DNA was discovered in the late 1860s by Friedrich Miescher, its helical structure in the form of three-dimensional right double helix B-DNA or dsDNA with two strands was elucidated only in 1953 by American biologist James Watson and English physicist Francis Crick. Other forms of DNA molecules are the short A-DNA, Z-DNA with left helix, stretched S-DNA, overtwisted P-DNA, and a single stranded ssDNA. The dsDNA structure of all species comprises the double helical chains, each coiled around the same axis, and each with a helical pitch of 3.3–3.4 nm and radius of 1–1.1 nm. The backbone of the DNA strand is made from alternating phosphate and sugar residues. The sugar in DNA is 2-deoxyribose. The discovery of the structure of DNA was the beginning of a process that has transformed the foundations of biology and accelerated the developments of the genetic engineering.

The sugars joined by phosphate groups form phosphodiester bonds between the third and fifth carbon atoms of adjacent sugar rings. These asymmetric bonds mean that a strand has a direction. The nucleotides in a double helix have one strand, with the direction of the bonds opposite to their direction in the other strand (strands are antiparallel). The asymmetric ends of DNA strands are called 5′ and 3′, with the 5′ end having a terminal phosphate group, and the 3′ end having a terminal hydroxyl group. One major difference between DNA and RNA is the sugar, with the 2-deoxiribose in DNA being replaced by the pentose sugar ribose in the RNA.

The genetic information is held within genes, and the complete set of this information in an organism is called the genotype. A gene is a unit of heredity, and is a region of DNA. Genes contain an open reading frame. Within a gene, the sequence of bases along a DNA strand defines a messenger RNA sequence that then defines one or more protein sequences. The genetic code consists of three letter words, called codons (the nucleoside triphosphates) that form a sequence of three nucleotides (e.g., ACT, CAG, TTT, where (A) is adenine, (G) is guanine, (T) is thymine, and (C) is cytosine).

Twin helical strands form the dsDNA backbone. Another double helix may be found tracing the spaces, or grooves, between the strands. As the strands are not symmetrically located with respect to each other, the grooves are unequally sized. The major grove is 2.2 nm wide, and the minor groove is 1.2 nm wide. As a result, proteins, such as transcription factors, usually have contacts to the sides of the bases exposed in the major groove.

DNA nanotechnology uses the unique molecular recognition properties of DNA and other nucleic acids to create self-assembling branched DNA complexes with the useful properties. DNA is thus used as a structural material, rather than as carrier of biological information. Recently, the first living organism has been created with artificial DNA.

The elastic properties of DNA are essential for its biological functions. They control its stretching, bending, and twisting, as well as the induction of structural modification in the molecule. These can affect its interaction with the cell machinery. The mechanical flexibility of DNA plays a key role in all of its cellular functions, including folding, packaging, regulation, recombination, replication, and transcription. The connections between DNA's molecular interactions and its mechanical properties are, therefore, of considerable interest to biologists and biophysicists.

From the very beginning, abstraction and modeling has played a significant role in the research on DNA. These models gave rise to mathematical concepts and techniques for the study of DNA configurations at microscopic and mesoscopic levels. Today, we can determine the structure of short DNA fragments, their stretching, bending, and twisting with nanometer and picoNewton resolution, with the application of atomic force microscopy (AFM) and with even higher measurement accuracy, up to femtometer and zeptogram, by AFM with higher oscillation modes. The number of DNA research publications is huge because their results have a large impact on the development of molecular biology and medicine.

This book is dedicated to the analysis of existing DNA molecule mechanical models, and the development of more comprehensive advanced mechanical models of DNA elasticity.

In Chapter 1, we discuss the mechanical properties of DNA molecules, and more than twenty effective mechanical models of DNA elasticity, based on discrete statistical mechanics, atomistic, continuum, and approximation methods. Among them, there are such models as freely jointed chain (FJC), wormlike chain (WLC), helical wormlike chain (HW or HWLC), and discrete persistent chain (DPC) models that played an important role in the development of the scientific approach to the study of DNA elasticity. We also discuss dynamic models of DNA, its persistence length features and structural phases, such as dsDNA (B-form), S-DNA (stretched one), ssDNA (single stranded), and P- DNA (overwind). Most of these models were introduced at the end of 1990s and the beginning of this century. The important Kratky-Porod model was introduced as early as 1949. These models play a significant role in the understanding and evaluation of DNA elasticity properties. But our analysis of these models shows that they have some limitations, specifically for the applicable force ranges with difference for small (less than 10 pN) and comparatively large forces (but less than 65 pN), and even some unfeasible results in calculations. Therefore, we concluded that the development of a more comprehensive explicit functional helicoidal model (EFHM or, briefly, EHM) that covers all stretch ranges from the beginning to the transition of dsDNA to S-DNA, overwinding options, non-linear elasticity and nonlinear thermomechanical fluctuations is very important. Modeling of ssDNA was done by A.V. Tkachenko with zigzag (ZZM), and by C. Storm and P.C. Nelson with DPC models, and may require application of the coiled ribbons (EHMR) model. In this chapter, we also discuss DNA dynamics with different pulling speeds, traveling waves, models of molecular length fluctuations, polymer dynamics and hydrodynamics, the persistence length definition, polymer material properties used for DNA modeling, limitations of micro object measurement accuracy, size and mass DNA conversion, and the DNA technical applications.

Chapter 2 focuses on the discussion and development of the experimental research methods for determining the mechanical properties of DNA, including force application by solutions flow, stretching micropipette, glass needles, optical and magnetic tweezers, the measurement of force and displacements by atomic force microscopy (AFM), and small-angle X-ray scattering interference measurements of the DNA molecular length fluctuations. Special attention in this chapter is paid to more sensitive atomic force microscopy with micro-nanocantilevers. In addition, the micromanipulation technique is also considered, including different types of cantilevers, and effective flexure hinges with cylindrical, elliptical, parabolic, hyperbolic, segmented, and any computer approximated concave profiles by an application of the reliable reverse conformal mapping.

Chapter 3 focuses on atomic force microscopy with higher vibration modes and sensitivity, up to zeptogram and femtometers. AFM calibration, effective increase of cantilever spring constants ratio, quality factor influence, sensitivity to additional mass, specifics of the V-shaped cantilever (with end extended mass) features at higher mode oscillations are considered. The contact dynamics at AFM tapping mode and an influence of measurand object profile surface roughness is also under consideration. The sensitivity of AFM with higher mode oscillations allows us to get increased resolution in mass, force, and dimensions measurements. Bimodal AFM is especially effective in measurement of biological objects and small dimensions in nanometer range, where interaction forces depend on the Hamaker and Lifshitz constants. However, the contact deformation can be larger at higher cantilever modes of vibration, as a result of large spring constants. Internal and external damping, ambient influence (temperature, vibration, acoustic noise, electromagnetic field, etc.) on the measurement and calibration accuracy (including information criterion of expanded uncertainty negligibility), and methods of measuring system protection are also discussed.

The kinematics of an elastic helicoidal beam and its physical realization in the thin pretwisted (helicoidal) micro-nanostrips – an important part of a new explicit functional helicoidal model of DNA, are considered in Chapter 4. This chapter shows the mathematical dependencies of the elastic helicoids parameters on their specific rigidities, stress, and fatigue conditions. We found, on this basis, the nonlinear relations in the pretwisted (helicoidal) strip's motion transformation, and its effective cross-section configurations. The parameters of the quasi-helicoids with shells (particular coiled ribbon) are also analyzed. We developed these relations using the principles and methods of continuum mechanics, because the connection between the motion and corresponding stress tensors, elastic, shear moduli, and Poisson's ratio plays a very important role in our helicoidal model of the DNA. The mathematical apparatus of this chapter is necessary in order to develop the functional relations of new DNA explicit helicoidal models presented in Chapter 6 and Chapter 7. Developed displacement-loading relations of helicoidal sensors are even wider for further DNA modeling in different environments and applications.

Chapter 5 is dedicated to the dynamics and vibration of elastic helicoids, their periodic and transient functions. We study the influence of large and small impedance, elastic longitudinal links, and internal friction (damping) asymmetry in the helicoidal multivibrators. State of stress and sensitivity to additional mass are also considered. This study shows that the relaxation of helicoids with different speed follows with less uncertainty than at a dynamic stretching. Limits of the transmission entropic fluctuations, flexural wave propagation, and break speed are also considered.

Mathematical relations, developed in Chapter 4 and Chapter 5, are essential to the structuring of the new explicit helicoidal models of the DNA molecule.

The main linear and nonlinear features of the DNA molecule elasticity are defined in Chapter 6, with the application of the explicit functional helicoidal model (EHM), where its Young's and shear moduli, conditional Poisson's ratio, persistence length, twist-stretch coupling, and overwinding options are evaluated. Calculation results are in good agreement with the experimental data for the certain DNA parameters. Problems of linear and nonlinear thermomechanical molecular length fluctuations of the DNA are carefully discussed, on the basis of EHM application. The dynamics of force application influence on the DNA stretching features, including the pulling speed factor, are also considered in this chapter.

Models for the transition of dsDNA to S-DNA and ssDNA at stretching are evaluated and developed in Chapter 7. We show that practical application of the discrete persistence chain (DPC) model is not a simple solution to the transition of B-DNA to S-DNA evaluation, as a result of the significant number of parameters used, the complexity of their estimation, and their complex mathematical relations. The theoretical approach of the quasi-reverse WLC transition B-DNA to S-DNA model does not take into account important aspects of the problem, such as the possible dependence of the parameters on the base pair (bp) sequence, the possibility of denaturation, and the effect of ionic strength of the medium. It also ignores the coupling of the chain elongation with torsion energy, that is, it assumes that the chain is free to rotate. Representation of B to S forms of DNA transition by a sequence of linear springs is useful for an approximation of the average experimental data at dsDNA and ssDNA stretching, but does not reveal a real mechanism of the dsDNA and ssDNA molecules' elasticity and overstretching behavior. It uses an imaginary combination of "multiple soft and hard linear springs" that does not exist in a real DNA molecule structure. Besides, the number of required constants depends on the accuracy of experimental data, and may not all be known *a priori*.

Calculation results with explicit functional helicoidal (EHM) model correspond well to the experimental data for all provided estimations and increase of the middle plateau by transformation with the unit singular function. This comprehensive model is simpler than other known models that reflect the transition from dsDNA to S-DNA or ssDNA. The widely known freely jointed chain (FJC), wormlike chain (WLC), and helical worm like chain (HW or HWLC) models do not represent this option directly at all. The EHM model works in a large range of stretching forces, and applied torques and, in the case of flexural waves, transition. Possible coiled ribbon applications for the single stranded ssDNA modeling (EHMR) at strong stretching and corresponding stress evaluation are also studied, in comparison with the ZZM and DPC chains. The EHM model can be included in eWLC and eFJC DNA extensible chain models, as well. The latter option is represented in detail. We also study in this chapter the buckling of a DNA molecule, and DNA mechanical stability. As a result, we have found that EHM, the explicit helicoidal model, is more comprehensive in comparison with other known mechanical models of the DNA, not only in the mathematical representation of main static and dynamical features of its elasticity and transitions, but also in the adequate presentation of this molecule's helical structure and geometric parameters.

This book reflects the author's experience, and represents his ideas and opinions. Results of many other researchers in this field are also discussed in the book.

Of course, all these assumptions and the explicit helicoidal model's application to DNA mechanics may be experimentally more thoroughly verified, and theoretically developed in more detail. However, the closeness of biomechanics and precision nanoelastic mechanics is evident.

Although every effort has been made to eliminate errors, experience dictates their inevitable presence in such comprehensive work. When detected, the author would be grateful for guidance on errors or omissions. Extensive references are included to enable the reader to verify discussed problems and solutions. We should also note that, in the preparation of this book, reasonable efforts have been made to publish reliable data and information, but the author cannot assume responsibility for the validity of all referenced materials, or for the consequences of their inappropriate use.

The author hopes that this book will help to develop new and more advanced mechanical models of DNA elasticity, and provide effective means for the better understanding of DNA mechanical features and functions.

Acknowledgments

The author would like to acknowledge the kind permission of Springer Publishing to reprint many figures and tables that were published in the author's earlier book *Structural Synthesis in Precision Elasticity* (Tseytlin, 2006). The author also thanks the Biophysical Journal, the American Physical Society, RightsLink Copyright Clearance Center, the Nanotechnology Journal (IOP), the Biopolymer Journal, ZAMM (J. Wiley and Sons Journal of Applied Mathematics and Mechanics), the Review of Scientific Instruments (AIP Publishing of the American Institute of Physics, AIPRights), PNAS (National Academy of Sciences USA), the American Chemical Society (Macromolecules), and the International Society of Automation (ISA) for permissions to reprint certain figures from their previous publications. Special appreciation goes to Professor R. Pelcovits of Brown University, for very helpful editorial assistance. The author expresses great gratitude to the publisher's editors, whose help in the index preparation and publication of this book was essential.

Chapter 1

DNA Molecules Mechanical Properties and Models

1.1 MECHANICAL PROPERTIES

The mechanical properties of the deoxyribonucleic acid (DNA) molecule play an important role in its function in genetic transition, and its interaction with ribonucleic acid (RNA). DNA elasticity includes stretching, bending, twisting, and the dynamics of molecular motion. The elastic properties of the DNA molecule are very important to its physiological behavior, such as wrapping around nucleosomes, packing inside bacteriophage capsids, and interacting with proteins.

The DNA molecule consists of a chain of base pairs (bp). Two nucleotides on opposite complementary DNA or RNA strands that are connected via hydrogen bonds form a base pair (Fig. 1.1): adenine (A) forms a base pair with thymine (T), and guanine (G) forms a base pair with cytosine (C). In RNA, thymine is replaced by uracil (U).

Among the main mechanical features and properties of the DNA molecule and its models are the following:

- Contour length L, L_c,
- End-to-end distance L_R, R_c, elongation $z = \Delta L$,
- Cross-section dimensions: width b_o, height h_o, diameter d of the outer contour circumference, width b_i, height h_i of the inner contour,
- Cross-section area A_c, A_{csec},
- Helix parameter $r_{p0} = k_{p0} = 2\pi/S_0$, where S_0 is a helix pitch,
- Spring constants j, j_θ, j_s, pulling force F, F_q,
- Persistence length l_p, A_{bp} ,
- Kuhn segments $b_k = 2l_p$,
- Twist-stretch coupling g_p,
- Stretch modulus E_{stm}, $E_{str} = EA_c$, Young's modulus E, shear modulus G,
- Poisson's ratio ν_p,
- Specific gravity ρ_γ,
- Ultimate tensile strength (UTS) σ_t, yield stress σ_e, endurance limit σ_{en}.

The smallest force at the molecular level is the Langevin force, due to thermal shocks from the environment. On the bacterium level (or a small bead), these shocks correspond to a force of about 0.03 pN applied for the average time

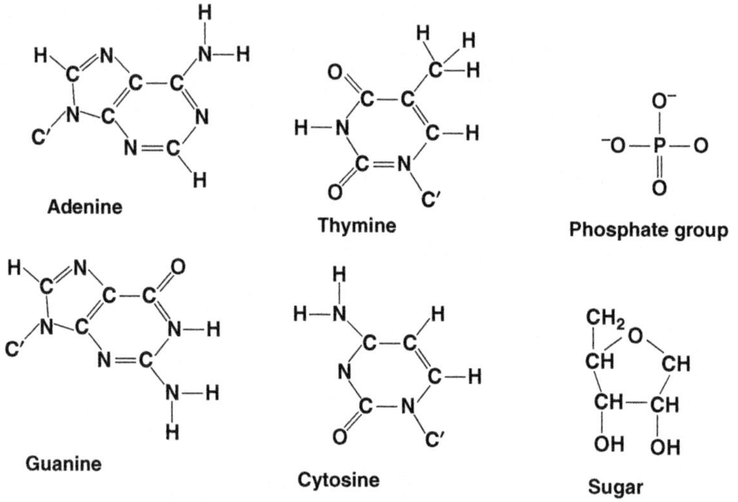

FIGURE 1.1 Basic building blocks of dsDNA: adenine, guanine, thymine, cytosine, phosphate group, sugar. *(After Chen et al., 2012, reprinted with permission.)*

of 1 s (Allemand et al., 2003). Much larger forces ($F \approx 6$ pN) are developed by molecular motors in order to displace submicron-sized objects (e.g., organelles). GC-rich and AT-rich sequences have different stickiness. GC-rich regions required a force of 15 pN (20 pN from Rief et al., 1999) to separate strands, whereas AT-rich regions required only 10 pN (Marko, 1997). The largest forces encountered at the molecular level, about 1000 pN, are associated with the rupture of a covalent bond. In nucleic acid studies, the interesting range of forces varies between 0.1 pN, at which molecules respond, and 100 pN, at which structural transitions associated with the rearrangement of the phosphodiester backbone have been observed (Allemand et al., 2003). Studies of a supercoiled DNA at its simultaneously twisting and pulling showed gradual stiffening of the entropic elasticity with twisting, and force plateaus appeared at force of only about 4 pN (Strick et al., 1996; Marko, 1997). Mechanical force at the molecular level is involved in the action of many enzymes. This is the case for processes of replication or transcription in which enzymes translocate with respect to DNA. Such translocation occurs unidirectionally, over long segments of DNA, and the enzymatic machinery has to develop a force against a number of impediments: the disruption of complementary base pairs, the possible attachments of the DNA or the enzymes to cellular components, structural proteins that coat the DNA, topological constraint, and viscous friction (Essevaz-Roulet et al., 1997). Long before the enzymes associated with replication were known, a simple model had been considered, in which the mechanism of unwinding the strands during replication is coupled to the rotation of the molecule as a whole. A molecular configuration with a **Y** shape had been proposed, in which the vertical part is the parent helix, and the two arms are the separated strands, which get replicated. For typical molecular interactions involving biomolecules, the forces are small (sub-picoNewton to tens picoNewton). For these reasons, sensitive measuring devices such as optical tweezers, soft microneedles, or micronanolevers of atomic force microscopy are used. In many protein-DNA complexes, the double helix is severely deformed from the classical Watson-Crick B-form (Watson and Crick, 1953) by being bent, stretched, and untwisted. A bacterial protein Rec.A has nonspecific binding to dsDNA that lengthens the dsDNA by a factor of 1.5 (Léger et al., 1998). Results from micromanipulation experiments indicate that Rec.A binds strongly to stretched DNA, as a result of spontaneous thermal stretching fluctuations that play a role in its binding to DNA. This is of a broad relevance to DNA-protein interactions. Structural proteins that bind DNA have nonspecific DNA-protein interactions. These proteins organize the DNA into a compact chromatin structure. In eukaryotes, this structure involves DNA binding to a complex of small basic proteins that are called histones, while in prokaryotes multiple types of proteins are involved (Sandman et al., 1998; Dame, 2005). The histones form a disk-shaped nucleosome complex that contains two complete turns of double-stranded DNA. Other nonspecific DNA-binding proteins in chromatin include the high-mobility group proteins that bind to a bent or distorted DNA (Thomas, 2001). These proteins are important in binding arrays of nucleosomes, and arranging them into the larger structures that make up chromosomes (Grosschedl et al., 1994). All this, again, shows the importance of the elasticity of DNA in its role in replication and transcription.

The sequence of three DNA or RNA nucleotides that corresponds with a specific amino acid or stop signal during protein synthesis is a codon. DNA and RNA molecules are written in a language of four nucleotides; meanwhile, the language of proteins includes 20 amino acids. Codons provide the key that allows these two languages to be translated into each other. Each codon corresponds to a single amino acid (or stop signal), and the full set of codons is called the genetic code. The genetic code includes 64 possible permutations, or combinations, of three-letter nucleotide sequences that can be made from the four nucleotides.

The mathematical difference between a single (ssDNA) and double (dsDNAS) helix is explained in *Physics of DNA* by Podgornik (2011), as follows. A continuous single helix is a spatial curve with a constant angle between a spatial direction and local tangent $\theta' = 0$, as well as having constant curvature and tortuosity (torsion). The simplest mathematical method to estimate tortuosity is the arc-chord ratio: the ratio of the length of the curve (L) to the distance between the ends R_c of it: $\tau = L/R_c$. The arc-chord ratio equals 1 for a straight line, and is infinite for a circle.

A continuous double helix has two helical curves inscribed on the same cylinder of radius R_a. A parametric description of these two helical curves is given by

$$r_1(t) = (R_a \cos t_1, R_a \sin t_1, r_s t_1) \text{ and } r_2(t) = (R_a \cos t_2, R_a \sin t_2, r_s(t_2 + \Delta\psi),$$

TABLE 1.1 Parameters of Different DNA Structures

DNA	Handedness	$\Delta\psi$	N	Ω_r (°)	ΔS_r [nm]	θ (°)
A	Right	0	11	31.1	0.26	34.8
B	Right	0.375	10.5	36.1	0.34	37.4
Z	Left	0	12.0	−30.0	0.37	−58.8

Note: $\Delta\psi$ is the shift angle of two helices, N is number of residues per turn, Ω_r is the rotational twist per residue, ΔS_r is the rise per residue, θ is the calculated helical pitch angle.
Source: Adapted from Olsen et al. (2010) with permission; with corresponding notations from Podgornik (2011).

where r_s is the helix factor, and the angle $\Delta\psi$ represents the angular shift between the two helical lines, in cut with a plane perpendicular to the central axis of the helices. If this angle is different from π, then the two helices make an asymmetric pattern with a well-defined major and minor "grooves," meaning that, in the direction of the long axis of the helix with radius R_a, the separation of helical lines are uneven. This unevenness of separation has an important meaning for the structural properties of the double helical DNA that is shown in Table 1.1.

Another double helix may be found tracing the spaces, or grooves, between the strands. As the strands are not symmetrically located with respect to each other, the grooves are unequally sized. The major grove is 2.2 nm wide, and the minor groove is 1.2 nm wide. As a result, proteins like transcription factors usually make contact with the sides of the bases exposed in the major groove (Fig. 1.2).

The elastic properties of DNA are essential for its biological function. They control its stretching, bending, and twisting, as well as the induction of structural modification in the molecule. These can affect its interaction with the cell machinery. The mechanical flexibility of DNA plays a key role in all of its cellular functions, including folding, packaging, regulation, recombination, replication, and transcription. The connections between DNA's molecular interactions and its physical properties are, therefore, of considerable interest to biologists and nanoengineers.

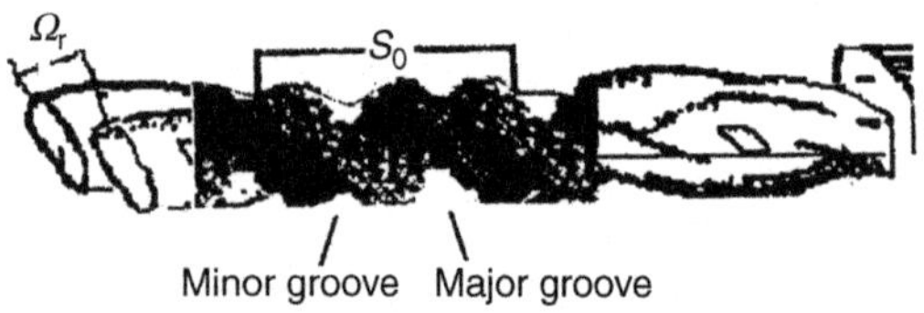

FIGURE 1.2 Conditional image of double helix B-DNA (dsDNA) molecule with two-shifted strands at angle Ω_r (Table 1.1) and helix pitch S_0.

1.2 DISCRETE, FLEXIBLE CHAINS, AND ATOMISTIC MODELS: WLC, FJC, DPC, WLRC, HW, ZZM

Mod 1. The wormlike chain (WLC) model (Type 1, Table 1.2) is based on a flexible rod where the derivatives of radius-vector $\boldsymbol{r}(S)$ with respect to the arc length S correspond to the tangent vector $t(S)=\dfrac{dr(S)}{dS}$ and the curvature $k(S)=\dfrac{dt}{dS}=\dfrac{d^2r(S)}{dS^2}$. Here, we have widely used (Marko and Siggia, 1995) formula $F=\dfrac{k_BT}{A_{bp}}\left[\dfrac{1}{4(1-z/L)^2}-\dfrac{1}{4}+\dfrac{z}{L}\right]$ for dependence of the applied force F and corresponding dsDNA molecule extension z at its contour length L. This formula includes the persistence length A_{bp} = l_p, Boltzmann's constant k_B = 1.3806488 $\times 10^{-23}$ J/K, and absolute temperature T in K. Note that $k_BT \approx 4.1$ pN·m at 20°C (293.15 K) temperature. This expression has the nice property of reducing to the exact solution at $F < 10$ pN, but may differ by as much as 10% for $z/L \cong 0.5$ (Bouchiat et al., 1999; Bustamante et al., 1994). It is clear, however, that at $z/L = 1$ (with division-by-zero), this formula will give a nonfeasible result with $F = \infty$.

Vučemilović-Alagić (2013), claims that this model is a numerical solution to the Kratky–Porod model (Yamakawa, 1976; Krajnc, 2012).

Another result of the Kratky–Porod model is presented in the following formula:

$$R_g^2=\langle R\rangle^2=\left(b_k\sum_{i=1}^{N}t_i\right)^2\approx 2Nb_k\xi_T=2L_0\xi_T,$$

where $L_0 = Nb$ is the contour length, $\xi_T = l_p = b_k/2$ is the persistence length, b_k is the Kuhn sector, $\boldsymbol{t}_i = \bar{t}$ is an orientation vector, $\langle t_i \bullet t_j\rangle = e^{-b(i-j)\xi_T}$, and bending energies of successive segments are

$$\varepsilon_{KP}=-\frac{B_b}{b_k}\sum_{i=2}^{N}\bar{t}_i\bullet\bar{t}_{i-1}=-\frac{B_b}{b_k}\sum_{i=2}^{N}\cos\theta_i,$$

where θ_i is the angle between successive orientation vectors and B_b is the bending modulus.

A critical value of tension $F_c=k_BT/l_p=80\,\text{fN}$ (80×10^{-15} N) separates regions of low and high tension. At this critical tension, a wormlike chain extends 45% of its contour length. For applied tensions much smaller than F_c, entropic effects dominate, and the kink energy can be calculated by a second-order expansion of the force-free equations. For tensions much larger than F_c, the kink energy is primarily determined by the enthalpic cost of bending

TABLE 1.2 DNA Molecules Elasticity Models with Statistical Mechanics and Interpolation

No.	Type	Scheme	Formula	Limit	Sources
1	Inextensible worm-like chain (WLC) for dsDNA	Fig. 1.3a	$F=\frac{k_BT}{A_{bp}}\left[\frac{1}{4\left(1-\frac{z}{L}\right)^2}-\frac{1}{4}+\frac{z}{L}\right]$, $z/L=x$ Uncertainty up to 10%	$F<10$ pN	Marko and Siggia (1995); Bustamante et al. (1994); Chen et al. (2012)
2	Modified eWLC	Fig. 1.3b	$F=\frac{k_BT}{A_{bp}}\left[\frac{1}{4\left(1-\frac{z}{L}+\frac{F}{E_{str}}\right)^2}-\frac{1}{4}+\frac{z}{L}-\frac{F}{E_{str}}\right]$	$F<10$ pN	Wang et al. (1997); Chen et al. (2012)
3	WLC polynomial enhanced interpolation, uncertainty 0.1% modified	Fig. 1.3a	$F=\frac{k_BT}{A_{bp}}\left[\frac{1}{4\left(1-\frac{z}{L}\right)^2}-\frac{1}{4}+\frac{z}{L}+\Sigma\alpha_i\left(\frac{z}{L}\right)\right]$ $\alpha_2=-0.5164228; \alpha_3=-2.737418;$ $\alpha_4=16.07497; \alpha_5=-38.87607;$ $\alpha_6=39.49944; \alpha_7=-14.17718$ $F=\left(\frac{k_BT}{A_{bp}}\right)\left[\frac{1}{4(1-l)^2}-\frac{1}{4}+l+\Sigma\alpha_1(l)^i\right],$ $l=z/L-F/E_{str}$	$F<10$ pN	Bouchiat et al. (1999); Chen et al. (2012)
4a	eWLC for dsDNA at larger force	Fig. 1.3b	$z=L\left(1-\frac{1}{2}\left(\frac{k_BT}{FA_{bp}}\right)^{1/2}+\frac{F}{E_{str}}\right)$	$F\leq65$ pN $E_{str}=500-1500$ pN	Odijk (1995); Smith et al. (1996); Chen et al. (2012)

4b	Helical WLC	–	$\rho = k_0/(k_0^2 - \tau_0^2)$-radius of helix $h_p = S_0 = 2\pi\tau_0/(k_0^2 - \tau_0^2)$-pitch of helix where k_0 is the curvature, τ_0 is the torsion	$<S^2>/x$ Mean-square radius of gyration/repeats x	Yamakawa (1997, 1999), for details, see Section 7.9, Swigon (2009)
5	WLC for larger force	–	$\frac{<z>}{L} = 1 - \frac{1}{2}\left[\frac{LF}{k_BT} - \frac{1}{32}\right]^{-1/2} + \frac{F}{E_{str}}$	0.1 pN < F<65 pN	Moroz and Nelson (1997)
6	Freely jointed chain (FJC) for ssDNA	Fig. 1.3c	$\left\langle \frac{z}{L} \right\rangle = \coth\left(\frac{Fb_k}{k_BT}\right) - \frac{k_BT}{Fb_k} = £\left(\frac{Fb_k}{k_BT}\right)$ $F = \frac{k_BT}{b_k}£^{-1}(z/L)$; Langevin function $£ = \frac{1}{3}z - \frac{1}{45}z^3 + \frac{2}{245}z^5 - \frac{1}{4725}z^7K$	$b_k = 0.17-1.7$ nm $b_k = 2A_{bp}$ as the Kuhn segment	Khokhlov et al. (1994)
7	Extensible freely jointed chain (eFJC) for ssDNA	–	$z = L\left[\coth\left(\frac{2A_{bp}^{ss}F}{k_BT}\right) - \frac{k_BT}{2A_{bp}^{ss}}\right]\left[1 + \frac{F}{E_{str}}\right]$ A_{bp}^{ss} is a persistence length for ssDNA	$b_k = 0.17-1.7$ nm $b_k = 2A_{bp}^{ss}$ the Kuhn segments	Smith et al. (1996) Kiang et al. (2010)
8a	DPC, Overstretching	Fig. 1.3d	$\left\langle \frac{z}{L} \right\rangle = k_BT\frac{d\Omega}{dF}$ $\Omega = b^{-1}\max_{\omega} \ln Y(\omega_1, \omega_{-1})$ $\ln Y(\boldsymbol{\omega}_1, \boldsymbol{\omega}_{-1}) = \frac{b^B}{2}(P + R + \sqrt{(P-R)^2 + 4Q^2})$ For detail, see Chapter 7	$b^S = \varsigma_s b^B$, $\varsigma_s > 1$, $\beta\varsigma_s < 1$, $A^S = \beta\varsigma_s A^B$ $F < 140$ pN	Storm and Nelson (2003a,b) $A^{B,S}$ stiffness, β-dimensionless ς_s is the elongation factor

(Continued)

TABLE 1.2 DNA Molecules Elasticity Models with Statistical Mechanics and Interpolation (*cont.*)

No.	Type	Scheme	Formula	Limit	Sources
8b	"Zigzag" model of ssDNA-ZZM (DPC version)	–	$\frac{z}{L}=\frac{b^*(F)}{b^*(0)}x(F)$ is the stretching curve with renormalization (see Chapter 7) $l_p = b^*(0)f^*/k_BT \approx Jb/k_BT$ $b = 0.6$ nm for ssDNA and bond length b which depends on stretching force F	ssDNA strongly stretched For detail see Chapter 7	Tkachenko (2007); Chen et al. (2012)
9	Mesoscopic coupled discrete wormlike chain-Ising model	–	Extension for ssDNA $z = L\left(1-\frac{1}{\sqrt{F^2+4k_bF}}\right)$ for $F>>1/k_b$. For B to S-DNA transition, Na_B is the contour length, $\frac{z}{a_BN}=\left(1+\frac{F}{E_B}-\frac{1}{2\alpha_B}\right)\varphi_B+\gamma\left(1-\frac{1}{2\alpha_S}\right)\varphi_S + \frac{\langle\sigma_i\sigma_{i+1}\rangle-1}{4}\left(\frac{1}{2\alpha_B}\frac{k_b-k_{bS}}{k_{bS}+F/2+\alpha_B}+\frac{\gamma}{2\alpha_S}\frac{k_S-k_{bS}}{k_{bS}+\gamma(F/2)+\alpha_S}\right)$	ssDNA stretching dsDNS-S-DNA overstretching	Manghi et al. (2012)
10	Generalized wormlike rod chain (WLRC) Asymmetric elasticity (ZZO) Plectonemic transition	Fig. 1.3e	$Tw=\frac{1}{2\pi}\int_0^L \tau\, ds$, $L = L_b\langle\cos\varphi\rangle$ $\tau=\frac{\tan\varphi}{R};\ \frac{ds}{ds_b}=\cos\varphi.$ $\frac{(F-p^2/(4A_b))A_b}{k_BT}=\left((1-z/X)^{-2}-1\right)+z/X$	<10 pN	Zhou and Lai (2005); Fain et al. (1997); Podgornik (2011); Chen et al. (2012)

a wormlike chain into its kinked shape. The average number of kinks per unit length (Wiggins et al., 2005) is equal to

$$\varsigma_{\mathrm{ki}} = \frac{4\pi k_j}{l_{\mathrm{ki}}} e^{-\epsilon},$$

where k_j is the spring constant, $k_j l_{ki} = l_p$ is the persistence length, $1/l_{ki}$ is a density of kinkable sites, and $\in$ is the energy penalty to realize the kink state (Dauxois, 1991). The average kink number for an unconstrained chain with length L equals $<m_k> = \varsigma_{ki} L$, and ς_{ki} is the density of kinks.

By defining a dimensionless force, $\overline{\overline{\overline{F}}} = F/F_c$, and setting $\alpha = \cos[(\pi - \theta)/4]$, it is possible to calculate (Blumberg et al., 2005) a closed-form expression for the kink molecule energy $\ni_{\mathrm{kink}}$:

$$\ni_{\mathrm{kink}} = 4k_B T\sqrt{\overline{\overline{\overline{F}}}}\left(1 + \frac{3}{2\alpha^2(1+\alpha)\overline{\overline{\overline{F}}}^{3/2}}\right)^1 (1-\alpha),$$

where α is the angle between the two pieces of DNA entering the loop.

Mod 2. The modified extensible WLC (eWLC) model (Type 2, Table 1.2) does not have the drawback of a nonfeasible force value at relative extension equal to 1; however, it is also valid only for small applied forces $F < 10$ pN. Besides, the formula is an implicit one.

Mod 3. A more accurate model for dsDNA extension in a stretch field $F < 10$ pN with an uncertainty of 0.1% is developed (Type 3, Table 1.2) with polynomial enhanced interpolation (Bouchiat et al., 1999). However, this solution becomes nonfeasible again with $F = \infty$ at $z/L = 1$. The modified polynomial solution does not have this problem, but the formula becomes implicit.

Mod 4a, 5. Extensible WLC models for larger stretch forces F < 65 pN are presented as Type 4a and 5 in Table 1.2 where E_{str} is a stretch modulus. Figure 1.3b shows a stiff chain under a gradually increasing tension: undulation-dominated elongation, the crossover, and elasticity-dominated regime.

Mod 4b. The helical wormlike HW (Type 4b, Table 1.2) chain (HWLC) generalizes the wormlike chain model by accounting for the twisting deformation of DNA (Shimada and Yamakawa, 1984; Yamakawa, 1997, 1999; Swigon, 2009).

However, WLC, eWLC, and HWLC models do not exhibit the transitions from dsDNA to S-DNA and ssDNA at the stretching of dsDNA.

Mod 6, 7. Freely jointed chain (FJC) and extensible freely jointed chain (eFJC) models are effective for single stranded ssDNA molecules (Types 6 and 7 in Table 1.2).

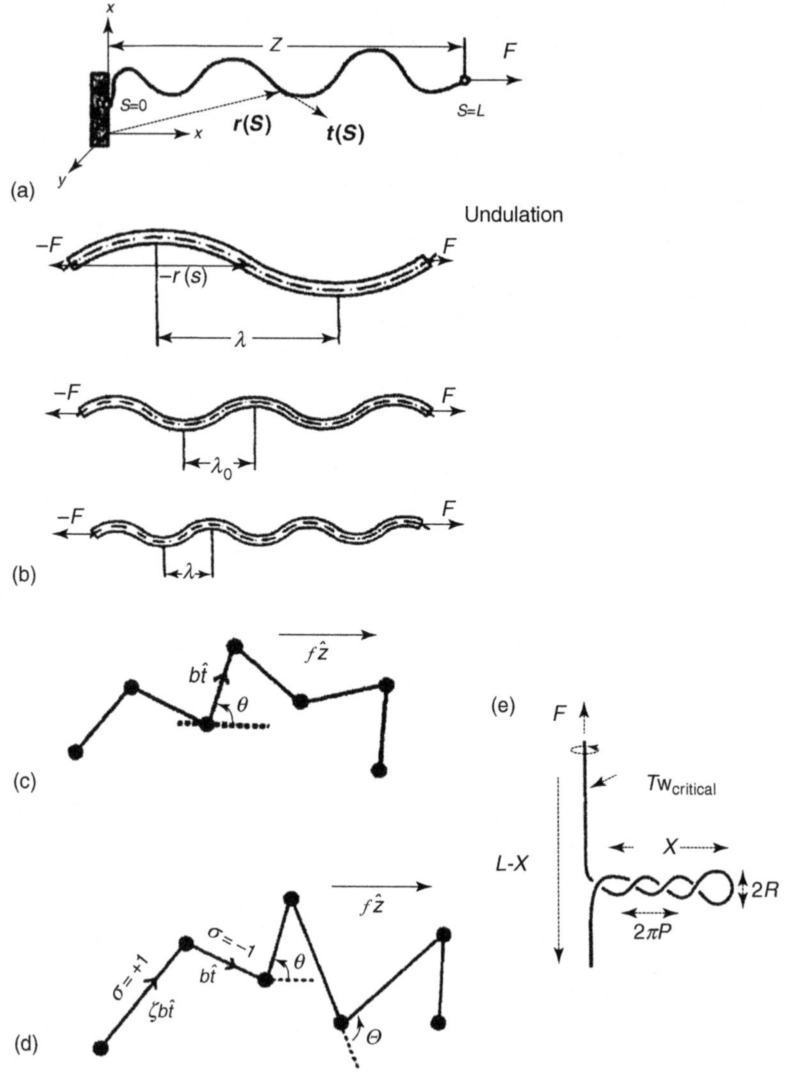

FIGURE 1.3 DNA models' schemes for Table 1.2 with references. (a) A WLC as a continuum elastic medium with position $r(S)$ and tangent $t(S)$ vectors where each is a function of the contour length arc coordinate S; (b) extensible wormlike chain with undulations, at the crossover, elasticity dominated regime; (c) the freely jointed chain which consists of identical segments of length b, joined together by free hinges, the configuration is fully described by the collection of oriented vectors $\{\hat{t}\}$. $\{\theta_i\}$ denotes the angle between $\hat{t}_i$ and the fixed direction $\hat{z}$ of the applied force; (d) the discrete persistent chain, viewed as a FJC with additional term in the energy proportional to the square of the polar angle θ between successive segments; and (e) shows a plectonemic transition. *(Part (b) after Odijk, 1995, reprinted with permission.)*

Mod 8a. The discrete persistent chain (DPC) (Type 8a, Table 1.2) exhibits the overstretching transition of dsDNA to *S*-DNA. Figure 1.3d illustrates this model. Each segment carries a discrete variable σ that takes the values ±1. $\sigma = +1$ means the segment is in the B state, and $\sigma = -1$ for the S state. The factor by which a segment elongates when going from B to S (Storm and Nelson, 2003a,b) is called ς_s, that is, $b^S = \zeta_s b^B$ with $\zeta_s > 1$. The bend stiffness parameter A^B is assigned to B-DNA and $A^S = \beta_{sn}\zeta_s A^B$ to S-DNA. β_{sn} is a dimensionless parameter with $\beta_{sn}\,\zeta_s < 1$. A least-square fit of this DPC model solution (Chapter 7) gives the overstretching results that are similar to the typical chart presented in Figure 1.3e.

Mod 8b. Tkachenko (2007) developed the Zigzag model (ZZM) of strongly stretched ssDNA, with a stretching force up to 1000 pN. This model (Type 8b, Table 1.2) is a version of DPC with renormalization of the moment of inertia, namely, $J^* = J/2$.

Mod 9. Mesoscopic models of DNA stretching under force F (Type 9, Table 1.2) are built as coupled discrete wormlike chains. In these models, k_b is a dimensionless bend modulus, and E_B is the stretch modulus in the B state. The formula in this model for larger force and transition from B-DNA to S-DNA has several specific parameters (Manghi et al., 2012). We will present here only some of them, because they are, and the whole approach to this problem, still complicated in the mesoscopic model. The authors recommend the following procedure: (i) fit the low force regime ($0 \le F \le 20$ pN) by fixing the known B-DNA bending modulus $k_b = 147$; (ii) The stretching modulus E_B is then determined by fitting the whole data before the transition ($0 \le F \le 60$ pN); (iii) The ratio $\gamma_m = a_S/a_B$ and the bending modulus of the S form, $k_{b,S}$, are fixed by fitting only the data after transition. Here,

$$a_{B,S} = \sqrt{k_{b,S}F + \left(\frac{F}{2}\right)^2}\,;\ \varphi_{B,S} = \frac{1}{2}\left(1 \pm \frac{\sinh \mu_0}{\sqrt{\sinh^2 \mu_0 + e^{-4J_0}}}\right).$$

The values of the parameters μ_0 and J_0 fix the position and the width of the transition, respectively. The calculation results are similar to Figure 1.3e that is common to this type of transition.

Finally, we see that the mesoscopic model is not a simple one, and does not explain the nature of the transition. It does not have the DNA molecule's twist feature.

Mod 10. The wormlike rod chain, WLRC (Type 10, Table 1.2) features asymmetric elasticity, including twist (*Tw*) in an elastic ribbon. In this model, s_b is the arc length of the backbone of the DNA molecule, R_{bb} is the half-length of the lateral distance between two back bones, L_b is the total contour length of the back bones, and φ is the folding angle. This model works at small-applied forces and small twist, where the twisted ribbon is in a stable state. The application of larger forces can cause plectonemic transition with bend constant A_b (Fig. 1.3e). Podgornik (2011) showed that extension of plectoneme, $L-L_p$, is the following function of the turns imposed on the chain:

$$b(Lk) = \left(1 + \frac{2K_c \sin\alpha^2}{C\cos 2\alpha}\right)L_c - \frac{4\pi R_p}{\sin 2\alpha}Lk,$$

where K_c is the elastic modulus for the curvature, C is elastic modulus for twisting, L_c is the chain contour length, $Lk = Tw + Wr$ (twist and writhe) is the linking number, R_p is the radius of the plectoneme helix, and α is the angle between the local direction of the helix and the average direction of the plectoneme.

We note that asymmetric elasticity is usually applied to materials with discrete structure, such as individual particles and molecules (Pal'mov, 1964).

Table 1.3 shows the DNA molecules' atomistic, continuum, and approximation elasticity models.

Mod 11. The coarse-grained wavelet-base model (Type 1, Table 1.3) uses an expression for continuum strain energy W where $\alpha(s)$ is a transition function considering the fraction of B-form DNA during the DNA stretching process; $W^1$is the strain energy function derived from the coarse-grained model (Fig. 1.4a,b) and W^2 is a phenomenological strain energy function representing the behavior of the S-form DNA, and reflecting the rapid strain hardening after transition.

$$W^2(I_1) = -\frac{G}{2}J_m \ln\left(1 - \frac{I_1 - 3}{J_m}\right),$$

where G is the shear modulus, I_1 is the principal invariant of the Green deformation tensor, and J_m is a parameter taking into account the limiting molecule extensibility.

The corresponding predicted force versus extension curve (Fig. 1.4c) is similar for both atomistic and coarse-grained simulation. This model does not include twist in the stretching process, and predicts a negative Poisson's ratio. The latter requires the usage of exotic materials, and excludes the option of experimentally observed overwinding in the model with helically wrapped elastic core (Type 3, Table 1.3). However, Baumann et al. (1997) also considered options for a negative Poisson's ratio.

An advanced coarse-grained DNA model was introduced by Korolev et al. (2014), in order to model DNA in nucleosome complexes. It consists of two types of sites: DNA core ("D") beads, each representing two base pairs, and "P" beads, representing phosphate groups (Fig. 1.5a). The internal DNA structure was maintained by four potentials (D–D, D–P, P–P, and P–P across the minor grove) and three angular potentials (D–D–D, P–P–P, and P–D–P). All these potentials are of harmonic type, with the selected parameters shown in

TABLE 1.3 DNA Molecules Atomistic, Continuum, and Approximation Elasticity Models

No.	Type	Schematic	Formula	Limit	Sources
1	Coarse-grained wavelet-base	Figs. 1.4a,b, and 1.5a	$W = \alpha(s)W^1 + (1-\alpha(s))W^2$ $\alpha = 1$ at $s < 1.2; \alpha < 1$ at $s > 1.2$ $W^{1,2}$-strain energy B-S DNA	400 pN $\nu_p < 0$	Chen et al. (2012) Korolev et al. (2014)
2	Clustered atomic coarse grained, Ionic medium I_{st}	Fig. 1.4a,b,c	$A_{bp}^{ds} = A_{bpo}^{ds} + \frac{0.0324}{I_{st}}$, $A_{bp}^{ss} = A_{bpo}^{ss} + \frac{0.03797}{I_{st}}$ -persistence length ds and ss DNA	dsDNA/ssDNA transition $A_{bpo}^{ds;ss} = 50;1.5$ nm	Yuan et al. (2005) Ambia-Garrido et al. (2010) Baumann et al. (1997) Han et al. (2006)
3	Elastic rod wrapped helically with wire	Core rod radius $r_r = 0.924$ nm Fig. 1.5b Diameter $\phi\#2 <. \phi\#1 = 2r_r$	$\alpha_c = \arctan(\sqrt{\nu_p}) \approx 0.62$ rad $x = \frac{L}{E_{str}}\left(F - \frac{g_p\theta}{L}\right)$;	Overwinding Option $\nu_p = 0.5$; $F \leq 30$ pN $S_{01} < S_o$-pitch	Gore et al. (2006)
4	Helical Spring-Kirchhoff-Love rod	Fig. 1.5c	$z = \frac{(K_1k_u^4 + K_3k_u^2\tau_u^2)}{K_1K_3(k_u^2 + \tau_u^2)^3}F + O(F^2)$ $\theta = \frac{(K_1 - K_3)k_u^2\tau_u}{K_1K_3(k_u^2 + \tau_u^2)^2}F + O(F^2)$	Linear to F, Over-winding option with nonlinear version is possible	Duričković et al. (2013)
5a	Multistep approxi-mation	Fig. 1.5e	$\xi = n_b l_b \left[z_b(F) + \frac{F}{E_{bstr}}\right] + (1-n_b) l_s \left[z_s(F) + \frac{F}{E_{sstr}}\right]$	$l_b = 0.35$ nm $l_s = 0.56$ nm for dsDNA and ssDNA <150 pN	Cizeau et al. (1997)

(*Continued*)

TABLE 1.3 DNA Molecules Atomistic, Continuum, and Approximation Elasticity Models (*cont.*)

No.	Type	Schematic	Formula	Limit	Sources
5b	Multiple linear spring approximation	Figs. 1.5f, and 7.1	$F = k_0 x + (a_1 + b_1(x - x_1))S_1 + (a_2 + b_2(x - x_2))S_2$; $S_i = \frac{1}{2}(\tan\beta_i(x - x_i) + \tanh\beta_i x_i)$ S_i are the hyperbolic tangent functions; i=1, 2. Detail in Chapter 7.	Multiple soft and hard linear springs F<120 pN $a_i.b_i$-constants; k_0, k_n -spring constants; n = 1, 12, 2, 3.	Xiao (2013)
6a	Pretwisted nanostrip (helicoidal), EHM explicit functional helicoidal linear and nonlinear model	Fig. 1.5d	See Chapters 6, 7. Main features: S_0 is the pitch, L_0 is a contour length, l_p persistence length, twist-stretch coupling g_p, overwinding, linear-nonlinear length fluctuation, transition dsDNA to S-DNA, options of buckling, dynamics	dsDNA-S-DNA, ssDNA<140 pN All linear and nonlinear mechanical, thermomechanical functions	Tseytlin (2011–2014)
6b	Flexible nanohelicoid with a particle	Fig. 1.5g	See Chapters 3, 6, 7	m_g –mass of a particle	Tseytlin (2011–2013)
6c	Flexible helicoid with two split single strands-coiled ribbons (EHMR)	Fig. 1.5h	See Chapters 3, 7	Coiled ribbons with 100–800 pN stretch force	Tseytlin (2011–2013)
6d	Flexible asymmetric implicit helicoid model (AEHM)	Fig. 1.5i	Imitation of asymmetric double helicoid (see Table 1.1) with two adjacent helicoidal strips shifted at angle $\Delta\psi$	Equivalent to the model 6a in most cases	–
7	Model Y (YM)	Y	Vertical stem corresponds to the dsDNA and the arms correspond to two strands of ssDNA.	–	Levinthal and Crane (1956)

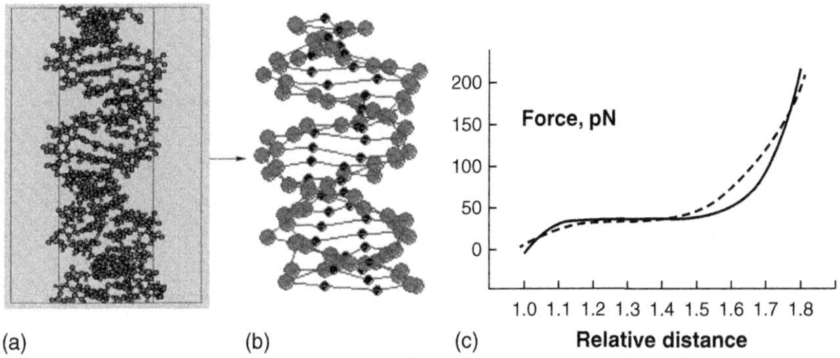

FIGURE 1.4 Graining of DNA molecule. (a) Fine-scale atomistic model, (b) corresponding coarse-grained model, and (c) force-extension curve with solid line for atomistic simulations and dashed line for grain simulations. *(After Chen et al., 2012, reprinted with permission; Table 1.3 with references.)*

Figure 1.5a. This model allows the evaluation of the flexibility (persistence length and torsional persistence length) of the DNA nucleosome core particles (NCP) array, as a function of monovalent salt (concentration of NaCl plus concentration of DNA counterions).

Mod 12. The Clustered Atomic coarse-grained model (Type 2, Table 1.3) includes the dependence of the persistence length of dsDNA and ssDNA on the ionic strength of the medium, which equals the ion concentration for monovalent ions, like Na^+. Initial persistence length $A_{bpo}^{ds} = 50\,nm$ for dsDNA, and initial persistence length $A_{bpo}^{ss} = 1.481\,nm$ for ssDNA. In this model, the DNA molecules are considered immersed in a saline solution with ionic medium I_{st}. This model does not allow for twist of the molecule upon stretching.

1.3 CONTINUUM AND APPROXIMATION MODELS

Mod 13. A model consisting of an elastic rod wrapped helically with wire (Type 3, Table 1.3) overwinds when stretched by force $\leq$ 30 pN, and its helix angle is less $\alpha_c = 35.5$ (angular degrees) similar to what was found in the DNA experiments (Gore et al., 2006). Meanwhile, the authors did not find the reason for the force limitation. We note that this model includes only a linear version of DNA molecule stretching x and may have external friction between the core rod and the wrapping wire. The model, however, has real stretching modulus $E_{str} = 1100$ pN, and twisting angleθ with twist-stretch coupling $g_p = 90$ pN·nm. Meanwhile, we have found the reason for that force limitation at overwinding in our EHM model (**Mod 16**), but with a somewhat different critical helix angle $\alpha = 49.0$ (angular degrees). Olsen et al. (2011) also found the critical angle of overwinding that is equal to $\alpha = 39.4$ (angular degrees). The geometrical investigation in this latter case is based on the study of double helices modeled as flexible tubes with hard walls. The dsDNA molecule has initial helix angle $\alpha \approx 27.3$ (angular degrees)

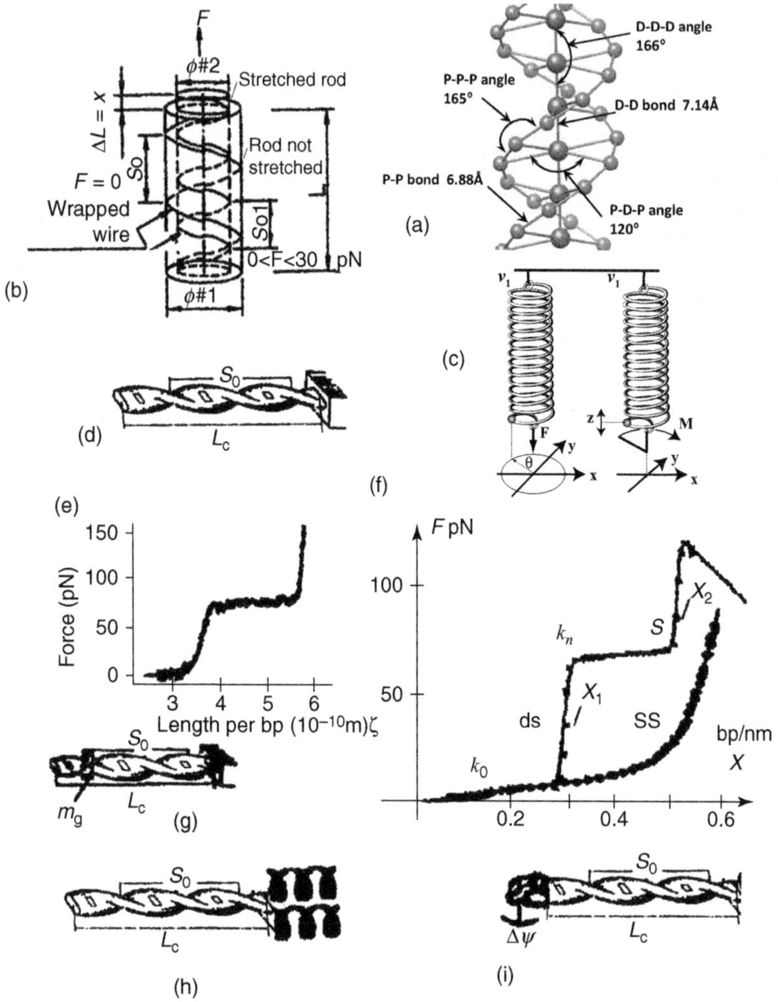

FIGURE 1.5 DNA models' schemes for Table 1.3 with references. (a) Coarse-grained (Korolev et al., 2014), (b) our presentation of elastic rod wrapped helically with wire and stretched, (c) helical spring-Kirchhoff-Love rod (Durîković et al., 2013; reprinted with permission) with applied axial tension ($v1$) and torque ($v2$) without of the tension force, (d) explicit helicoidal strip, EHM, (e) multistep approximation (Cizeau and Viovy, 1997; reprinted with permission), (f) fourfold chart for multiple soft and hard linear springs (Xiao, 2013; reprinted with permission), (g) flexible nanohelicoid with a particle, (h) explicit helicoidal model with coiled ribbons, EHMR, and (i) asymmetric explicit helicoidal model, AEHM.

that is less than any of the found theoretical critical angles. Therefore, the DNA molecule overwinds at the beginning of its stretching in all cases.

Mod 14. A helical spring-rod model (Type 4, Table 1.3) shows how a wrench applied to a helical spring gives rise to equilibrium: ($v1$) increment the rotation angle θ when an axial tension is applied with no torque, and ($v2$) extension or contraction of the spring as a result of an axial torque M loading with no applied force F. In the equations for the extension z and rotation θ, the following parameters are used: k_u and τ_u are a constant curvature and torsion of the unstressed helical configuration; K_1 and K_3 are the bending and torsional stiffness of the rod. Both equations are linear, and cannot provide a comprehensive solution of the dsDNA to S-DNA transition. However, authors (Duričković et al., 2013) show that their equations reduce to quadratic conditions on the curvature and torsion that allows them to move beyond the linear analysis, and predict the overwound and unwound configurations of the rod. No numerical results for this situation are presented.

Mod 15 a, b. Models Type 5a and 5b (Table 1.3) are multistep approximations that describe with certain accuracy the experimental results, but do not show the real cause of such behavior of the DNA molecule at stretching.

Model 5a (Table 1.3) combines (Cizeau et al., 1997) several previous approaches that describe particular regimes of stretching, that is, the entropic elasticity of the DNA considered as a WLC, the chain elongation, and the statistical model for the transition of DNA from the B state to the S state. Each base pair is assumed to be in one of two states: the regular state B and extended state S, with rises per base pair l_b and l_s, respectively. The longitudinal elastic stretch modulus in the B state equals E_{bstr} and in S state equals E_{sstr}. The Kuhn length in B state is assumed to be equal to 35 nm. The equation for mean distance per base pair $\zeta_{bp}(z)$ versus force F in Table 1.3 contains parameters n_b as a fraction of bp in the B state, and z_b and z_s, which recommended to define from WLC equation (see model Type 1 in Table 1.2), but that equation works only for $F \leq$ 10 pN. The presented analysis for this model is interesting, but the cause of such DNA behavior is not explained.

In the model Type 5b (Table 1.3), DNA elastic nonlinearity is explained (Xiao, 2013) as a multiple smooth combination (Fig. 1.5f) of soft and hard linear springs with constants: for the initial soft spring, for hard spring with high rigidity, for soft overstretching spring with low rigidity, for the second hard spring with high rigidity, and a breaking one with negative slope. Besides, the first and second hardening extensions X_1 and X_2 lie in the middle of the two hard spring functions. At the midpoint of overstretching with soft spring part comes the overstretching force, and, at the peak, emerges the maximum force. This approximation has constants that are obtained from test data. The model does not show a real mechanics of DNA nonlinearity, and may be used only for experimental data approximation (see Chapter 7, Fig. 7.1).

Mod 16 a, b, c, d. The explicit functional helicoidal model (**EFHM**, or briefly **EHM**, Type 6a, Table 1.3) is a more comprehensive and comparatively simple one, at least for the simulation of elasticity of single molecule manipulation (SMM) and short double stranded DNA molecule; this answers many questions about features and mechanical and thermomechanical linear and nonlinear functions of the DNA molecules, including stretch-twist coupling, overwinds options, nonlinear length fluctuations, transition from dsDNA to S-DNA and to ssDNA – as will be shown in Chapters 6 and 7 of this book. But this model is based on the linear and nonlinear mathematical apparatus of continuous mechanics for precision pretwisted (helicoidal) strip nanosensors that we discuss in Chapters 4 and 5. Model type 6b has a conjugated particle m_g similar to a protein or another substance connected with the DNA molecule. Model 6c (**EHMR**) has coiled ribbons after the split of dsDNA to two single strands (ssDNA). Model 6d imitates the asymmetric double helix with shift angle$\Delta\psi(\Omega_r)$ between two adjacent helicoidal strips (AEHM).

Mod 16c is similar to **Y** model (Levinthal and Crane, 1956) mentioned in Essevaz-Roulet et al. (1997). In that model (Type 7, Table 1.3), the vertical stem corresponds to the dsDNA, and the arms correspond to two strands of ssDNA.

If the ends of the molecule are allowed to rotate freely under traction, then the structure of S-DNA is ladderlike, and can be considered as an unwound double helix. S-DNA may be significant biologically, since it allows easier access to the base pairs for the transcription process (Ahsan et al., 1998).

1.4 DYNAMICS AND FLUCTUATION

Dynamic models of DNA can be divided (Swigon, 2009) into two groups – those based on theories of elastic rod, and those based on theories of polymer dynamics.

Elastic rod dynamics includes fast motion, internal friction, traveling waves, hydrodynamics, and thermal fluctuations.

Mod 17. In Type 1 model (Table 1.4), $\dot{\boldsymbol{C}}$ and $\dot{\boldsymbol{I}}$ are the linear and angular momentum of the cross-section, the dot stands for the time derivative, $\boldsymbol{F}'$ and $\boldsymbol{M}'$ are the contact forces and moments applied to the cross-section at s arc-length point by material with arc-length greater than s, and $\boldsymbol{c}_f$ and $\boldsymbol{m}_f$ are external forces and moments (Swigon, 2009). Brownian dynamics has been employed in the study of DNA tumbling and twisting.

Mod 18. In Type 2 model (Table 1.4), $\sigma_L^2(n)$ represents variance of molecule's length, $nC(1)$ reflects the chain length, $C(m)$ include correlations of the helical rise in dinucleotide separated by (m−1) steps. Term $C(1)$ is positive-definite. Mazur (2009) shows possible differences in thermomechanical fluctuations between variance $\sigma_L^2(n)$ for counter length, and variance σ_{LR}^2(n) for end-to-end distance.

Mod 19a, b. In polymer dynamics models (Table 1.4, Type 3a; Fig. 1.6c), a DNA molecule is replaced by a collection of rigid spheres that are connected by elastic linear springs along their length, with the addition of nonlinear coupling between masses (m) of each chain, simulating the bending and twisting rigidity of DNA (Forinash et al., 1991). At high hydrodynamics resistance of the solvent, the molecule is assumed to move by diffusion, resulting in Brownian type dynamics. The lifetime of base pair, that is, the time during which it stays

TABLE 1.4 Elastic Rod Dynamic Models

No.	Type	Schematic	Formula	Limit	Sources
1	Elastic rod	–	$\dot{C} = F' + c_f$ $\dot{I} = M' + x' \times F + m_f$	–	Swigon (2009)
2	Accordion bellows DNA length fluctuation	Fig. 1.6a	$\sigma_L^2(n) = nC(1) + \sum_{m=2}^{n} C(m)$	–	Mazur (2009)
3a	Two-dimensional nonlinear	Fig. 1.6c	$H = \sum_n \frac{1}{2} m\left(\overset{*}{u}{}_n^2 + \overset{*}{v}{}_n^2\right) + \frac{1}{2} k[(u_n - u_{n-1}) + (v_n - v_{n-1})^2 V(u_n - v_n - 1)$	H is a Hamiltonian	Forinash et al. (1991) Muto et al. (1990); Peyrard et al. (2008)
3b	Traveling waves	Bell-shaped and periodic	$dy/d\xi = A\varphi^3 + B\varphi^2 + C\varphi$; $d\varphi/d\xi = y$	A,B,C-constants, see Chapter 7	Ouyang et al. (2014) Muto et al. (1990)
4	Pretwisted microstrip with asymmetric damping, puling speed	Fig. 1.6b	See Chapters 6, 7 $S_{1,2}$ are the stretch values	–	see Chapters 6 and 7
5	Thermomechanical DNA length fluctuation	SAXSI Fig. 1.6e,d	See Chapter 2, Chapter 6	800 bp	Mathew-Fenn et al. (2008a, 2008b, 2009)
6	Pulling-speed-dependence	Stem-flower (Trumpet) At $v > v_0^c$ Fig. 1.7	$v_0^c = k_B T / (\eta_0 R_0^2) = R_0 / \tau_z^0$, where $R_0 \approx \sqrt{L l_p}$; $\tau_z^0 = \eta_0 R_0^3 / k_B T$ L-contour length, l_p is the persistence length, η_0 is solvent viscosity	τ_z^0 is Zimm relaxation time	Lee and Thirumalai (2004)

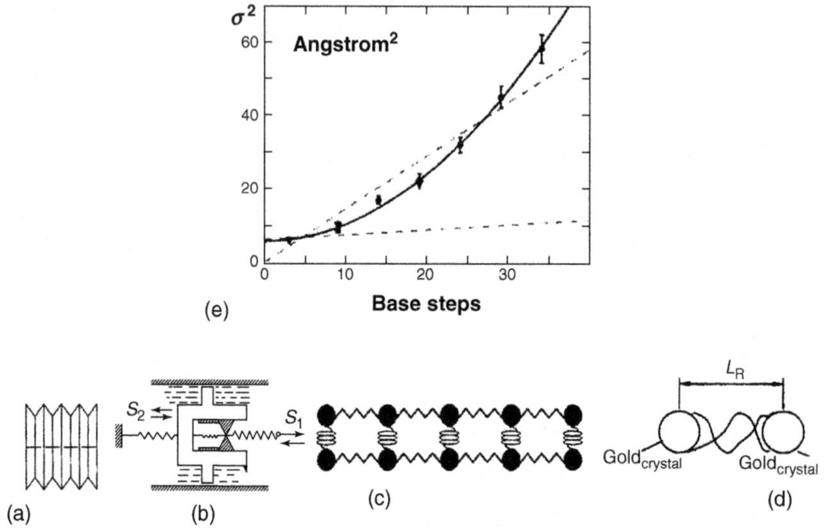

FIGURE 1.6 Dynamic models' schematics for Table 1.4 with references. (a) Accordion bellow, (b) pretwisted strip with damping, (c) double chain model, (d) thermomechanical fluctuation model, (e) thermomechanical fluctuation function (Mathew-Fenn et al., 2008b). *(Part (c) after Forinash et al., 1991, reprinted with permission.)*

closed, is only of the order of few milliseconds (Peyrard et al., 2008). Experiments show that fluctuations, known as the "breathing of DNA," are highly localized, and may open a single base pair, while the adjacent ones stay closed. The following relations are typical for the double-chain nonlinear model and traveling wave model:

$\xi = (x\sqrt{\rho / E} - \sqrt{2}t)$; $u(x, t) = \varphi(\xi)$; $u(x, t)$ is difference of longitudinal displacements of the bottom and top strands, t is time, $x = (u_n + v_n)/\sqrt{2}$, and $y_n = (u_n - v_n)/\sqrt{2}$, v represents the difference of the transverse displacements of the bases from their equilibrium positions along the direction of the hydrogen bond that connects the two bases of the base pair (Table 1.4, Type 3a, b, and Sec. 7.8).

Mod 20. Asymmetric dynamic model DAEHM (Type 4, Table 1.4) has different force-stretch functions at fast and slow stretching and relaxation.

Mod 21. Mean nanocrystal separation distance and distance variance for DNA duplex measured in model Type 5 (Table 1.4) with one type of solution (Mathew-Fenn et al., 2008a) and under different ionic composition in solutions are shown (Mathew-Fenn et al., 2009).

Mathew-Fenn et al. (2008a) measured the mean and variance of the end-to-end length for a series of DNA double helices in solutions, using small-angle x-ray scattering interference (SAXSI) between gold nanocrystal balls. In the absence of applied tension, DNA is at least one order of magnitude softer than measured by single-molecule stretching experiments. The data rule out the

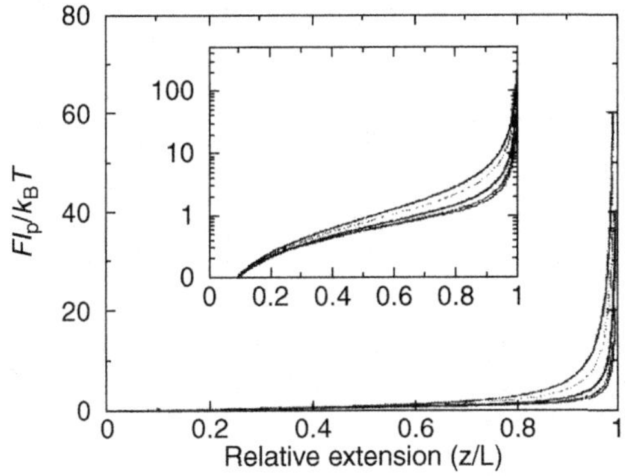

FIGURE 1.7 Theoretically determined dynamical (F, z) curves for WLC at various pulling speeds. *(After Lee et al., 2004, reprinted with permission.)*

conventional elastic rod model. The variance in end-to end length follows a quadratic dependence on the number of base pairs, indicating that DNA stretching is cooperative over more than two turns of the DNA double helix.

Mod 22. Significant nonuniform effects can be discerned upon pulling the molecule only when the pulling speed exceeds v_0^c (Type 6, Table 1.4). The value of v_0^c at water viscosity $\eta \sim 1$ cP (1 centipoise = 1 mPa·s, millipascal second) for typical parameters of DNA ($L \sim 10$ μm, $l_p \sim 50$ nm) is $v_0^c \sim 800$ μm/s. The typical pulling speeds used in experiments is $v_0 = (1\text{–}10)$ μm/s, at least two orders-of-magnitude smaller than v_0^c (Lee and Thirumalai, 2004). However, nonequilibrium response may be observed in force-extension profiles of longer DNA molecules.

As described by Brouchard-Wyart (1995), nonuniform profiles (trumpet and stem-flower profiles) manifest themselves when the pulling speed reaches $v^c(F) = k_BT/\eta_0 b_f^2 > v_0^c$, where b_f is the transverse size of the chain.

The chain relaxation time under tension is $\tau_z^f = \eta_0 b_f^3 / k_BT$. Below the threshold pulling speed, the chain confirmation can be approximated as a uniform cylinder. The values of the pulling speeds in Figure 1.7 change in order $v_0 =$ (0.5, 1, 2, 3) b/t_0 from bottom to top (after Lee and Thirumalai, 2004). The persistence length $l_p = 20b$ at $N = 100$, and the bond length $b = 2.5\text{–}2.8$ nm. The insert in Figure 1.7 is in a linear-log scale.

More generally, experimental mechanical overstretching and subsequent force-induced melting of the DNA double helix can also lead to different results, depending on the details of the attachment of the molecule to pulling tip and gold base. If the molecule is attached to the substrate by only one of its two strands, and is picked up by the AFM tip at the complementary strand (i.e., the molecule is attached to the substrate and the AFM tip with both 3′ ends and both of its 5′ ends), the connection will be lost upon force-induced melting of the double helix. In other cases, permanent conversion to single-stranded DNA occurs upon overstretching. Here, it is possible that single-stranded breaks (nicks) in the

DNA strand that melts off, as well as detachment of this strand from the AFM tip and the gold substrate, facilitate complete detachment of the strand, and render the recombination of the double helix upon relaxation of the molecule impossible. In some cases, only a part of the molecule recombines to its double-helical conformation, indicating that only a part of the strand that melts off remains connected to the substrate or the AFM tip. In most relaxation traces, however, reannealing can be performed up to several hundred times with the same molecule. Here, the whole strand that melts off must remain attached to one of the two opposing surfaces, or, in the presence of a nick, it must remain attached to both surfaces. Finally, if the molecule is attached by both strands to both surfaces (tip and substrate) and no single-strand breaks is present in the molecule, no conversion to ssDNA should be possible, and the hysteresis between extension and relaxation traces should be small (Lee and Thirumalai, 2004).

In contrast to the B-S transition, where only internal rearrangement of the double helix, such as restacking of the base and partial unwinding of the helix, are taking place, the two strands have to separate upon melting (Clausen-Schaumann et al., 2000). The molten strand presumably rotates around the stretched strand, and then it is likely to assume a random coil conformation with possible intrastrand hairpins. Upon relaxation, this sequence in one way or another has to occur in reverse order. This typically gives rise to a large hysteresis, and often even requires nearly complete relaxation of the molecule before reannealing of the double-helix is observed. Figure 1.8 shows the difference in the stretching and relaxation behaviors of dsDNA in relation to fast motion and waiting time.

Three fast stretching and relaxation cycles on Figure 1.8a obtained from a piece of dsDNA in 150 mM NaCl, 10 mM Tris (pH 8), 1 mM EDTA (ethylenediaminetetraacetate) at 20°C. In the first cycle, the molecule is stretched and immediately relaxed again in a 112 ms. In the second cycle, a waiting time of 100 ms is introduced at the fully extended position, and ~25% of the molecules are converted to the single-stranded conformation, as can be seen from hysteresis in the B-S plateau. If the waiting time at the fully extended position is increased to 500 ms (the third cycle), the entire piece of DNA is converted to ssDNA, as can be seen from the lack of B-S plateau in the relaxation curve.

Three fast deformation cycles of dsDNA (in the same buffer as in Fig. 1.8a) are shown on Figure 1.8b. Here, the molecule is first stretched to the fully extended position, where it is allowed to rest for several seconds to ensure complete melting of the double-helix. Then the molecule is relaxed and immediately stretched again in 112 ms. During this fast cycle, ~30% of the molecules are able to recombine to the double-helical conformation, as can be seen from the partial appearance of a B-S plateau in the extension curve.

If a resting time of 100 ms is introduced in the relaxation position, ~60% of the strands are able to recombine, and if the resting time is increased to 1 s, the full B-S plateau reappears, indicating that 100% of the DNA is able to

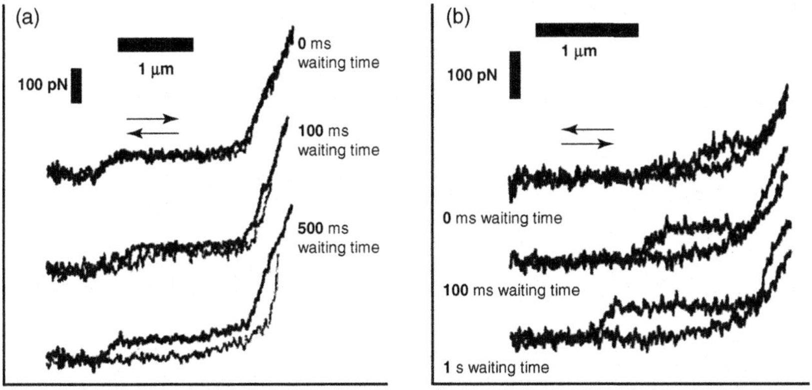

FIGURE 1.8 Fast stretches and relaxation cycles with waiting time (a) and fast stretching with waiting time (b) of dsDNA. *(After Clausen-Schaumann et al., 2000, reprinted with permission.)*

recombine to the double-helical conformation. Zhou and Liu (2013) studied the stability of kinked DNA with a generalized rod model.

The separation of double-stranded DNA into single-stranded DNA is fundamental to DNA replication in living organisms. In equilibrium, DNA will separate when the free energy of the separated ssDNA is less than that of the ds-DNA (Danilowicz et al., 2003). In most studies of DNA separation, the strands are separated by increasing the temperature of the sample until the DNA melts. In living organisms, however, DNA separation is not thermally driven. Rather, enzymes and other proteins force the two strands apart. When the dsDNA is unzipped at a constant force, metastable states result in significant pauses, during which the number of separated base pairs does not change with time, despite the constant applied force.

In polymer dynamics models, a DNA molecule is replaced by a collection of rigid spheres (Mod 19a) connected by elastic linkages simulating the bending and twisting rigidity of DNA. The total energy of the chain is composed of stretching, bending, twisting, and electrostatic components. Because of hydrodynamics resistance of the solvent, the molecule is assumed to move by diffusion that results in a Brownian type dynamics (Swigon, 2009).

Migration of macromolecules across streamline in fluid flows has been observed by various researchers. The migration of ssDNA molecules has been investigated in a fully developed pressure-driven channel flow (Benke et al., 2011) at Reynolds number 0.075, based on the bulk velocity and channel height of 0.75 μm.

1.5 PERSISTENCE LENGTH FEATURES

Many of the known DNA models use the persistence length feature $A_{bp} = l_p$ that is a measure of the rigidity of a linear polymer backbone. Persistence length has a precise mathematical definition (Fig. 1.3a), namely,

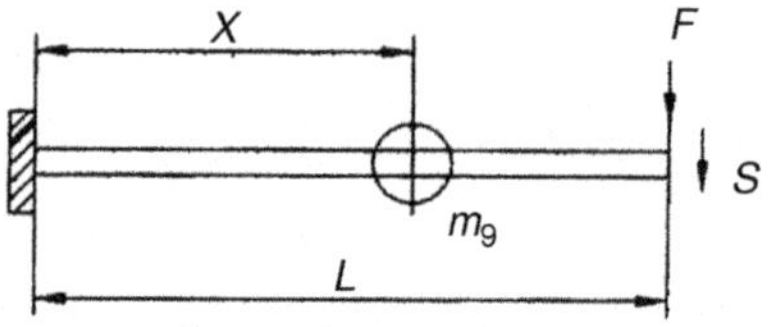

FIGURE 1.9 Flexible rod with a particle.

$$l_p = A_{bp} = \frac{-\Delta s}{\ln\left\langle \frac{\partial r}{\partial s}\Big|_{s_1} \bullet \frac{\partial r}{\partial s}\Big|_{s_2} \right\rangle}, \qquad (1.1)$$

where angular brackets denote the average of a fluctuating property. The persistence length should be independent of the distance $\Delta s = (s_2 - s_1)$ separating points s_1 and s_2 on the arc. The persistence length provides the means for understanding DNA's configuration, the shape of the DNA molecule as a function of contour length and external environments. The nonbending Kuhn segments b_k in *FJC* models have a length equal to twice the persistence length of semiflexible chain. Each Kuhn segment is assumed to be freely jointed with its neighbors. The length of a fully stretched chain is $L = Nb_k$ for the Kuhn segment chain where N is the number of Kuhn segments. The average end-to-end distance for this chain satisfying the random walk model is $\langle L_R^2 \rangle = Nb_k^2$.

Boal (2002) showed that the persistence length of rods and ropes can be expressed as follows:

$\xi_p = YJ/k_BT$, where in our notations $\xi_p = l_p = A_{bp}$, $Y = E$ *is* Young's modulus, J is a moment of cross section inertia, k_B is Boltzmann's constant, and T is the absolute temperature in K. We will use this expression for the persistence length in our DNA models.

The persistence length for the flexible rod with a particle (similar to the model 6b, Table 1.3) can be evaluated (Fig. 1.9) as follows:

$$l_p = K_b \frac{L^3(1 + m_g g q^2/F)}{3\sqrt{2}k_BT},$$

where $K_b = F/s$ is the bending constant of the rod, m_g is the mass of a particle, g is the free fall acceleration (acceleration of gravity), F is an applied force, and $q = x/L$.

DNA molecules have different values of persistent length: around 50 nm for dsDNA, 0.75–0.85 nm for S-DNA and ssDNA. Persistence length depends on the environment conditions, ionic strength of the medium (model Type 2 in Table 1.3), etc. Electrostatic interactions lead to a higher effective bending modulus of a semirigid chain (Tkachenko, 2006). This results in renormalization of the persistence length:

$$l_p^* = l_p + \alpha_e^2 r_s^2 / 4l_B, \tag{1.2}$$

where α_e is the Manning parameter that is the number of electron charges e per Bjerrum length. We recall that Bjerrum length is $l_B = e^2/D_e k_B T = 0.7$ nm, $r_s = 1/\sqrt{4\pi l_B n}$ is the Debye screening length, D_e is the dielectric constant of water. On length scales smaller than the electrostatic blob, $\xi_e = (l_p^2 l_B \alpha_e^2)^{1/3}$, the electrostatic effects may be neglected.

If the sodium ion concentration I in the used solution decreases from 1000 mM to 2.57 mM, the persistence length of dsDNA increases from 46 nm to 59 nm, while the elastic stretch modulus of dsDNA remains approximately constant (Wenner et al., 2002). General dependence of persistence length l_p^{ds} on the [Na$^+$] concentration I_{st} in this publication is given by the following expression:

$$l_p^{ds} \approx l_{p0}^{ds} + \frac{0.324}{I_{st}} \bullet \left(\frac{l_B}{\alpha_w} \right),$$

where $l_{p0}^{ds} = 50$ nm is the initial value of persistence length, l_B is the Bjerrum length, $\alpha_w = (1 - \delta K/K_0)^3$, δK is the electrostatic contribution, and K_0 is the limit of high salt (nonelecrostatic value).

Double-stranded DNA (dsDNA) bending has been investigated experimentally by transient electric birefringence, solution cyclization, single-molecule stretching, atomic force microscopy, nuclear magnetic resonance (NMR), X-ray crystallography, and small-angle X-ray scattering (SAXS). The majority of these experiments have borne out the predictions of the WLC model, in which bending energy increases quadratically with curvature (Kratky and Porod, 1949). The WLC is specified by a single persistence length parameter, the characteristic distance over which thermal bending occurs. Polymers shorter than their persistence length behave as elastic rods, and polymers significantly longer behave as random walks. Values for the persistence length of duplex DNA under roughly physiological conditions range from 44 nm to 55 nm, with variation due to salt concentration and sequences.

Despite the success of the WLC model, nearly all experiments on DNA flexibility remain limited to thermally accessible bending energies, leaving the regime of greater curvature comparatively unexplored. The range of bending angles over which the WLC model retains validity is a topic of longstanding and ongoing debate.

A recent review emphasized that the conditions favoring putative kinks in DNA are still poorly understood (Vologodskii and Frank-Kamenetskii, 2013). Crick first suggested that the bending energy may be significantly lower than the WLC prediction when the bend angle is large, giving rise to a stable kinking of DNA under sufficient imposed curvature. Consistent with this hypothesis, base pair (bp) DNA minicircles were found to be digested by single-strand-specific nucleases, suggesting that bending and/or torsional stress induce local structural disruptions. A recent study has used single-molecule Förster resonance energy transfer (FRET) to measure cyclization of subpersistence length linear dsDNA with complementary single-stranded overhangs. The observed cyclization rate exceeded the predicted rate from the WLC model. On the other hand, SAXS data on DNA as short as 42 bp were consistent with the WLC. The differing results may arise because the cyclization data were dominated by rare extreme conformational fluctuations, whereas the SAXS data were dominated by the most prevalent conformations in the ensemble.

Simplified Euler buckling studies of DNA (Fields et al., 2013) shows that the buckling length at constant compression force F for double-stranded DNA is equal to:

$$L_{\text{buckle}} = \pi\sqrt{k_{\text{ben}}/F}, \tag{1.3}$$

where $k_{\text{ben}} = l_{\text{p}} k_{\text{B}} T$ is the bending modulus, l_p is the persistence length, k_{B} is the Boltzmann constant, and T is the absolute temperature in Kelvin. Rods shorter than this length can support compression; rods longer cannot and instead adopt "elastica" bent confirmation. This estimate is based purely on continuum mechanics, and neglects thermal fluctuations, as well as all molecular details. However, at high ionic strength, dsDNA was found to be more bent than predicted by the WLC model. The free energy to form a kink (Fields et al., 2013) has two components:

$$G_{\text{kink}} = G_{\text{bubble}} + G_{\text{collapse}},$$

where G_{bubble} is the free energy cost (positive) to form a sharply bent region of locally melted DNA, and $\text{G}_{\text{collapse}}$ is the free energy gain (negative) by conformational relaxation in the rest of the molecular vise.

$G_{\text{bubble}} \approx -\text{G}_{\text{collapse}} \approx$ 15–20 $k_{\text{B}}T$ under the conditions where kinking was observed.

In addition, recent research shows that DNA molecules with length smaller than persistent length can also be effectively bent, or have extreme bendability (Vafabakhsh and Ha, 2012). However, these measurements reporting enhanced flexibility of DNA at short length scale were conducted in "unphysiologically" high sodium or magnesium, raising the question of whether these results apply under physiological conditions (Fields et al., 2013). DNA bending is an integral part of many biological processes, including genome storage, genetic recombination, mismatch repair, and transcriptional regulations. Quantitative descriptions of these processes require a detailed understanding of the energetics of DNA bending.

1.6 A, B, Z DNA FORMS, S- AND P-DNA PHASES

The possible existence of nucleic acids in different helical forms, especially the left-handed Z-DNA structure *in vitro*, has been well documented (see images of A, B, Z-DNA in Fig. 10.15 of Pierce, 2005; and Pray, 2008). A variety of conditions, such as water, salt concentration, temperature, presence of ethanol, and some specific proteins and methylation are known to influence the conformational state of DNA. A-DNA structure is short and wide with right helix direction, average base pairs per turn 11, distance between adjacent bases 0.26 nm, helix pitch 2.8 nm, and diameter of the cross section circumference 2.3 nm. B-DNA structure is long and narrow, right-handed with average base pairs per turn 10, diameter of the cross section circumference 1.9–2.1 nm, distance between adjacent bases 0.34 nm, and helix pitch 3.4 nm. Z-DNA structure is left-handed with 12 average base pairs per turn, distance between adjacent bases 0.37 nm, and diameter of the cross-section circumference 1.8 nm. Z-DNA does not have a major groove; minor groove is very narrow and deep. The B to Z helical transformation can be induced in supercoiled DNA by untwisting the macromolecule. Since these transformations are cooperative in nature, statistical mechanical methods offer a convenient approach to understanding the underlying physicochemical basis. An Ising statistical mechanical model with sequence assumption was developed. In the proposed Ising model (Ghosh, 1992), the helical states of DNA are considered at first approximation to exist in right-handed (R) and left-handed (L) forms in equilibrium with each other, that is, R↔L. Considering nearest neighbor interaction, the state of a complementary pair of nucleotides may be expressed by a matrix, where the relative weight factor of an R (right) state with a neighboring R state is 1, that with a neighboring L (left) state is σ_{L}, and the relative weight factor of an L state with a neighboring L state is S_{L}.

The eigenvalue of the matrix is given by:

$$\begin{vmatrix} 1-\lambda_m & \sigma_{\mathrm{L}} \\ \sigma_{\mathrm{L}} & S_{\mathrm{L}}-\lambda_m \end{vmatrix} = 0.$$

Hence, $\lambda_m = \frac{1}{2}\left\{(1+S_{\mathrm{L}}) \pm \sqrt{(1-S_{\mathrm{L}})^2 + 4\sigma_{\mathrm{L}}^2}\right\}$.

The partition function of the nucleotide chain of length N is given by $Z(N) = C\lambda_{\max} N$, where C is a constant.

Therefore, the fraction of nucleotide in the L state is:

$$n_{\mathrm{L}} = \frac{1}{N}\frac{\partial \ln Z(N)}{\partial \ln S_{\mathrm{L}}} = \frac{S_{\mathrm{L}}}{2\lambda_m}\left\{1 - \frac{(1-S_{\mathrm{L}})}{\sqrt{(1-S_{\mathrm{L}})^2 + 4\sigma_{\mathrm{L}}^2}}\right\}.$$

When the stretching force reaches 65–70 pN, a structural transition to a new DNA phase, called S-DNA, is observed. At this force, the length of the molecule overstretches abruptly by about 70%. Numerical simulations of the structure of dsDNA under a large stretch have suggested that S-DNA could be associated with two possible DNA structures (Allemand et al., 2003): a ladder structure in which the two backbones are antiparallel and perpendicular to the bases; or a double-helical structure with inclined bases. Our helicoidal model (Type 6, Table 1.3) represents transaction to the S-DNA phase directly (Chapter 7). Comparison of force-extension data for S-DNA and single-stranded DNA shows that S-DNA to be distinct from both double helix and single-stranded forms (Cocco et al., 2004). The simple thermodynamical model for tension-melted double stranded DNA indicates that the overstretching transition near 65 pN cannot be explained in terms of the conversion of double helix to noninteracting polynucleotide strands. However, the single-strand-like response observed in some experiments can be explained in the terms of "unpeeling" of large regions of one strand, starting from nicks on the original double helix. S-DNA becomes unstable to unpeeling at large forces at low ionic strength, or for weakly base-paired sequences. Unpeeling can preempt formation of S-DNA. Strongly base-paired regions generate large barriers that stabilize DNA against unpeeling. For long genomic sequences, these barriers to unpeeling cannot be kinetically crossed until the force exceeds approximately 150 pN.

Upon twisting the DNA molecule with torque of ≈20 pN·nm, an overwound DNA molecule adopts a new inside-out overtwisted structure called P-DNA (Allemand et al., 1998; Léger et al., 1999). An analysis of the force/extension curves for different degrees of supercoiling reveals that P-DNA is a double-helical structure with pitch of 2.6 bases per run. DNA supercoiling to toroidal stationary state was researched by Shi and Hearst (1994), on the basis of nonlinear Schrödinger equation. In statistical mechanics, the Zimm and Bragg (1959) model is a helix-coil transition model that describes helix-coil transactions of molecules, usually polymer chain.

Our helicoidal model (Type 6, Table 1.3) may successfully describe the transition to the P-DNA phase as well.

It is possible to unzip mechanically the double helix, putting apart the two strands of dsDNA. In this case, the typical stick-slip (or saw tooth) signal is observed. During the stick-slip phase, tension builds up as unwound strands are stretched, until the stored energy is large enough to unlock the base pair closest to the fork defined by the dsDNA and two single strands. This is followed by rapid slip phase, in which this energy is dissipated through the unlocking of many successive base pairs: tension decreases and extension increases as the unwound strands become longer. The maximum force observed in the stick phase (~15 pN) reflects the GC content of the adjacent region (Allemand et al., 1998). Results of melting of the double helix into separate strands upon heating and stocking of strands by quenching published by Ivanov et al. (2004). Qualitatively, the unpairing transition is relatively narrow in temperature, while stacking interactions display a broader transition. The experiments were provided in phosphate buffer saline at ionic strength of pH = 7.4 and 3.6. The stacking enthalpies are lower at lower pH.

It has been shown that the elastic response of ssDNA can be understood under a variety of conditions, and over a large force range using an extension of the freely jointed model. An ssDNA is modeled as a chain of charged (self-avoiding), slightly extensible segments of size $b_s = 1.6$ nm that are free to adapt any orientation.

RNA (ribonucleic acid) structure is similar to ssDNA, but it is a stable state. This is the main difference with dsDNA that is usually found in a double-stranded form. The lack of paired strand allows RNA to fold into complex, dimensional structures.

DNA polymerases are enzymes that proceed along an ssDNA template synthesizing the complementary strand. By simply binding to DNA, a protein may modify its local elastic properties (bending, twisting, stretching, or locally melting the molecule). As the tension is increased, the nucleosome DNA is released, thereby relaxing part of the tension. The resulting force-extension signal (Fig. 1.10) has stick-slip characteristic (saw tooth or zigzag) view, with each peak corresponding to the ejection of one nucleosome. In contrast to DNA polymerases, an RNA polymerase does not modify the extension of its template during transcription (Allemand et al., 2003).

For $F > 5$ pN, replication results in the shortening of the molecule by an amount of 0.1 nm/nucleotide at $F = 20$ pN. This shortening implies that the enzyme is performing work against the force. It can be slowed by applying a force and even stopped at a force of about 30 pN.

A single nucleosome is realized as the tension on the DNA increased (up to about 15 pN) thereby relaxing part of the tension. Topoisomerases are enzymes that control DNA supercoiling and are responsible for disentangling molecules during replication, DNA repair, and recombination.

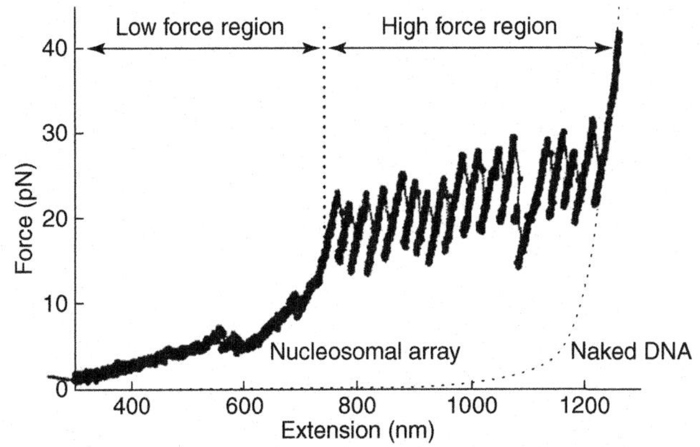

FIGURE 1.10 Releasing a nucleosome at DNA molecule tension. *(After Allemand et al., 2003, reprinted with permission.)*

1.7 POLYMER MATERIALS

Most researchers have assumed that DNA molecule has mechanical properties which similar to those of polymer materials (Kalpakjian et al., 2007; Fried, 1995). These materials include ABS (acrylonitrile, butadien, and styrene), acrilics, polyester, polymethyl, polycarbonate, etc. Their elastic modulus, E, is in the range from 0.14 GPA to 12 GPA, Poisson's ratio is in the range from 0.35 to 0.49, ultimate tensile stress (UTS) is approximately in the range from 14 MPa to 160 MPa, and yield stress is in the range from 7 MPa to 69 MPa. Zero Poisson's ratio has a cork.

Several polymers are effective (Walia et al., 2015) as a flexible substrate for metamaterials for sensors and modeling: polydimethylsiloxane with very low Young's modulus (7.5×10^{-4} GPa), polyimide (Kapton) with intrinsic Young's modulus of 2.5 GPa and strong adhesion to metal coatings, and polyethylene terephthalate with Young's modulus of 4.0 GPa.

Mechanical, optical, and electrical properties of polymers may be enhanced (particular significant increase of Young's modulus) by incorporation of nanoscopic particles (Sarkar and Alexandridis, 2015). Tensile test of polymer micro- and nanoscale fibers shows that their stress-strain relationships can be nonlinear (in some cases with strain in the third power). SU-8 is a negative epoxy-based photo-resist developed for the micro technology. It is a polymer that can be spin-coated to thickness ranging from below 1 μm to 1 mm and the final structures can be defined by UV lithography.

However, our research shows the option to estimate DNA molecule properties on the basis of its experimental data (Chapters 6 and 7) instead of using assumed average data of polymers without experimental verification.

RNA ladders composed of RNA molecular-weight size markers were initially developed by using the synthetic circle method to produce different-sized markers. This technique was improved upon by inventor Eric T. Kool (Stanford University) to use circular DNA vectors as a method for producing RNA molecular-weight size markers. Referred to as the rolling circle method, the improvements of this technique stem from its efficiency in synthesizing RNA oligonucleotides. From the circular DNA template, single-stranded RNA varying in length from 4 bp to 1500 bp can be produced without the need for primers and by recycling nucleotide triphosphate. DNA can also be synthesized from the circular template, adding to this technique's versatility. In comparison to runoff transcription, the synthetic circle method produces RNA oligonucleotides without the runoff.

1.8 DNA TECHNICAL APPLICATIONS

Elastic kinematic systems are now widely used in nanotechnology because they provide very low friction, do not require grease, and can have miniature structures. A nontraditional miniature elastic kinematic system can be made on the basis of DNA molecule (Bustamante et al., 2003). Outside of traditional DNA biology, the ease of synthesis and the well-characterized elasticity of DNA make it an ideal structure for stiff molecular handles to manipulate other molecules. Such handles have been used to mechanically unfold molecules of RNA. Tiny DNA tweezers can catch and release an object (Han et al., 2008) under control.

DNA self-assembly begins with the synthesis of single strand DNA molecules that self-assemble into macromolecular blocks called tiles. These tiles facilitate further assembly into larger structures known as DNA tiling lattices. In principle, DNA tiling assembles can form any two or three-dimensional pattern. These assembles can be used as programmable scaffolding to position molecular electronics and robotics with precision and specificity, facilitating the fabrication of complex nanoscale devices (Yan et al., 2003).

1.9 DNA SIZE AND MASS CONVERSION

Molecular weight, external, and internal damping in molecule is important for dynamics of the molecular processes. See: http://www.lifetechnologies.com/us/en/home/references/ambion-tech-support/rna-tools-and-calculators/dna-and-rna-molecular-weights-and-conversions.html

Molecular weight (M.W.) is usually measured in Daltons (1 Dalton = 1 atomic mass unit = 1.6605×10^{-27} kg).

M.W. (Table 1.5) is the sum of the atomic weights of the constituent atoms.

Exact M.W. of RNA: M.W. = $(An \times 329.2) + (Un \times 306.2) + (Cn \times 305.2) + (Gn \times 345.2) + 159$. *An*, *Un*, *Cn*, and *Gn* are the number of each respective nucleotide (nt) within the polynucleotide. Calculated M.W. is valid at physiological conditions (pH). Addition of "159" to the M.W. takes into account the M.W. of a 5′ triphosphate. Exact M.W. of ssDNA (e.g., Oligonucleotides): M.W. = $(An \times 313.2) + (Tn \times 304.2) + (Cn \times 289.2) + (Gn \times 329.2) + 79.0$. Addition of 79.0 to the M.W. takes into account the 5′ monophosphate left by most restriction enzymes. No phosphate is present at the 5′ end of strands made

TABLE 1.5 Size and Mass dsDNA, ssDNA, and RNA Conversions

	Mass in Daltons		
Size (nt)	dsDNA	RNA	ssDNA
20	12,306	6,569	6,153
100	60,898	32,200	30,449
300	182,378	96,309	91,189
500	303,858	160,409	151,929
1,000	607,558	320,659	303,779
1,800	1,093,478	577,059	546,739

by primer extension. Every base pair in a DNA molecule contributes about 600 Daltons to the molecular weight of a DNA molecule.

M.W. of dsDNA = (number of nucleotides × 607.4) + 157.9.

Anhydrous molecular weight of each nucleotide in DNA is:

A (adenine) = 313.21; T (thymine) = 304.2; C (cytosine) = 289.18; and G (guanine) = 329.21.

For the purpose of a rough estimate, the average weight of a DNA nucleotide (in salt solution) is 325 Daltons.

In accordance with quantum aspects (Landau and Lifshitz, 1969), a solid body model is "microscopic" if its smallest size is less than $L_{\text{mic}} \leq \hbar v_s/(k_B T)$, where v_s is the velocity of sound in the body; $\hbar$ is the reduced Planck's constant; k_B is Boltzmann constant, and T is the absolute temperature.

L_{mic} at $T \leq 293$ K is comparable to an atom's diameter, thermal phonon wavelength, and to the thickness of very thin films or coatings on the solid body surfaces. It has been shown with quantum mechanics methods that the limit for the smallest measurable displacement is equal to $\Delta L \geq (\hbar\tau/m)^{1/2}$, where m is the levitation body's mass and τ is the measurement impulse duration. For example, at $\tau = 10^{-3}$ s and $m = 10^4$ g $\Delta L \geq 10^{-19}$ m (Tarbeev et al., 1983).

1.10 DEFLECTION AT EQUILIBRIUM

Boltzmann constant, widely used in this book, is equal to $k_B = 1.3806488 \times 10^{-23}$ m^{2}kg/s^{2}K and normal absolute temperature in Kelvin equals 293.15 K (corresponding to 20°C). Therefore, we use, as already mentioned earlier, in our calculations for regular conditions the value of $k_B T \approx 4.1$ pN·nm.

Macroscopic bodies' physical models should have the smallest size that is larger than microscopic size L_{mic}. In an equilibrium state, systems of macroscopic bodies and external physical medium have parameters that are approximately equal to their mean values. However, random small deflections from mean values take place as a result of thermodynamic fluctuations, even at the macroscopic equilibrium. These thermodynamic fluctuations have a standard deviation, which may be equal to:

$$\sigma_{X_a} = (k_B T_T X_a^2 / 2R_{v\,\text{min}})^{1/2}, \tag{1.4}$$

where $R_{v\,\text{min}}$ is the minimal work necessary for reversible small changes of the thermodynamic quantity X_a, k_B is Boltzmann's constant, and T_T is an absolute temperature. For a straight linear beam, the tensile stiffness equals EA_c/L, and the minimal deformation work is equal to:

$$R_{v\,\text{min}} = EA_c X_a^2 / 2L, \tag{1.5}$$

where E is Young's modulus of the beam's material, A_c is the area of the body's cross section, and L is the beam's length. Substitution (1.5) into (1.4) yields

$$\sigma_{X_a} = (k_B T_T L/EA_c)^{1/2}, \tag{1.6}$$

and subsequently, the limit of these fluctuations with normal distribution is equal to: $\Delta_{\varepsilon M} = 3\sigma_{X_a}$. From this formula, it is evident that increasing the length and decreasing the cross-section area of the body yields increasing statistical fluctuation of the beam's length. Fluctuations at room temperature can practically be larger than 10^{-10} for steel body, and would also be higher for materials with smaller elastic modulus. The expanded uncertainty U_{th} of thermal fluctuations at Gauss distribution with confidence level of 0.95 are equal to $U_{th} = 2\sigma_{X_a}$. Therefore, we have a factor 4 in the numerator under square root after substitution of the expression for σ_{X_a} from (1.6) as follows:

$$U_{th} = (4K_B T_T X_a^2 / 2R_{v\,min})^{1/2}. \tag{1.7}$$

These values will limit the accuracy of small objects dimensions measurement.

REFERENCES

Ahsan, A., Rudnick, J., Bruinsma, R., 1998. Elasticity theory of the B-DNS to S-DNA transition. Biophys. J. 74, 132–137.

Allemand, J.-F., Bensimon, D., Lavery, R., Croquette, V., 1998. Stretched and overwound DNA form a Pauling-like structure with exposed bases. Proc. Natl. Acad. Sci. USA 95, 14152–14157.

Allemand, J.-F., Bensimon, D., Croquette, V., 2003. Stretching DNA and RNA to probe their interactions with proteins. Curr. Opin. Struct. Biol. 13, 266–274.

Ambia-Garrido, J., Vairub, A., Pettit, B.M., 2010. A model for Structure and thermodynamics of ssDNA and dsDNA near a surface: a coarse grained approach. Comput. Phys. Commun. 181 (12), 2001–2007.

Baumann, C.G., Smith, S.B., Bloomfield, V.A., Bustamante, C., 1997. Ionic effects on the elasticity of single DNA molecules. Proc. Natl. Acad. Sci. USA 94, 6185.

Benke, M., Shapiro, E., Drikakis, D., 2011. Mechanical behavior of DNA molecules-elasticity and migration. Med. Eng. Phys. 33, 883–886.

Blumberg, S., Tkachenko, A.V., Meiners, J.-C., 2005. Disruption of protein-mediated DNA looping by tension in the substrate DNA. Biophys. J. 88 (3), 1692–1701.

Boal, D., 2002. Mechanics of the Cell. Cambridge University Press, Cambridge, UK.

Bouchiat, C., Wang, M.D., Allemand, J.-F., Strick, T., Block, S.M., Crquette, V., 1999. Estimating the persistence length of worm-like chain molecule from force-extension measurements. Biophys. J. 76, 409.

Brouchard-Wyart, F., 1995. Polymer-chains under strong flows-stem and flowers. Europhys. Lett. 47, 171–174.

Bustamante, C., Marko, J.F., Siggia, E.D., Smith, S., 1994. Entropic elasticity of λ-phage DNA. Science 265, 1599.

Bustamante, C., Bryant, Z., Smith, S.B., 2003. Ten years of tension: single-molecule DNA mechanics. Nature 421, 423–427.

Chen, J.-S., Lee, C.-H., Teng, H., Wang, H., 2012. Atomistic to continuum modeling of DNA molecules. In: Advances in Soft Matter Mechanics. Springer, Berlin, Heidelberg, (Chapter 1).

Cizeau, P., Viovy, J.L., 1997. Modeling extreme extension of DNA. Biopolymers 42, 383–385.

Clausen-Schaumann, H., Rief, M., Tolksdorf, C., Gaub, H.E., 2000. Mechanical stability of single DNA molecules. Biophis. J. 78, 1997–2007.

Cocco, S., Yan, J., Léger, J.-F., Chatenay, D., Marko, J.F., 2004. Overstretching and force-driven strand separation of double-helix DNA. Phys. Rev. E 70, 011910.

Dame, R.T., 2005. The role of nucleoid-associated proteins in the organization and compaction of bacterial chromatin. Mol. Microbiol. 56 (4), 858–870.

Danilowicz, C., Coljee, V.W., Bouzigues, C., Lubensky, D.K., Nelson, D.R., Prentiss, M., 2003. DNA unzipped under a constant force exhibits multiple metastable intermediates. Proc. Natl. Acad. Sci. USA 100 (4), 1694.

Dauxois, T., 1991. Dynamics of breather modes in a nonlinear "helicoidal" model of DNA. Phys. Lett. A 159 (8–9), 390–395.

Duričković, B., Goriely, A., Maddocks, J.H., 2013. Twist and stretch of helices explained via the Kirchgoff-Love rod model of elastic filaments. Phys. Rev. Lett. 111, 108103.

Essevaz-Roulet, B., Bocklemann, U., Hesiot, F., 1997. Mechanical separation of the complementary strands of DNA. Proc. Natl. Acad. Sci. USA 94, 11935–11940.

Fain, B., Rudnick, J., Östlund, S., 1997. Conformation of linear DNA. Phys. Rev. E 55, 7364.

Fields, A.P., Meyer, E.A., Cohen, A.E., 2013. Euler buckling and nonlinear kinking of double-stranded DNA. Nucleic Acids Res. 41 (21), 9881–9890.

Forinash, K., Bishop, A.R., Lomdahl, P.S., 1991. Nonlinear dynamics in double-chain model of DNA. Phys. Rev. B 43 (13), 10743.

Fried, J.R., 1995. Polymer Science and Technology. Prentice Hall PTR, Upper Saddle River, NJ.

Ghosh, S., 1992. Ising model for B-Z transition in supercoiled DNA. Bul. Math. Biol. 54 (5), 727–732.

Gore, J., Briant, Z., Nollman, M., Le, M.U., Cozzarelli, N.R., Bustamante, C., 2006. DNA overwinds when stretched. Nature 442, 836.

Grosschedl, R., Giese, K., Pagel, J., 1994. HMG domain proteins: architectural elements in the assembly of nucleoprotein structures. Trends Genet. 10 (3), 94–100.

Han, C.H., Chou, C.Y., Wu, C.J., Chiang, K.N., 2006. Investigation of the ssDNA backbone molecule mechanical behavior using atomistic-continuum mechanics method. In: Technical Proceedings of the 2006 NSTI Nanotechnology Conference and Trade Show, vol. 2. pp. 321–324. NSTI-Nanotech (Chapter 4).

Han, X., Zhou, Z., Yang, F., Deng, Z., 2008. Catch and release: DNA tweezers that can capture, hold, and release an object under control. J. Chem. Soc. 130 (44), 14414.

Ivanov, S., Zeng, Y., Zocchi, G., 2004. Statistical mechanics of base staking and pairing in DNA melting. Phys. Rev. E Stat. Nonlin. Soft Matter Phys. 70 (5 (Pt1)), 051907, arXiv.org/ftp/cond-mat/papers/0411/0411563.pdf

Kalpakjian, S., Schmid, S., Schmid, S.R., 2007. Manufacturing Processes for Engineering Materials, fifth ed. Prentice Hall, Upper Saddle River, NJ.

Khokhlov, A.R., Grosberg, A.Y., Pande, V.S., 1994. Statistical Physics of Macromolecules (Polymers and Complex Materials). American Institute of Physics, New York, USA.

Kiang, C.H., Chen, W.S., 2010. Using mechanical forces to study nano-bio materials properties. In: Proceedings of the Eighteenth Annual International Conference on Composites or Nano Engineering (ICCE-18), Anchorage, Alaska, 4–10 July, 2010.

Korolev, N., Luo, D., Lubartsev, A., Nordenskiöld, L., 2014. A coarse-grained DNA model parameterized from atomic simulations by inverse Monte Carlo. Polymers 6, 1655–1675.

Krajnc, M., 2012. Elasticˇna teorija DNA: raztegovanje in topljenie DNA, University of Lubljana, Faculty of Math and Phys., Seminar, Avgust 6, Mentor Prof. Dr. Podgornik, R. (in Slovenian).

Kratky, O., Porod, G., 1949. Röntgenuntersuchung gelöster Fadenmoleküle. Rec. Trav. Chim. Pays-Bas. 68, 1106.

Landau, L.D., Lifshitz, E.M., 1969. Statistical Physics. Pergamon Press, Oxford, UK.

Levinthal, C., Crane, H.R., 1956. On the unwinding of DNA. Proc. Natl. Acad. Sci. USA 42, 436–438.

Lee, N.-K., Thirumalai, D., 2004. Pulling-speed-dependent force-extension profiles for semiflexible chains. Biophys. J. 86 (5), 2641–2649.

Léger, J.F., Robert, J., Bourdieu, L., Chatenay, D., Marko, J.F., 1998. RecA binding to a single double-stranded DNA molecule: A possible role of DNAS conformational fluctuations. Proc. Natl. Acad. Sci. USA 95, 12295–12299.

Léger, J.F., Romano, G.A., Sarkar, J., Robert, L., Bourdieu, D., Chatenay, Marko, J.F., 1999. Structural transitions of a twisted and stretched DNA module. Phys. Rev. Lett. 83 (5), 1066–1069.

Manghi, M., Destainville, N., Palmeri, J., 2012. Mesoscopic models for DNA stretching under force: new results and comparison with experiments. Eur. Phys. J. E 35, 110.

Marko, J.F., 1997. Twist and shout (and pull): molecular chiropractors undo DNA. Proc. Natl. Acad. Sci. USA 94, 11770–11772.

Marko, J.F., Siggia, E.D., 1995. Statistical mechanics of supercoiled DNA. Phys. Rev. E 52, 2912.

Mathew-Fenn, R.S., Das, R., Harbury, P.A.B., 2008a. Remeasuring the double helix. Science 322, 446–449.

Mathew-Fenn, R.S., Das, R., Silverman, J.A., Walker, P.A., Harbury, P.A.B., 2008b. A molecular ruler for measuring quantitative distance distribution. PLoS One 3, e3229, Supplementary materials section.

Mathew-Fenn, R.S., Das, R., Fenn, T.D., Schneiders, M., Harbury, P.A.B., 2009. Response to comment on "Remeasuring the double helix". Science 325, 538-c.

Mazur, A.K., 2009. Analysis of accordion DNA stretching revealed by the gold cluster ruler. Phys. Rev. E 80, 010901, (R).

Moroz, J.D., Nelson, Ph., 1997. Torsional directed walks, entropic elasticity, and DNA twist stiffness. Proc. Natl. Acad. Sci. USA 94 (26), 14418–14422.

Muto, V., Lomdahl, P.S., Christiansen, P.L., 1990. Two-dimensional model for DNS dynamics: longitudinal wave propagation and denaturation. Phys. Rev. A 42 (12), 7452–7458.

Odijk, T., 1995. Stiff chains and filaments under tension. Macromolecules 28, 7016.

Olsen, K., Bohr, J., 2010. The generic geometry of helices and their close-packed structure. Theor. Chem. Acc. 125, 207–215.

Olsen, K., Bohr, J., 2011. The geometrical origin of the strain-twist coupling in double helices. AIP Adv. 1, 012108.

Ouyang, Z.Y., Zheng, S., 2014. Traveling wave solutions on nonlinear dynamical equations in a double-chain model of DNA. Abs. Appl. Anal. 2014, Article ID 317543, 5 pages.

Pal'mov, V.A., 1964. Fundamental equations of the theory of asymmetric elasticity. PMM 28 (3), 401, (in Russian).

Peyrard, M., Cuesta-Lopez, S., James, G., 2008. Modeling DNA at the mesoscale: challenge for nonlinear science? Nonlinearity 21, T91–T100.

Pierce, B., 2005. Genetics: A Conceptual Approach, second ed. W.H. Freeman, New York, NY.

Podgornik, R., 2011. Physics of DNA, www.fmf.uni-lj.si/~podgornik/.../physics-of-DNA.

Pray, L.A., 2008. Discovery of DNA structure and function: Watson and Crick. Nat. Edu. 1 (1), 100 http://www.nature.com/scitable/topicpage/discovery-of-dna-structure-and-function-watson-397

Rief, M., Clausen-Scaumann, H., Gaub, H.E., 1999. Sequence-dependent mechanics of single DNA molecules. Nature Struct. Biol. 6 (4), 346–349.

Sandman, K., Pereira, S., Reeve, J., 1998. Diversity of prokaryotic chromosomal proteins and the origin of the nucleosome. Cell. Mol. Life Sci. 54 (12), 1350–1364.

Sarkar, B., Alexandridis, P., 2015. Block copolymer-nanoparticle composites: structure, functional properties, and processing. Prog. Polym. Sci. 40, 33–62.

Shi, Y., Hearst, J., 1994. The Kirchhoff elastic rod, the nonlinear Schrödinger equation, and DNA supercoiling. J. Chem. Phys. 101 (6), 5186–5200.

Shimada, J., Yamakawa, H., 1984. Ring-closure probabilities for twisted wormlike chain. Macromolecules 17, 689.

Smith, S.B., Cui, Y., Bustamante, C., 1996. Overstretching B-DNA: the elastic response of individual double-stranded and single-stranded DNA molecules. Science 271, 795.

Storm, C., Nelson, P.C., 2003a. Theory of high-force DNA stretching and overstretching. Phys. Rev. E 67, 051906.

Storm, C., Nelson, P.C., 2003b. The bend stiffness of S-DNA. Europhys. Lett. 62 (5), 760.

Strick, T.R., Allemand, J.-F., Bensimon, D., Bensimon, A., Croquette, V., 1996. The elasticity of a single supercoiled DNA molecule. Science 271, 1835–1837.

Swigon, D., 2009. The mathematics of DNA structure, mechanics and dynamics. In: Benham, C.J., Harvey, S., Olson, W.K. (Eds.), Mathematics of DNA Structure, Function and Interactions. Springer, New York, USA.

Tarbeev, Iu.V., Tseytlin, Y.M., et al.,1983. Metrology of small length. Investigations in the Field of Linear and Angular Measurement. Ĕnergoatomisdat, Leningrad, (in Russian).

Thomas, J., 2001. HMG1 and 2: architectural DNA-binding proteins. Biochem. Soc. Trans 29 (Pt 4), 395–401.

Tkachenko, A.V., 2006. Electrostatic effects in DNA stretching. Phys. Rev. E 74, 041801.

Tkachenko, A.V., 2007. Elasticity of strongly stretched ssDNA, arXiv: q-bio/0603027v1[q-bio.BM] 12 Mar.

Tseytlin, Y.M., 2011. An effective model of DNA like structure: with length fluctuation nonlinearity. AIP Adv. 1, 012118.

Tseytlin, Y.M., 2012. Flexible helicoids, atomic force microscopy (AFM) cantilevers in high mode vibration, and concave notch hinges in precision measurements and research. Micromachines 3, 480.

Tseytlin, Y.M., 2013. Functional helicoidal model of DNA molecule with elastic nonlinearity. AIP Adv. 3, 062110.

Tseytlin, Y.M., 2014. DNA molecule elastic nonlinearity: a functional helicoidal model. ZAMM 94 (6), 505–508.

Vafabakhsh, R., Ha, T., 2012. Extreme bendability of DNA less than 100 base pairs long revealed by single-molecule cyclization. Science 337 (6096), 1097.

Vologodskii, A., Frank-Kamenetskii, M.D., 2013. Strong bending of the DNA double helix. Nucleic Acids Res. 41 (14), 6785–6792.

Vučemilović-Alagić, N., 2013, Kratky-Porod model. University of Split, Faculty of math and natural science. http://www-f1.ijs.si/~rudi/sola/KratkyPorodmodel.pdf

Walia, S., Shah, C.M., Guturut, P., Nili, H., et al.,2015. Flexible metasurfaces and metamaterials: A review of materials and fabrication processes at micro- and nano-scales. Appl. Phys. Rev. 2, 011303, 15 p.

Wang, M.D., Yin, H., Landick, R., Gelies, J., Block, S.M., 1997. Stretching DNA with optical tweezers. Biophys. J. 72, 1335–1346.

Watson, J.D., Crick, F.H., 1953. Molecular structure of nucleic acids: a structure of deoxyribose nucleic. Acid. Nature 171 (4356), 737–738.

Wenner, J.R., Wiliams, M.C., Rouzina, I., Bloomfield, V.A., 2002. Salt dependence of the elasticity and overstretching transition of single DNA molecules. Biophys. J. 82, 3160.

Wiggins, P.A., Phillips, R., Nelson, P.C., 2005. Exact theory of kinkable elastic polymers. Phys. Rev. E 71, 021909-1-19.

Xiao, H., 2013. DNA elastic nonlinearities as multiple smooth combination of soft and hard linear springs. ZAMM 93, 50–56.

Yamakawa, H., 1976. Statistical mechanics of wormlike chain. Pure Appl. Chem. 46, 135–141.

Yamakawa, H., 1997. Helical Wormlike Chains in Polymer Solutions. Springer, Berlin.

Yamakawa, H., 1999. A new framework of polymer solution science. The helical wormlike chain. Polymer J. 31 (2), 109–119.

Yan, H., Park, S.H., Finkelstein, G., Reif, J.H., LaBean, T.H., 2003. DNA-templated self-assembly of protein arrays and highly conductive nanowires. Science (301), 1882–1884.

Yuan, C.-A., Han, C.-N., Chiang, K.-N., 2005. Investigation of the sequence-dependent dsDNA mechanical behavior using clustered atomic-continuum method. NSTI-Nanotech. 2, 561.

Zimm, B.H., Bragg, J.K., 1959. Theory of the phase transition between the helix and random coil polypeptide chain. J. Chem. Phys. 31, 526–531.

Zhou, Z., Lai, P.-Y., 2005. On the asymmetric elasticity of twisted dsDNA. Physica A 350, 70.

Zhou, X., Liu, J., 2013. Stability analysis of kinked DNA with generalized rod model. Phys. E 47, 152–156.

Chapter 2

Force Application, Measurement, and Manipulation Accessories

2.1 STRETCHING MICROPIPETTE, GLASS MICRONEEDLES, AND HYDRODYNAMICS

Two counter-propagating laser beams "trap" a bead in the middle of a specially designed fluid chamber (Fig. 2.1). The other bead is held by a micropipette introduced inside the chamber. The micropipette is moved to increase or decrease the tension in the DNA molecule that bridges the two beads. The environment around the molecule can be altered by exchanging the solution inside the chamber. Hydrodynamic coupling between DNA and the surrounding moving fluid was studied by Perkins et al. (1995, 1997). See also Noeth et al. (2014) for integrated cantilever-based flow sensors with tunable sensitivity, for the inline monitoring of flow fluctuations in microfluidic systems.

Glass microneedles have been used to pull nucleic acids anchored at their other end on to a solid surface. The bending of the microneedle yields a direct measure of the force that can be as small as 1 pN, and as large as desired. The typical dimensions of a tapered glass microneedle (Essevaz-Roulet et al., 1997) are as follows: shank diameter 1 mm, tip diameter 1 μm, and tapered length 1 cm. The needles' stiffness varies from needle to needle, and should be calibrated. The tip of a treated microneedle sticks readily to streptavidin-coated microbeads. After measurement, the beads can be removed by mechanical shock, or by using the meniscus effect: taking the microneedle out and then putting it back into the liquid. Calibration of the microneedle is achieved using a magnet and recording the time-dependent position of the bead on video with stroboscopic illumination, during the acceleration toward the magnet. The local velocity v is proportional to the force on the bead, in accordance with Stokes' formula: $F = 6\pi\eta R v$, where η is the viscosity of the buffer and R is the radius of the bead. The microneedle used in research by Essevaz-Roulet et al. (1997) had a stiffness of 1.7 pN/μm ± 20%.

Yamakawa (1997, p. 210) estimated the friction force and intrinsic viscosity of circular DNA in solutions against its molecular weight. Effective AFM tip radius determination can be done by using monodispersed silicon oxide nanoparticles (Efimov et al., 2014).

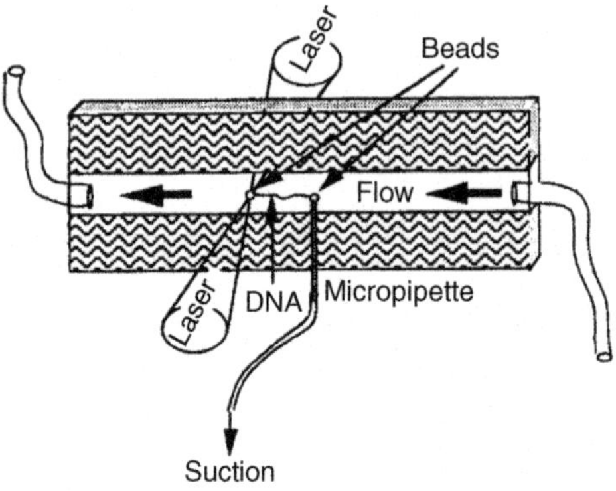

FIGURE 2.1 Schematic diagram of exchanging chamber. *(After Baumann et al., 1997, reprinted with permission.)*

2.2 OPTICAL TRAP AND TWEEZERS

Optical tweezers are formed when a laser beam is focused by a high numerical aperture microscope objective. The large electromagnetic field gradient generated at the focal point can be used to trap a micron-sized glass or plastic bead with a force F varying between 0.1 pN and 100 pN. Because the light intensity and position can be quickly modulated, multiple traps can be achieved with a single laser beam (Allemand et al., 2003). The time resolution of this technique is a few tens of a millisecond. Figure 2.2 shows a schematic diagram of the optical trap and detection system, where the solid lines represent the path of the Nd laser and the dashed lines represent the illuminating light from a tungsten lamp. Optics include lenses (L_1, L_2), mirrors (M_1, M_2), a dichroic filter D, and a microscope objective (L_3). The laser beam can be shifted in the plane of the specimen (O) by deflecting it by two orthogonally mounted acousto-optic modulators (AO), and by moving the mirrors M_1, M_2. The illuminating light is split between a video camera (VC) and a quadrant photodiode detector (QD) that indicates the position of the trapped bead. The feedback loops can be closed by feeding the output signals from the quadrant detector into the driver circuits for the voltage-controlled oscillator. PZT is a piezoelectric ceramic (lead – *plumbum*, in Latin, zirconite, titanate). A block diagram of electronic circuits has pulse inputs to open and close feedback loops (Simmons et al., 1996, for more detail). The forces F_x and F_y acting on the bead are equal to

$$F_x = k_{\mathrm{Tj}}\left(x_{\mathrm{T}} - x_{\mathrm{B}}\right) - b_{\mathrm{f}}\frac{dx_{\mathrm{B}}}{dt},$$

$$F_y = k_{\mathrm{Tj}}\left(y_{\mathrm{T}} - y_{\mathrm{B}}\right) - b_{\mathrm{f}}\frac{dy_{\mathrm{B}}}{dt},$$

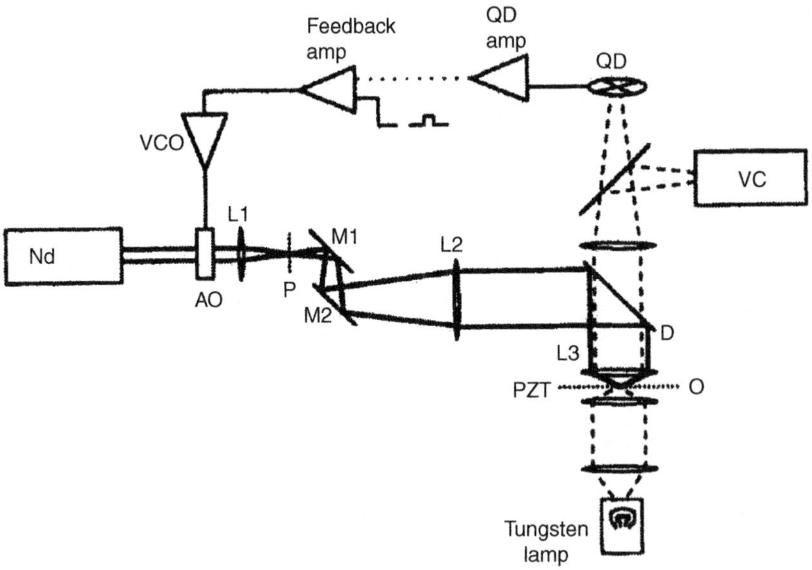

FIGURE 2.2 Optical trap with detection system. Nd, laser; L_1, L_2, lenses; M_1, M_2, mirrors; D, dichroic filter; L_3, microscope objective; O, specimen; AO, acousto-optic modulators; VC, VCO, video camera; and QD, quadrant photodiode detector. *(After Simmons et al., 1996, reprinted with permission.)*

where k_{Tj} is the stiffness of the trap; b_f is the damping factor; x_B, y_B are the signals after the quadrant detector; and x_T, y_T are the signals after feedback amplifier in the detection VCO system.

Another example of stretching the DNA with optical tweezers is shown by Wang et al. (1997).

It was previously difficult to provide rotational motion with the optical tweezers. However, recently developed single beam optical vortex tweezers can overcome this limitation (Gecevičius et al., 2014). Bodensiek et al. (2013) developed a high-speed vertical optical trap for the mechanical testing of living cells at picoNewton (pN = 10^{-12} N) forces.

In this case, the probe is held in a tightly focused laser beam that effectively replaces the AFM cantilever (Section 2.5). The noise level is less than 1 pN. The trapping occurs due to a balance between optical radiation pressure and the direction of the intensity gradient in the focused field. The vertical trap maintains all of the advantages of AFM, but adds the ability to apply very low forces to the sample.

2.2.1 The Radiation Pressure of Light

Relations. Let us recall from classical and quantum physics that the pressure of light p_{lt} can be expressed by the following formula:

$$p_{lt} = \langle w \rangle (1 + R_{ref}) \cos^2 i_{\varphi}, \tag{2.1}$$

where $<w> = 2\pi\hbar n_0 \nu_{lgt}$ is the mean value of the spatial density of incidence light, R_{ref} is the coefficient of light reflection, i_φ is the incident angle of light on the flat surface, n_0 is the concentration of photons in incidence light, $2\pi\hbar$ is Planck's constant, and $\nu_{lgt} = \frac{cn_{ref \cdot i}}{\lambda_{vac}}$ is the frequency of light in a medium. In the latter expression, the velocity of light c = 299,792,458 m/s, λ_{vac} is the wavelength of light in vacuum, and $n_{ref.i}$ is the refractive index of the medium.

2.2.2 Experimental Test

Jones (1988) carried out, in the 1950s, very careful research on light pressure in dispersive media, with the help of a special precision elastic installation, based on the thin torsional suspension of a light mirror. He concluded that, within the limits of his measurements' accuracy (standard uncertainty of order 0.05%), the pressure exerted by a steady beam of light on a mirror in an optically dense medium is directly proportional to the phase refractive index of the medium, and that it is independent of the group velocity ratio and the plane of polarization in oblique incidence, at least up to angles of 20°. We can see that the main conclusion of Jones' fine experiments about direct proportionality between the light's pressure and the refractive index of the medium correspond to Eq. (2.1), after substituting certain expressions for the spatial density $<w>$ and light frequency ν_{lgt}, such that

$$p_{lt} = 2\pi\hbar n_0 c n_{ref} / \lambda_{vac}. \tag{2.2}$$

In a pioneering work, Braginsky and Manukin (1977) demonstrated that radiation pressure acting on the mirrors of a Fabry–Pérot cavity leads to a measurable coupling between the optical and mechanical properties of the resonator. In recent years, submicron micromanipulators driven by light have been developed and used. Observation and characterization of an optical spring are shown by Sheard et al. (2004). It is possible to identify double-stranded and single-stranded DNA with circular dichroism spectroscopy by the absorption of left and right circular polarized light.

2.3 SMALL-ANGLE X-RAY SCATTERING INTERFERENCE (SAXSI)

The scattering interference ruler complements existing molecular ruler techniques in several ways (Mathew-Fenn et al., 2008). It provides calibrated distances and the ability to record complex distributions. Another unique property is that it yields an instantaneous snapshot of a molecular ensemble because the nuclei are slow, with respect to the electronic transitions that produce scattering interference. In this technique, gold nanocrystals are attached to the ends of DNA double helices ranging from 10 to 35 base pairs in length, via 3′ thiol linker. These nanocrystals are clusters of 75 gold atoms with electron density that is enough for a strong scattering of X-rays. Interference between the X-rays scattered from the DNA generated a scattering profile that can then be Fourier transformed in order to give the mean distance between each cluster, from center to center. The range of the scattering interference ruler can be extended beyond the 40–170 Å distances. The short end of the range is limited by the diameter of the nanocrystal probe, although distances much smaller than the probe diameter can be resolved. In principle, nothing limits the long end of the range (Mathew-Fenn et al., 2008).

However, in practice, longer distances are associated with broader distributions that cause a more rapid decay of the scattering interference signal with the scattering angle. Longer distances also produce higher frequency oscillations, so that the interference pattern should complete many cycles before disappearing into the noise. The problem is that, for sufficiently long distances, all of the measurable oscillations occur below the low angle cut-off of a typical small-angle X-ray scattering instrument. The solution to this problem is to decrease the low angle cut-off. This can be accomplished by changing the optical geometry of the existing X-ray beamline, improving the signal-to-noise of scattering interference data in proportion to the nanocrystal volume. An important application of the X-ray ruler will be the measurement of distance distribution within proteins.

2.4 MAGNETIC TWEEZERS

Optical and magnetic tweezers can be contrasted in how they apply force to probe particles. Whereas optical tweezers typically trap a probe particle in a potential well, magnetic tweezers apply a constant force to the probe particle (Lansdorp et al., 2013). Magnetic tweezers use the gradient of the horizontal magnetic field generated by small magnets in order to pull vertically on a magnetic bead (Smith et al., 1992) anchored to a surface by a DNA molecule. The force is obtained by measuring the amplitude of the Brownian fluctuations of the bead. The force varies typically between 0.01 pN and 100 pN. The time resolution of the technique is limited by the viscous drag of the sensor (magnet bead), and is of the order of 0.1 s. Magnetic tweezers offer a natural force clamp. Another useful feature of magnetic tweezers is that they offer a simple and noninvasive way to twist the anchored molecule simply by rotating the magnets around the vertical axis.

Optical tweezers are widely used for the micromanipulation of cells or biomolecules. However, they do not allow easy rotary motion (until the recent development shown in Gecevičius et al., 2014). Magnetic tweezers overcome that drawback. In the magnetic tweezers shown in Figure 2.3, electromagnets are coupled to a microscope-based particle tracking system through a digital feedback loop (Gosse and Croquette, 2002). Magnetic beads are first trapped in a potential well of stiffness 10^{-7} N/m, where they can be manipulated in three dimensions, at a speed of 10 μm/s, and rotated along the optical axis at

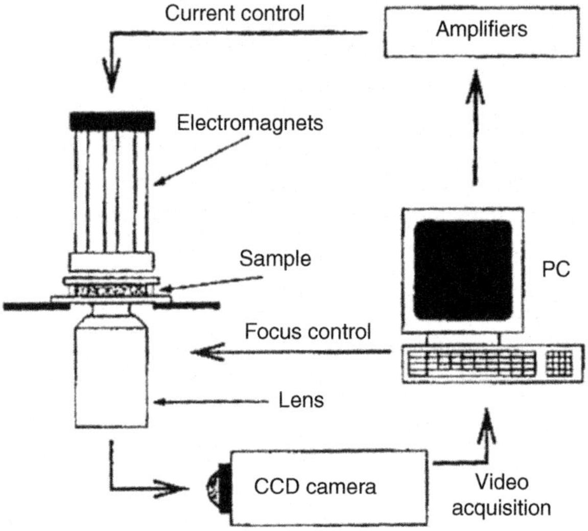

FIGURE 2.3 General magnetic tweezers setup. *(After Gosse and Croquette, 2002, reprinted with permission.)*

a frequency of 10 Hz. This apparatus can work as a dynamometer, relying either on usual calibration against the viscous drag, or complete calibration using Brownian fluctuation.

A measured vertical stretching force ranging from 50 fN to 20 pN was applied to a DNA molecule between a magnetic particle and a glass surface. The cell containing the magnetic particles in solution is held on the stage of an inverted microscope (lens in Fig. 2.3). The cell is a small capillary tube with a rectangular section. A system of six vertical electromagnets with their pole pieces arranged in a hexagonal pattern is placed just above the capillary tube. Parallel light illuminates the sample through a 2-mm diameter aperture, located at the center of the hexagon. An *xyz* translation stage allows the accurate positioning of the electromagnets, with respect to the optical axis of the objective. During micromanipulation, magnetic particles are located with nanometer accuracy, by video analysis.

The bead positioning software determines the DNA extension ΔL_{DNA} and the particle transverse fluctuations δx. Using the equipartition theorem, the vertical magnetic force F_{mag} can then be evaluated (Velthuis et al., 2010) through the formula

$$F_{\mathrm{mag}} = k_{\mathrm{B}} T \Delta L_{\mathrm{DNA}} / <\delta x^2>.$$

The high-speed magnetic tweezers (Lansdorp et al., 2013) have the ability to resolve particle position to within 1 Å at 100 Hz, and to measure the extension of 1566 bp DNA with 1 nm precision at 100 Hz, in the presence of thermal noise. The requirement for sufficient light at high frames rates is fulfilled with a super luminescent diode (SLD), and replacing the conventional 60 Hz CCD with a high-speed CMOS camera. The probe particles are typically tethered to the surface of a glass coverslip, and are subjected to forces between 0.1 pN and 50 pN. Video-based images of probe particles are sent to a computer processor to convert them into three-dimensional positions. As a result, the high-speed magnetic tweezers resolve accurately DNA hairpin dynamics that were unresolvable with the previous generations of magnetic tweezers. These new tweezers have a noise below 0.6 Å in the three dimensions, at a bandwidth of 100 Hz. Yu et al. (2014) developed a force calibration new standard for magnetic tweezers.

Chen et al. (2015) developed magnetic tweezers with high permeability electromagnets for fast actuation of magnetic beads.

2.4.1 Magnetic Influence Field

Sources and action. The sources of a magnetic field are (i) closed electric currents and (ii) magnetized bodies. In their action, the latter are equivalent to a collection of elementary magnets, or the elementary circuits of current, called dipoles. The mechanical action with the attractive force Q_{mag} of the magnetic field is determined by the magnetic inductance B_{mag} as $Q_{mag} = 4.06\, B_{mag}^2 A_{mag}$, or by a magnetic flux Φ_δ in the gap, as $Q_{mag} = 4.06\, \Phi_\delta^2 / A_{mag}$, where A_{mag} is the area through which the magnetic flux Φ_δ flows. In the engineering industry, the typical sources of significant magnetic fields are magnetic plates and other magnetic grippers: cubes, blocks, stand's bases, faceplates, chucks, jigs, etc. Magnetic plates are also a source of heat, equal to $Q_{hT} = 0.24\, I_E^2 R_r t$, where I_E is electrical current and R_r is the chain resistance.

2.4.2 Magnetostriction

High saturation strain has such materials (Joshi, 2000) as Ni (-50×10^{-6}, Curie temperature 630 K), Fe (-14×10^{-6}, 1040 K), $SmFe_2$ (-2340×10^{-6}, 690 K), $TbFe_2$ (2630×10^{-6}, 700 K), and $Tb_{0.6}Dy_{0.4}$ (6300×10^{-6}, 75 K). Here, Sm is samarium, Tb is terbium, and Dy is dysprosium. All of these chemical elements belong to the lanthanum group. In recent times, the giant magnetostrictive film material, such as Terefenol-D, was developed. A commercially available material is the Metglas ribbon (www.metglas.com).

2.5 ATOMIC FORCE MICROSCOPY

Stretching a single DNA by optical tweezers and magnetic beads is effective only in the case of relatively large, micrometer-size beads spheres, and long DNA chains of several tens of kilo base pairs.

The invention in 1986 of the scanning force microscope, also called the atomic force microscope (AFM), has equipped biologists with a powerful new tool of

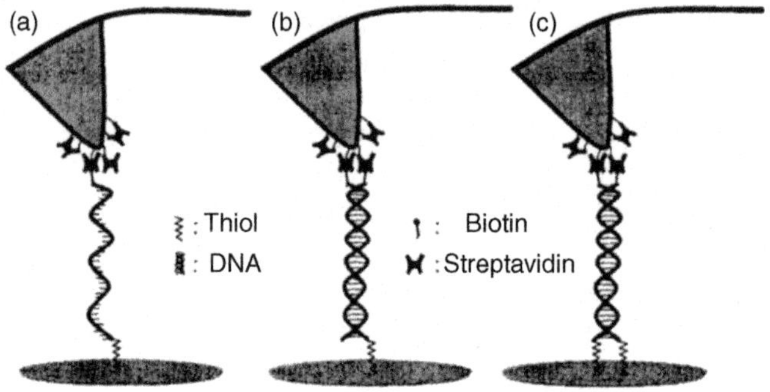

FIGURE 2.4 Schematic view of DNA stretching by streptavidin-modified AFM tips. An ssDNA modified with 3′-biotin and 5′-thiol (a), one strand of a dsDNA modified with 3′-biotin and 5′-thiol and one end of the other strand, modified with biotin, while the other end left free (b), and both strands are modified with 3′-5′-thiols and 5′-3′-biotins (c). The thiolated ends are attached to the Au-surfaces, while the biotinylated ends are captured by streptavidin tips. *(After Nguyen et al., 2010, reprinted with permission.)*

structural characterization of protein–nucleic acid interactions (Bustamante and Rivetti, 1996). AFM can measure very small forces (in subpicoNewtons), small displacements (in subnanometers), and even provide chemical identification of individual surface atoms (Sugimoto et al., 2007).

The AFM uses a small cantilever to pull on a molecule (DNA, protein) bound at its other end to a surface. The applied force is measured through the deflection of the cantilever, and ranges from 5 pN to a few nanoNewtons. Small cantilevers have very good spatial and time resolutions.

Recently, AFM, with the capability of acquiring measurements in the picoNewton range, has been widely used to measure both inter- and intramolecular forces directly. Using these techniques, the structural transition and the stiffness of even small fragments of the dsDNA molecules can be measured (Nguyen et al., 2010). Figure 2.4 shows the schematics for the stretching of ssDNA (a) that is modified with 3′-biotin and 5′-thiol, one strand of a dsDNA that is modified with 3′-biotin and 5′-thiol, and one of the other strands is modified with biotin, while the other strand is left free (b), and (c) where both strands are modified with 3′-5′-thiols and 5′-3′-biotins. The latter is double fixed. The flat surface is covered with bare gold at flatness within 1 nm. AFM, in these cases, is functioning in the tapping mode.

Clausen-Schaumann et al. (2000) studied the mechanical stability of the single DNA molecule with AFM as a function of stretching velocity.

2.5.1 Micro-Nano-Cantilevers

Used in AFM, micro–nanocantilevers are of different shapes: (i) rectangular (Fig. 2.5a); (ii) triangular (Fig. 2.5b), (iii) V-shaped with the triangular (Fig. 2.5c) and trapezoid (Fig. 2.5d) recesses, double-taper, and wedge type.

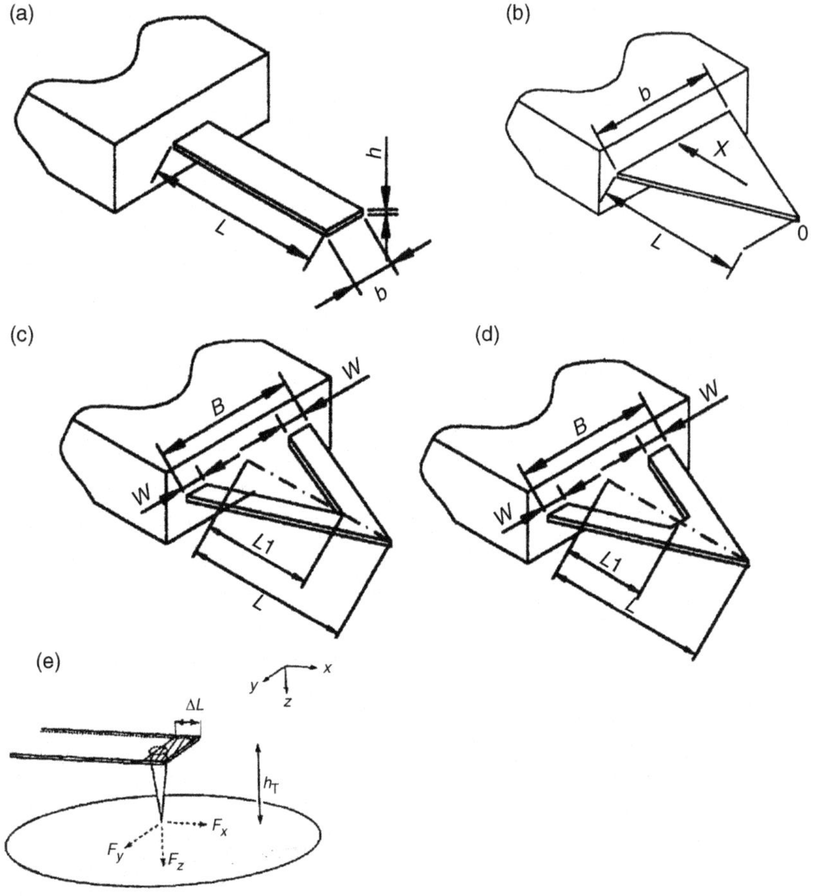

FIGURE 2.5 Cantilevers with different structures. (a) Rectangular; (b) triangular; (c), (d) V-shaped; and (e) with probe.

These cantilevers are usually used in the static regime, and at vibration with natural frequency. Let us estimate the natural frequency of solid cantilevers.

Triangular cantilevers' natural frequency. This can be estimated by Rayleigh's method, with application of a force to the movable end of the cantilever, in which case the flexure curve is represented by the following formula:

$$X(x) = \frac{QL^3}{6EJ}\left(\frac{x^3}{L^3} - \frac{3x}{L} + 2\right),$$

where L is the length of the cantilever (Fig. 2.5b), x is the longitudinal coordinate with the origin at the movable end, and the flexural moment of sectional inertia equals $J = b(x)h^3/12$, where $b(x) = bx/L$ and h is the thickness of the cantilever.

In accordance with Rayleigh's method, the square of the natural frequency of the triangular cantilever is equal to:

$$f_0^2 = \frac{1}{(2\pi)^2} \frac{Q^2L^3/3EJ}{\int_0^L \rho_\gamma b \frac{x}{L} h(QL^3/6EJ)^2 \left(\frac{x^3}{L^3} - \frac{3x}{L} + 2\right)^2 dx} = \frac{68.571EJ}{(2\pi)^2 \rho_\gamma bhL^4}$$

because the integral $\int_0^L I_f = \int_0^L \frac{x}{L}\left(\frac{x^3}{L^3} - \frac{3x}{L} + 2\right)^2 dx = 0.175L$. If one assumes that the radius of gyration $r_g^2 = J/bh = h^2/12$ of the triangular cantilever cross-sectional area is constant, we obtain the following expression for the fundamental resonant frequency of the solid triangular cantilever:

$$f_0 \approx \frac{h}{0.837\pi L^2}\sqrt{E/\rho_\gamma} \approx 0.38\frac{h}{L^2}\sqrt{E/\rho_\gamma}. \tag{2.3}$$

A similar expression may be found in Weaver et al. (1990), where the following formula is presented for the cantilever beam with variable cross-section form:

$$f_1 = \frac{\alpha_j r_g}{2\pi L^2}\sqrt{E/\rho_\gamma}. \tag{2.4}$$

Here, r_g is the radius of gyration of the built-in section, α_j is a constant depending on the shape of the beam and on the mode of vibration. If the variation of the cross-sectional area and of the moment of inertia along the x axis can be expressed in the forms

$$A_c = A_0\left(1 - c_j\frac{x}{L}\right) \text{ and } J = J_0\left(1 - c_j\frac{x}{L}\right),$$

typical for the triangular cantilever at $c_j = 1$ with x being measured from the built-in end, then the quantity $\alpha_j = 7.16$. Substituting this quantity and expression for the constant r_g in the previous formula (2.4), we obtain the natural frequency of the triangular cantilever that equals

$$f_1 = \frac{h}{0.967\pi L^2}\sqrt{E/\rho_\gamma} \approx 0.33\frac{h}{L^2}\sqrt{E/\rho_\gamma}. \tag{2.5}$$

Expression (2.3) differs within 15% from the presumably more accurate result (2.5). This difference may be caused by the insufficiently accurate approximation of the elastic curve used in Rayleigh's method. The numerical coefficients on the right-hand side of these expressions are close to the coefficient 0.31 that has been chosen for the curve fitting the plot to the results obtained for the triangular cantilever by FEA, with ANSYS software, where the experimental results, the analytical plot, and the ANSYS plot are matched within 10%. If the cantilever has a tapering trapezoid form, then the corresponding quantity c_j and the natural frequency of the cantilever are smaller than for the triangular one with the sharp, free end. The interested reader can find the fundamental frequency coefficients for spring-hinged cantilevers and single- and double-tapered beams in the publication by Magrab (1979).

2.5.2 Cantilever Preferred Shape

V-shaped cantilevers. These cantilevers have a fundamental frequency intermediate between those of rectangular and triangular cantilevers. We can find many publications devoted to this subject because V-shaped cantilevers are widely used in AFM. The simplest theoretical model describes a V-shaped cantilever by two parallel rectangular cantilevers. Some authors showed good agreement between this model and experimental data. But this model does not reflect the specifics of V-shaped cantilevers' geometry. Other publications are devoted to more complicated and presumably more accurate models: (i) "correction factor" with numerical integration (Butt et al., 1993) and (ii) "variational" representation (Chen et al., 1994, 1995). Let us recall that a proper calculation of the cantilever's fundamental frequency is very important for the synthesis of a precision system in AFM. It should be noted, however, that not all of the more complicated models demonstrate notable increase in accuracy and most of them are difficult to use for structural synthesis. Therefore, we develop a simple model for a V-shaped cantilever that reflects the specifics of its geometry and provides satisfactory accuracy. This model may be built by Rayleigh's method, on a similar basis as we did for a solid triangular cantilever because the radius of gyration remains constant for all cross sections in the model (Fig. 2.5c,d). The main parameters of a V-shaped cantilever that have influence on its fundamental frequency are thickness h, overall length L, recess length L_1, Young's modulus E, and density ρ_γ of material. The following formula reflects all of these parameters:

$$f_0^2 = \frac{1}{(2\pi)^2}\frac{F\dfrac{FL^3}{3EJ}}{\rho_\gamma Bh\int\limits_0^{L-L_1} I_\mathrm{f} + \int\limits_{L-L_1}^{L}\rho_\gamma B\left(1-\dfrac{L_1}{L}\right)h\left(\dfrac{FL^3}{3EJ}\right)^2\left(\dfrac{x^3}{L^3}-\dfrac{3x}{L}+2\right)^2 dx},$$

where

$$\int\limits_0^{L-L_1} I_\mathrm{f} = \int\limits_0^{L-L_1}\frac{x}{L}\left(\frac{x^3}{L^3}-\frac{3x}{L}+2\right)^2 dx = 0.175(L-L_1) \text{ and}$$

$$\int\limits_{L-L_1}^{L}\left(\frac{x^3}{L^3}-\frac{3x}{L}+2\right)^2 dx = 0.943L_1.$$

Hence, after substitutions and cancellations, we finally obtain a quasiuniversal and simple enough formula for the fundamental frequency of V-shaped cantilevers, as follows (Tseytlin, 2006):

$$f_0 = \frac{h}{2\pi L^2}\sqrt{\frac{E}{\rho_\gamma}\frac{1}{(0.175+0.768L_1/L)}}. \tag{2.6}$$

One can use this quasiunified formula for a solid triangular cantilever, if we assume that $L_1 = 0$, and for the rectangular cantilever if we assume that $L_1 = L$. The calculation results on the basis of Eq. (2.6) agree to within a few percent with the experimental data and more complicated theoretical methods presented by other authors.

The use of V-shaped microcantilevers has been recommended in AFM application, as they were assumed to be less susceptible to lateral forces than rectangular ones. However, Sader (2003), has found in his study that V-shaped cantilevers are generally more prone to the effects of lateral forces than rectangular AFM cantilevers. It is an important and interesting conclusion that should be related to the moments from the lateral forces because the lateral spring constants k_x and k_y in Sader's formulae depend on the height h_T of a probe with tip (Fig. 2.5e). Later, however, Sader and Green (2004), on the basis of regular formulae for the bending of rectangular beams and wedges (Timoshenko, 1955), have found that the comparison of the in-plane deformation of rectangular and V-shaped cantilever plates reveals the superiority of a V-shaped cantilever that has a higher spring constant in this direction. Let us recall that, unlike the wedge, a V-shaped cantilever has some cut-off (Fig. 2.5c) that requires solving a special differential equation of the cantilever bending, as follows:

$$y''(x) = \frac{12M}{EhB^3}\left\{\left(1-\frac{x}{L}\right)^3 - \left(1-\frac{x}{L_1}\right)^3\left(1-\frac{2w}{B}\right)^3\left[1-U(x-L_1)\right]\right\}^{-1}.$$

The unit-step function $U(x-L_1) = 0$ at $x < L_1$ and is equal to one, at $x \geq L_1$. If the cut off is assumed to be absent, $w = B/2$, and the second member in braces becomes equal to zero because $2w/B = 1$. Only in this case, the wedge model will be an exact one. We should also note that in-plane bending of the thin rectangular strip can cause its buckling.

Another aspect of the problem that should be considered is the part of the cantilever that remains in the space beyond probe (shaded area in Fig. 2.5e). This part is assumed to be a rigid one, but it has a certain surface and mass that will be many times larger for a rectangular cantilever, than for a V-shaped cantilever at the same distance ΔL from the axis of the tip to the end of the cantilever. This may cause a larger damping in the ambient air, especially in a dynamic AFM. Note that ΔL cannot be equal to zero in any cantilever's shape, due to the finite diameter of the probe base. It may conditionally be considered zero for the position of clamped spheres, for calibration purposes.

Frequency. A V-shaped cantilever (Fig. 2.5c) with a tip has fundamental resonant frequency often expressed by the formula

$$f_0 = \frac{0.162\,h}{L^2}\sqrt{E/\rho_\gamma}$$

and the following representation, $f_0 = \frac{1}{2\pi}\sqrt{k_{ef}/m_{ef}}$, where $k_{ef} = (EJ/3L^3) - F'$ is the effective bending spring constant (Arutyunov and Tolstikhina, 2002), F' is the force gradient normal to the surface, $m_{ef} = m_c + n^*m_{cn}$ is the effective mass, m_c is the probe mass, m_{cn} is the cantilever mass with mass factor n^*. The resonance frequency is altered upon varying the force gradient F' because the force gradient affects the effective rigidity. If, for example, there is an attractive force, $F' > 0$, the resonant frequency is reduced. The resonant frequency is increased for $F' < 0$. One can thus record any change in f_0 to measure the force gradient in the interaction between the probe and the surface. Let us note that, in the previously mentioned reference, the accepted value of mass factor $n^* = 0.24$ is the same as for a rectangular cantilever because many authors assume that V-shaped cantilever frequency response is similar to the response of the corresponding two parallel positioned rectangular cantilevers (Butt et al., 1993). However, it was found in experimental studies, and on the basis of a developed "variational" method (Chen et al., 1994, 1995), that, for commercial V-shaped cantilevers, n^* ranges from 0.137 to 0.2. We will investigate this problem in more detail.

2.5.3 Effective Mass Factor

Let us recall that the resonance frequency f_0 of the cantilever may be presented through its bending spring constant K_{ct} and its effective mass, as follows, $f_0 = \frac{1}{2\pi}\sqrt{\frac{K_{ct}}{n^*m_{cn}}}$, where m_{cn} is the cantilever's mass and n^* is a mass factor that depends on the cantilever's shape. The mass of the V-shaped cantilever may be presented as $m_{cn} = Lh\,\rho_\gamma[\tfrac{1}{2}B(1-\zeta_L)^2 + 2w\zeta_L]$ and the mass factor n^* can be found from the equation $n^* = K_{ct}/[m_{cn}(f_0/2\pi)^2]$. Substituting the expressions for spring constant, mass of cantilever, and its fundamental frequency found through (2.6), we obtain, after simple cancellations, the following quasiunified formula for the V-shaped cantilever's mass factor that is equal to:

$$n^* = \frac{2(0.175 + 0.768\zeta_L)}{\left[4\zeta_L + (1-\zeta_L)^2 B/w\right]\left[\frac{6(1-\zeta_L)^2}{B/w} + 2\zeta_L\left(\zeta_L^2 - 3\zeta_L + 3\right)\right]}.$$

This expression for the rectangular cantilever with $\zeta_L = 1$ yields a commonly used value, $n^* = 0.24$.

If $\zeta_L < 1$ for the triangular cantilever with a trapezoidal recess, then the mass factor n^* becomes significantly smaller than 0.24.

TABLE 2.1 Fundamental Frequency of Silicon V-Shaped Cantilevers

Parameter	Data			
L, μm	190	200	190	200
L_1, μm	87.5	115	87.5	115
B, μm	165	164	165	165
w, μm	33.3	20	33.5	20
h, μm	0.6	0.6	0.3	0.3
*K, N/m	0.094	0.05	0.012	0.006
*f_0 (exp.), kHz	31.50	25.80	17.46	13.10
*f_0 (var.), kHz	34.00	25.80	17.30	12.90
f_0 (2.6), kHz	31.65	26.38	15.82	13.20
δf_0 (2.6, exp.), %	0.5	2.3	−9.4	0.8

All dimensional notations correspond to Fig. 2.5c,d.
*Data for these parameters: silicon modulus of elasticity $E = 179$ GPa, its density $\rho_\gamma = 2330$ kg/m^3, experimental values f_0 (exp.), and the calculation results of the "variational" method f_0 (var.) are the same as presented by Chen et al. (1994); f_0 (2.6) represents the calculation results by formula (2.6), and δf_0 (2.6, exp.) = 100 [f_0 (2.6) − f_0 (exp.)]/f_0 (exp.).
Source: After Tseytlin (2006), reprinted with permission.

We assume that the "variational" method, as well as the "correction factor" method (Butt et al., 1993), are effective but not easily used and require a numerical integration. Therefore, we prefer a more tractable approach. Let us recall that we derived a simple formula (2.6), for the calculation of the fundamental frequency for the straightforward design of cantilevers. Table 2.1 presents the comparison of the results of the fundamental frequency calculation by the different methods and experimental data for silicon V-shaped cantilevers.

As we can see, the largest differences between the experimental data and calculation results by formula (2.6) do not exceed 9.4%. There is no influence of the width parameters (B, w) on the fundamental frequency of the V-shaped cantilever. This, of course, follows from the assumed model with the constant cross-sectional radius of gyration, and is clearly manifested in formula (2.6).

Composite cantilever. Formula (2.6) can be used for the calculation of the composite cantilevers as well, but, in this case, one should estimate the composite's modulus of elasticity and the average density (Lange et al., 2002) of the composite.

For example, if a V-shaped cantilever (Fig. 2.6a) of silicon nitride, with thickness $h_{si} = 0.54$ μm, Young's modulus $E_{si} = 150$ GPa, and density $\rho_{\gamma si} = 3{,}100$ kg/m^3 is coated with gold film ($E_{gf} = 80$ GPa, $\rho_{\gamma gf} = 19{,}300$ kg/m^3, and thickness $h_{gf} = 0.04$ μm), then its composite elastic modulus E_c can be estimated from the following expression:

$$E_c = E_{si} + E_{gf} J_{gf} / J_{si},$$

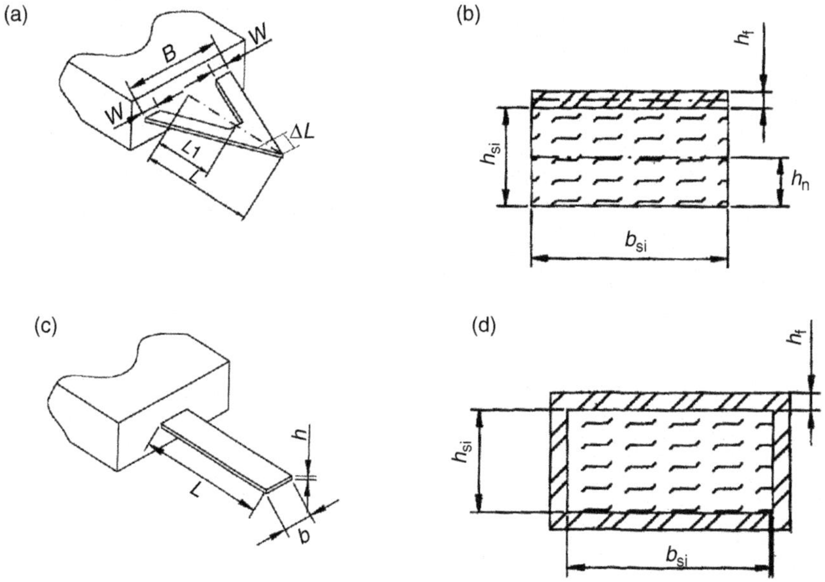

FIGURE 2.6 Composite microcantilevers. V-shaped (a) with film layer (b) and rectangular (c) with film cover (d).

where J_{gf} and J_{si} are the moments of inertia with respect to the common axis of the cantilever's cross section, E_{gf} and E_{si} are the apparent Young's moduli of gold film and silicon nitride, respectively.

The position h_n of the neutral layer (Fig. 2.6b) for a composite beam with the film layer may be estimated similar to the centroid or center of mass, such that

$$h_n = \frac{E_{si}h_{si}\dfrac{h_{si}}{2} + E_{gf}h_{gf}\left(h_{si} + \dfrac{h_{gf}}{2}\right)}{E_{si}h_{si} + E_{gf}h_{gf}} = \frac{E_{si}h_{si}^2 + E_{gf}(2h_{gf}h_{si} + h_{gf}^2)}{2(E_{si}h_{si} + E_{gf}h_{gf})},$$

and, therefore,

$$h_n/h_{si} = \frac{E_{si}h_{si} + E_{gf}(2h_{gf} + h_{gf}^2/h_{si})}{2(E_{si}h_{si} + E_{gf}h_{gf})}.$$

After substituting the numerical data for all parts of this equation, one can find that $h_n = 0.52\ h_{si} \approx ½h_{si}$. We will obtain a similar result if we suppose that the thin nanostructured gold film has slightly different Young's modulus that is equal to 69.1 GPa (Salvadori et al., 2003).

Let us also recall that the moment of inertia with respect to different parallel axes differs by product of cross-sectional area, and a square of the distance between the respective axes. Hencc, we have the following relationships with respect to the neutral axis in the middle of the silicon layer $J_{si} = \frac{bh_{si}^3}{12}$; $J_{gf} = \frac{bh_{gf}^3}{12} + bh_{gf}\left(\frac{h_{si}+h_{gf}}{2}\right)^2 = \frac{b}{12}\left(4h_{gf}^3 + 3h_{gf}h_{si}^2 + 6h_{si}h_{gf}^2\right)$; and

$J_{gf}/J_{si} = 3\frac{h_{gf}}{h_{si}}\left[1 + 2\frac{h_{gf}}{h_{si}} + \frac{4}{3}\frac{h_{gf}^2}{h_{si}^2}\right]$.

Therefore, after substituting all of the constituents' data, one finds that $E_c \approx 170.5$ GPa for the silicon nitride cantilever coated with the gold film. The average density $\rho_{\gamma c}$ of this composite cantilever is equal to:

$$\rho_{\gamma c} = \frac{h_{si} \times \rho_{\gamma si} + h_f \times \rho_{\gamma gf}}{h_{si} + h_f} = 4196.6\ \text{kg/m}^3.$$

Table 2.2 shows the calculated fundamental frequency for composite V-shaped cantilevers of silicon nitride coated with gold film (Fig. 2.6a,b) by the following formula for the composite cantilever:

$$f_{0c} = \frac{h_{si}}{2\pi L^2}\sqrt{\frac{E_c}{\rho_{\gamma c}} \frac{1}{0.175 + 0.768 L_1/L}}. \tag{2.7}$$

TABLE 2.2 Fundamental Frequency of V-Shaped Composite Cantilevers of Silicon Nitride Coated with Gold Film

Parameter	Data		
L, μm	202	195	119
L_1, μm	116	145	61
B, μm	205	198	121
W, μm	40.2	21.2	24.1
h (Si_3N_4), μm	0.54	0.54	0.54
h_f (gold film)	0.04	0.04	0.04
*K, N/m	0.058	0.034	0.17
*f_0 (exp.), kHz	18.25	14.88	50.28
*f_0 (var.), kHz	18.35	16.26	51.93
f_0 (2.7), kHz	17.1	16.65	51.19
δf_0 (2.7, exp.), %	−6.3	11.8	1.8
δf_0 (2.7, var.), %	−6.8	2.4	−1.4

All dimensional notations correspond to Fig. 2.6a,b.
*Data for these parameters, silicon nitride (Si_3N_4) modulus of elasticity E_{si} = 150 MPa, its density $\rho_{\gamma si}$ = 3,100 kg/m^3, gold film modulus of elasticity E_{gf} = 80 GPa, its density $\rho_{\gamma gf}$ = 19,000 kg/m^3, experimental values f_0 (exp.) and the calculation results of the "variational" method f_0 (var.) are the same as presented by Chen et al. (1994); f_0 (2.7) represents the calculation results by formula (2.7) that is the version of formula (2.6) for composite cantilever, δf_0 (2.7, exp.) = 100[f_0 (2.7) − f_0 (exp.)]/f_0 (exp.); and δf_0 (2.7, var.) = 100[f_0 (2.7) − f_0 (var.)]/f_0 (var.).
Source: After Tseytlin (2006), reprinted with permission.

If the composite rectangular cantilever is coated uniformly on all its sides (Fig. 2.6c,d), then the position of its neutral axis remains in the center of the cross section, and the composite Young's modulus can be estimated by the following expression: $E_c = E_{si} + 2E_{g(p)f}\left[\frac{J_f}{J_{si}} + \frac{(h_{si}+2h_f)^3 h_f}{b_{si}h_{si}^3}\right]$, where h_{si} and h_f are the thickness of the cantilever and the coating film, respectively; E_{si} = 162 GPa for a silicon cantilever, E_{gf} = 69.1 ± 2.6 GPa for a gold film, and E_{pf} = 139.7 ± 2.7 GPa for a platinum film; in addition, $J_f = \frac{b_{si}h_f^3}{12} + b_{si}h_f\left(\frac{h_{si}+h_f}{2}\right)^2$ is the moment of inertia for the film cross section on one wide side of the cantilever, $J_f / J_{si} = h_f\left[h_f^2 + 3(h_{si}+h_f)^2\right]/h_{si}^3$, and b_{si} is the width of the cantilever without coating. The average density of this composite cantilever can be estimated as follows:

$$\rho_{\gamma c} = \rho_{\gamma si} + 2\rho_{\gamma f}\frac{h_f}{h_{si}b_{si}}(b_{si} + h_{si} + 2\,h_f),$$

where $\rho_{\gamma si}$ = 2.33 g/cm^3 and $\rho_{\gamma f} = \rho_{\gamma gf}$ = 19.32 g/cm^3 for gold film and $\rho_{\gamma f} = \rho_{\gamma pf}$ = 21.44 g/cm^3 for the platinum film are the density of the cantilever and the film materials, respectively.

A comparison of the resonance frequency calculations with formula (2.7) and experimental data (Salvadori et al., 2003) for a cantilever coated with gold or platinum f_c and uncoated f_u is shown in Table 2.3, and Figure 2.7a,b, where the experimental data are given by circles, and the solid curve shows the theoretical results as a function of the gold (Fig. 2.7a) and platinum (Fig. 2.7b) film thickness h_f, respectively. The silicon cantilever dimensions, in the case of gold coating, are presented in Table 2.3, and, for the platinum coating, they are as

TABLE 2.3 Uncoated (f_u) and Coated (f_c) Cantilever Resonance Frequency

h_f (nm)	* f_c (exp.) (kHz)	f_c (2.7) (kHz)	100|f_c (exp.) − f_c (2.7)|/f_c (exp.) (%)
18.78 ± 0.4	300.849	298.4	0.53
28.55 ± 0.25	294.537	293.61	0.13
39.68 ± 0.32	288.612	288.5	0.04
54.19 ± 0.51	283.135	282.32	0.29
61.62 ± 0.40	277.756	279.28	0.55

Si coated with Au, uncoated f_u (exp.) = 308,745; f_u (2.7) = 308.519 kHz. The silicon cantilever's for gold coating dimensions are L = 136.9 μm, h_{si} = 4.23 μm, b_{si} = 39.38 μm; and h_f is the thickness of gold coating.
Experimental data for f_c (exp.) resonance frequency are from Salvadori et al. (2003).
Source: After Tseytlin (2006), reprinted with permission.

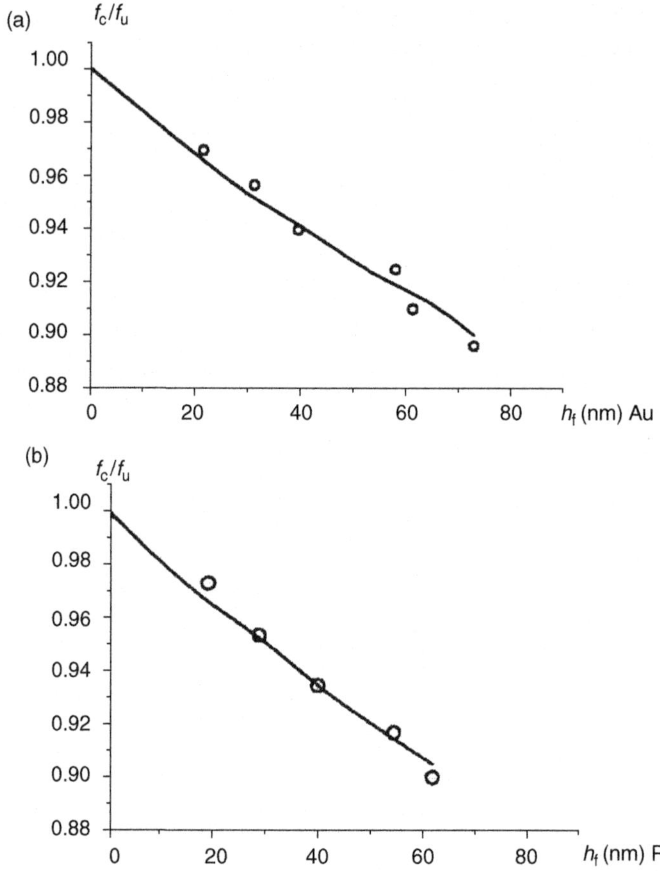

FIGURE 2.7 Theoretical and experimental charts of ratio for coated and uncoated cantilevers' frequencies. (a) Coated with gold film, and (b) coated with platinum film. *(After Tseytlin, 2006, reprinted with permission.)*

follows: $L = 138.8\ \mu m$, $h_{si} = 4.41\ \mu m$, and $b_{si} = 41.23\ \mu m$. The bias between the theoretical calculations by formula (2.7) and the experimental data is not more than 0.6% in all cases presented in Table 2.3, and in Figure 2.7a,b.

Young's modulus and density measurement of thin atomic layer deposited films can be estimated by using resonant vibrations in two directions: out of plane and in plane (Ilic et al., 2010).

Sensitivity to additional mass. Let us now estimate the shift of the resonant frequency for the system of a cantilever with an additional mass Δm clamped to its free end (probe, specimen, etc.). In this case, we have to represent the shift $\Delta f_0 = f_0(m_c) - f_0(m_c + \Delta m)$ of the resonance frequency $f_0(m_c)$ in the system with the cantilever that has mass m_c, and the resonance frequency $f_0(m_c + \Delta m)$ with the additional mass Δm attached to the cantilever. It is convenient first to express the difference of these frequencies' squares through the bending spring constant K_{ct} of the cantilever and the corresponding masses, as follows:

$$f_0^2(n^*m_c) - f_0^2(n^*m_c+\Delta m) = \left(\frac{1}{2\pi}\right)^2\left(\frac{K_{ct}}{n^*m_c} - \frac{K_{ct}}{n^*m_c+\Delta m}\right).$$

If $(\Delta m/m_c) \ll 1$, the following approximation is helpful:

$$f_0^2(n^*m_c) - f_0^2(n^*m_c+\Delta m) = \Delta f_0\left[f_0(n^*m_c) + f_0(n^*m_c+\Delta m)\right] \approx 2f_0\Delta f_0(n^*m_c).$$

Therefore, we can find the simplified expression for the shift of frequency Δf_0 for the rectangular cantilever, in the form:

$$2\pi\Delta f_0 = \frac{K_{ct}}{4\pi f_0}\frac{\Delta m}{n^*m_c(n^*m_c+\Delta m)} \quad \text{or} \quad \Delta f_0/\Delta m \approx 2.08 f_0/m_c.$$

A similar expression for a V-shaped cantilever has a larger numerical factor before the ratio f_0/m_c as a result of the lesser mass factor n^*. The effective mass factor for a doubly clamped prismatic beam is equal to 0.37.

2.5.4 The Largest Attainable Natural Frequencies

Old estimates. The natural frequency f_0 for a beam can be represented by the expression

$$f_0 \approx (1/2\pi)\sqrt{F/\delta_s m_{eq}}, \tag{2.7a}$$

where F is the force applied to the beam, δ_s is the sag, m_{eq} is the point-load mass equivalent to the elastic element's mass of unit length. The formulae presented in Table 2.4 for different elastic elements were obtained by substituting the expressions of δ_s and m_{eq} into (2.7a).

The stress σ_{max} is the largest acceptable stress for the material under working conditions: σ_e is the yield stress, σ_t is tensile (UTS) stress, or σ_{en} the endurance limit.

The last column in Table 2.4 shows the relationship between natural frequency f_0, minimal sag δ_{smin}, and maximum stress σ_{max} without any elastic element's geometrical dimensions figuring in these formulae. It is important to note that, in the previous reasoning, a complete utilization of the elastic element's strength characteristics was assumed.

A similar procedure was used for the round plane diaphragms subjected to a distributed load (pressure), whose formulae are also shown in row 4 of Table 2.4. Analysis of the formulae in Table 2.4 shows that the product of the natural frequency f_0 of the elastic element, and its minimal sag δ_{min}, does not depend on the element's size, but is completely determined by the selected material features. Moreover, this product is structurally the same for all studied

TABLE 2.4 Formulae for the Largest Attainable Natural Frequency

Structure's number, name, and load	Parameters of dimensions and loading*	Natural frequency, f_0	Product, $f_0 \times \delta_{smin}$*
1. Prismatic cantilever, concentrated force	h – thickness; b – width; L – length; F – end force	$(h/1.97\pi L^2)\sqrt{E/\rho_\gamma}$ **	$\sigma_{max}/(3\pi\sqrt{E\rho_\gamma})$
2. Prismatic cantilever, distributed load	q – distributed load	$(h/1.2\pi L^2)\sqrt{E/\rho_\gamma}$ *	$\sigma_{max}/(4\pi\sqrt{E\rho_\gamma})$
3. Doubly clamped beam, distributed load	q – distributed load	$(h/0.3\pi L^2)\sqrt{E/\rho_\gamma}$ †	$\sigma_{max}/(3.2\pi\sqrt{E\rho_\gamma})$
4. Clamped round plate with diameter $2R$ and distributed load	q – distributed load; h – thickness	$q = gh\rho_\gamma$; $(h/2R^2)\sqrt{E/\rho_\gamma} \times \gamma_f$, $\gamma_f = 1$, or ‡	$\sigma_{max}/(7.8\sqrt{E\rho_\gamma})$

*see Dolitskii and Fedorenko (1972), Tseytlin (2006), for more detail.
**This formula is based on the frequency equation cos αL cosh$\alpha L = -1$ with the first root $\alpha L = 1.875$.
†The formula is based on the frequency equation cosαLcosh$\alpha L = 1$ with the first root $\alpha L = 4.730$ (Panovko, 1968); if the beam has one fixed and one simply supported end as in the contact mode of AFM its frequency equation is tanαL = tanhαL with the first root $\alpha L = 3.927$, and the natural frequency of the simply supported beam increases with addition of axial tensile force Q_{ax} and decreases with the axial compression force $-Q_{ax}$ in accordance with (Timoshenko and Young, 1964) the following expression: $f_o = \frac{hn_f^2 L^{-2}}{0.702\pi}\sqrt{E/\rho_\gamma}\sqrt{1+\frac{Q_{ax}L^2}{n_f^2 EJ\pi^2}}$, where we accept $J = bh^3/12$ and n_f is the mode number.
‡Influence of surrounding air or liquid (Weaver et al., 1990) with density $\rho_{\gamma 1}$ reflects in the factor $\gamma_f = 1/\sqrt{1+\beta_f}$, where $\beta_f = 0.6689\frac{\rho_{\gamma 1}}{\rho_\gamma}R/h$.

elastic systems. The computation's results of f_0 for elastic elements made of the different materials (steel, beryllium bronze, titanium, and aluminum alloys), in the case when minimal sag $\delta_{smin} = 10$ μm, show that the natural frequencies are in the range from 50 kHz to 113 kHz. However, we should recall that the value of the minimal sag was selected, in this case, as a typical for the pressure electromechanical transducers of different types known in the 1970s. It should be noted that this value is a conditional one, and is different in new structures that have much higher fundamental frequencies. The largest natural frequency for the flat strings can be defined from the expression

$$f_n = 0.5n_f L^{-1}\sqrt{\{\sigma_e\}/\rho_\gamma}, \tag{2.7b}$$

where n_f is the harmonic's number and $\{\sigma_e\}$ is the permissible yield stress for the string's material. Then, at the minimum value of the string's length

$L = 1$ mm (this number is conditional with respect to the mesoscopic structure's options), the largest yield stress value $\{\sigma_e\} \approx 1800$ MPa of the elastic materials in the bulk state, and the density of the material $\rho_\gamma \approx 8\times10^{-6}$ kg/mm^3, we obtain $f_{n=1} \leq 230$ kHz.

Revisited estimates. Today, these options have made large progress in the micro- and nanotechnology field. The nanowire-based very-high-frequency electromechanical resonator is already fabricated with a natural (fundamental) frequency of 105.3 MHz (Husain et al., 2003). The resonator is built of suspended platinum nanowire, and has a diameter of $d_w = 2R_d = 43$ nm, length of $L = 1.3$ μm with a suspended mass of 40 fg, and a volume of 1.9×10^{-15} cm^3. This microresonator is among the smallest NEMS structures whose motion has been detected directly. Advanced sensing applications require both high responsivity and ultrahigh frequency operation. Among the most challenging of these problems are the attainment of subattoNewton, high frequency force sensing for magnetic resonance force microscopy, and the study of mechanical motion in the quantum regime. Let us recall that a platinum microstrip has the yield stress σ_e = 1,570 MPa, Young's modulus E = 169 GPa, and density ρ_γ = 21,090 kg/m^3. If we assume that the microresonator is a flat string, its maximum fundamental frequency estimated by (2.7b) is equal to: $f_0 = \dfrac{1}{2\times1.3\times10^{-6}}\sqrt{\dfrac{1570\times10^6\times9.8}{210900}} = 103.9$ MHz. This number is within 1.3% of the value measured by Husain et al. (2003) for the nanowire resonator's natural frequency. A very simple first modeling by Husain et al. (2003) of the response of this microresonator as an unstrained doubly clamped beam (Timoshenko and Young, 1964) yields the fundamental frequency

$$f_0 = \frac{22.4}{2\pi}\frac{R_d}{2L^2}\sqrt{\frac{E}{\rho_\gamma}} = 64 \text{ MHz},$$

which is much lower than the measured response. The authors (Husain et al., 2003) realized, however, that this is likely "due to the differential thermal contraction of the beam and the substrate, which should cause the beam to be under tension." Therefore, the correct modeling of this resonator's response is a string version with formula (2.7b) or with the formula for a beam under tension (see notes in Table 2.4). This example shows how important the correct theoretical model is for structural analysis and design. Indeed, the fundamental frequency for nanoresonators can be even higher (Section 3.7.2) because a whisker of many materials has much higher mechanical properties than in a bulk state.

Magnetomotive detection, in which an ac current drives a string in a transverse magnetic field, has been employed to actuate and read successfully NEMS resonators at frequencies up to, and exceeding, 1 GHz. The measured quality Q-factor of the microresonator is approximately 8500–8800 in vacuum, and decreases slightly with an increasing magnetic field from 1 T to 8 T. Previous studies have shown that quality factors measured for NEMS generally decrease with an increasing surface area–volume ratio, apparently indicating that surface processes contribute strongly to energy dissipation. Carr et al. (1999), for example, measured the quality factors of single-crystal silicon beams, and found that the Q-factor decreases from 3000 for beams with a surface–volume ratio $0.02\ \text{nm}^{-1}$ to a value 1000 for devices with the ratio of $0.06\ \text{nm}^{-1}$. The response of the platinum nanowire resonator, as the frequency sensor is swept both upward and downward, confirms the expected hysteretic nature in the nonlinear regime. In the simplest model for the phenomenon, the critical oscillation amplitude x_c above which instability occurs is dependent only upon the geometry, Q-factor, and Poisson's ratio v_p of the beam, and is given by the expression $x_c = \frac{h_{ge}\sqrt{2}}{\sqrt{0.528Q\left(1-v_p^2\right)}}$ (Landau and Lifshitz, 1969), where for round beam, $h_{ge} = d_w\sqrt{3/4}$ and d_w is the diameter of the beam; for the rectangular cross-section beam, $h_{ge} = h$ and h is the thickness of the beam at equal normal stress. The critical amplitude for the platinum microresonator evaluated on this basis is in the order of nanometers that pose a significant challenge for detecting the motion of nanowire and nanotube devices. See also Noeth et al. (2014) on integrated cantilever.

High frequency example. The natural frequency of a nanocantilever of chromium with the length $L = 2\ \mu\text{m}$, width $b = 150$ nm, thickness $h = 50$ nm, $E = 248$ GPa, and density of 7.19 g/cm^3, on the basis of the formula for version 2 in Table 2.4, is equal to 19.3 MHz, a value significantly larger than was calculated for version 2 of a mechanical system with regular sizes. The increased natural frequency, in comparison with the accurate solution for the rectangular cantilever, has a sharp-end wedge cantilever whose natural frequency one can estimate on the basis (Weaver et al., 1990) of the following formula: $f_0 = \frac{h_r/2}{0.65\pi L^2}\sqrt{E/\rho_\gamma}$, where h_r is the height of the wedge in the fixed end.

2.5.5 Spring Constants

Bending (flexural). Let us now estimate the spring bending constant of a V-shaped cantilever that has a triangular part on the free end, and two quasirectangular beams connecting the triangular part to the fixed base. If we apply a concentrated force F_c to the free end of the cantilever, each of its parts develops a flexural deflection that we can estimate separately and then total. One can find the deflection y_{tr} of the triangular part by integration of the following flexural differential equation,

$$y''_{tr} = \frac{12F_cL(1-x/L)}{EB(1-L_1/L)(1-x/L)h^3} = \frac{12F_cL}{EBh^3(1-L_1/L)},$$

where x is the V-shaped cantilever's longitudinal axis and B is its overall width (Fig. 2.5c,d).

The length of the V-shaped cantilever's triangular part equals $L_{tr} = L - L_1$. Therefore, $y_{tr} = \dfrac{6L^3F_c(1-L_1/L)^2}{EBh^3}$. Hence, the spring constant of a solid triangular cantilever equals $K_{tr} = \dfrac{EBh^3}{6L^3}$ because, in this case, $L_1 = 0$ (Sarid, 1997). The deflection of the quasirectangular part of the cantilever with its length L_1 consists of (i) the deflection $y_{rec}(F_c)$ under the action of force F_c, (ii) the deflection $y_{rec}(M_{F_c})$ induced by the corresponding bending moment $M_{F_c} = F_c(L-L_1)$, and (iii) the angular deflection $\vartheta(F_c, M_{F_c})$ of this part, causing an additional motion $y_{tr\vartheta} = \vartheta L_{tr}$ of the triangular part's end, as a solid body.

Let us recall that $y_{rec}(F_c) = \dfrac{2F_cL_1^3}{Ewh^3}$, $y_{rec}(M_{F_c}) = \dfrac{3F_c(L-L_1)L_1^2}{Ewh^3}$, and $y_{tr\vartheta} = \dfrac{6(L-L_1)^2L_1 + 3L_1^2(L-L_1)}{Ewh^3}F_c$.

Finally, we find the spring constant of the V-shaped cantilever with a trapezoidal or rectangular recess $K_{vsh} = F_c/[y_{tr} + y_{rec}(F_c) + y_{rec}(M_{F_c}) + y_{tr\vartheta}]$ as a result of all shown deflection constituents by the following expression,

$$K_{cvsh} = \frac{Ewh^3}{L^3}\left[\frac{6(1-\zeta_L)^2}{B/w} + 2\zeta_L(\zeta_{L^2} - 3\zeta_L + 3\right]^{-1}. \tag{2.8}$$

Here, we denote the ratio $L_1/L = \zeta_L$ for the more convenient presentation of this and other related formulae. If we assume that for the rectangular cantilever without a recess $\zeta_L = 1$ and $w = b/2 = B/2$, then its spring constant calculated with expression (2.8) equals the common value for the rectangular cantilevers, $K_{crec} = EBh^3/4L^3$. Expression (2.8) works for the triangular cantilever without a recess, as well, if we assume that in this case $w = b/2 = B/2$ and $\zeta_L = 0$. The spring constant of the trapezoidal part of the V-shaped cantilever with a recess may be similarly estimated on the basis of the following approximate formula,

$$K_{ctrap} = \frac{EBh^3m_B^3}{6L^3\left[1-(1-m_B)^2\right](1-m_B\zeta_L)^2}, \tag{2.9}$$

where $m_B = 1 - B_1/B$, $B_1/B \ll 1$, and B_1 is the width of the trapezoid's free smaller end. If we assume that this cantilever is a triangular one and does not have a recess, then $B_1 = 0$, $\zeta_L = 0$, $m_B = 1$, and the formula converges to the expression for the spring constant of a common triangular cantilever K_{tr} shown earlier. One can use the same spring constant formula for the triangular cantilever with the probe shifted by δ_L from its sharp end. In this case, $m_B = 1 - \delta_L/L$. A more accurate estimate of the spring constant for a trapezoidal cantilever or a triangular cantilever with a probe shifted by δ_L can be found on the basis of the following differential equation for the cantilever's free end, or the shifted probe displacement under applied force F_c,

$$y'' = \frac{12F_c(L-x-\delta_L)}{EBh^3(1-x/L)},$$

with the boundary-value conditions on the cantilever's fixed end $y' = 0$ and $y = 0$ at $x = 0$.

The solution of this equation leads to the following expression for the spring constant of the previously mentioned cantilevers:

$$K_{\mathrm{ctr}\delta} = \frac{F_{\mathrm{c}}}{y} = \frac{EBh^3}{6L^3\left(m_{\mathrm{L}}^2 L^2\right)}\left\{1 + \frac{2\left(1 - m_{\mathrm{L}}L\right)}{m_{\mathrm{L}}^2 L^2}\left[m_{\mathrm{L}}L + (1 - m_{\mathrm{L}}L)\ln(1 - m_{\mathrm{L}}L)\right]\right\}^{-1}, \quad (2.10)$$

where $m_{\mathrm{L}}L = 1 - \delta_{\mathrm{L}}/L$, and, for a V-shaped cantilever with a recess, the width B should be multiplied by $(1 - \zeta_{\mathrm{L}})$. Expression (2.10) is presumably more accurate than (2.9) but it is more unwieldy. This expression at $\delta_{\mathrm{L}} = 0$ converges to the common formula for the spring constant of a triangular cantilever. A similar unified formula for the spring constant of a trapezoidal cantilever without a recess has been derived by Sarid (1997), who showed that his formula is applicable to a V-shaped cantilever without a recess, where $B_1 = 0$, and for a rectangular cantilever, where $B_1 = B$. Unfortunately, Sarid's formula is also an unwieldy one, and is difficult to use. However, if we assume that $1 - m_{\mathrm{L}}L = \delta_{\mathrm{L}}/L \ll 1$, then at the approximation with the series expansion $\ln(1 - m_{\mathrm{L}}L) \approx -m_{\mathrm{L}}L$, the member of (2.10) in braces significantly simplifies, and becomes equal to $1 + 2(1 - m_{\mathrm{L}}L)$. Therefore, after substitution of the latter expression into (2.10), we have a more accurate equation than Eq. (2.9), and a useful approximation for the bending spring constant of a trapezoidal or triangular cantilever with the probe shifted by $\delta_{\mathrm{L}} = \Delta L$. This expression is as follows:

$$K_{\mathrm{ctr}\delta} \approx \frac{EBh^3}{6L^3(m_{\mathrm{L}}L)^2\left[1 + 2(1 - m_{\mathrm{L}}L)\right]}.$$

Sader et al. (2012), developed methods of spring constant calibration of atomic force microscope cantilevers of arbitrary shape. Sader et al. (2014) also studied the effect of cantilever geometry on the difference between dynamic (at natural frequency) and static spring constants. They found the ratio of the dynamic–static spring constants within 1.03 for a rectangular cantilever, 1.054–1.072 for V-shaped cantilevers, and 1.181 for a triangular equilateral one. The dynamic to static ratios are critical for the accurate implementation of the thermal noise method for AFM spring constant calibration.

Torsional. The torsional spring constant K_{ctors} for a thin rectangular cantilever may be found from the well-known expression $K_{\mathrm{ctors}} = GLbh^3/3$.

Calibration of an AFM rectangular microcantilevers' torsional spring constant may be provided on the basis of the following (Love, 1944) formula

$$K_{\mathrm{ctors}} = 3^{-1}\pi^2\rho_\gamma bh^3 Lf_{\mathrm{t,vac}}^2,$$

where $f_{\mathrm{t,vac}}$ is the fundamental frequency of the cantilever torsional vibration in vacuum. This formula, however, is of limited use, due to the difficulty in measuring both the cantilever mass ($\rho_\gamma bhL$) and the resonant frequency in vacuum. A simplified calibration procedure is considered in Jing et al. (2007).

Recent studies show that nanomechanical resonators can achieve fundamental resonance frequencies exceeding 1 GHz with quality factor Q in the range $10^3 \leq Q \leq 10^5$ in vacuum. However, experimental measurement of their Q-factor in air revealed very low single numbers. It is suggested that the Q-factor of the nanocantilever system can be improved by operating a nanocantilever at a high-resonance frequency, or by the active Q-control feedback method. Fan et al. (2012) have found that quality factor is dependent on the distance above snap-to-contact Z (μm) with linear relationship ($Q = 2.62Z + 13.17$) for distances greater than 0.03 μm. The cantilever's Q-factor is linearly dependent on the tip-surface distance that is far from the surface ($Z > 0.1$ μm), while closer to surface ($Z < 0.1$ μm), the sensitivity drastically increases. The snap-to-contact of the tip to the surface occurs because of capillary wetting attraction at the probe tip.

Common features. Cantilever-based scanning probe microscopy (AFM) techniques are among the most powerful approaches for imaging, manipulating, and measuring nanoscaled properties and phenomena. Let us recall that a very sensitive tunneling microscopy can work only with conducting and semiconducting materials mostly under ultrahigh vacuum, whereas the AFM can work with any kind of substances in air, gas, and liquid. This microscopy generates surface, molecules, and atom images by measuring small forces between the sample surface and the probe. Common forces detected using these techniques are van der Waals, Casimir, electrostatic forces, surface stress, etc. The effects of the electrostatic field and Casimir force on the cantilever vibration and damping are thoroughly studied by many researchers (Chumak et al., 2004).

Conventional microcantilevers are fabricated from isotropic and crystalline quartz, silicon, silicon carbide, and silicon nitride (Si_3N_4), using electron-beam or optical lithography, and have typical length, width, and thickness of ~125, ~35, and ~4 μm, respectively.

Cantilevers with probes that are integrated (the tip height is a few orders of magnitude larger than the radius of its curvature) and colloidal (the tip height is in the same order as the radius of curvature) should be calibrated for direct and lateral force action. The latter is important because the coefficient of friction between the polysilicon measurand and a silicon probe can be in the range of 0.19–0.35. This friction can cause a torsional displacement of the probe, stick-slip motion, and corresponding additional measuring uncertainties.

The cantilever's tips are made from tungsten, diamond, silicon, cobalt, samarium, cobalt-samarium permanent magnets, and iridium, for various purposes. New tips in scanning microscopy systems are based on growing carbon nanotubes and implanted diamonds. These carbon tips with high aspect ratio (up to 20:1 with the radius of curvature of 1 nm) are nearly ideal for AFM imaging because of their size and durability (exceptionally long wear of up to 200 times longer than silicon probes). The AFM is routinely used to measure a number of interfacial phenomena, including surface forces, friction, and adhesion (pull-off force of ~0.4 nN for hydrophobic tip at RH ≥ 30%, and ~4 nN for hydrophilic one). Of fundamental importance to these quantitative measurements is an accurate knowledge of the cantilever spring constant. Furthermore, by controlling the interaction geometry by attaching a particle to the end of the cantilever, it is possible to compare the measured interaction force with theory.

It is clear that the static AFM requires very soft cantilevers. The deflection of the cantilevers is most commonly measured by optical means, through laser heterodyne Doppler, Michelson, or Fabry–Pérot interferometer (threshold noise up to femtometer per $\sqrt{\text{Hz}}$) with spot size up from 3 μm, or by bouncing light rays off the cantilever. The detection in the latter system is provided by segmented or linear position sensing detectors (PSD), its calibration may be performed with the precision calibrated step sample, or by the inverse calibrated piezolever. The deflection of the piezoresistive cantilevers is usually measured by making them part of a Wheatstone bridge. Let us recall that silicon piezoresistive transducers are nonlinear, and require the use of confluent statistical analysis at their calibration.

If the cantilever in a variable-load AFM is tilted toward the specimen sample, its tip may have significant parasitic displacement along the cantilever axis direction. The deflection of the heated cantilever can be measured by the cantilever thermal response (Park et al., 2007).

Cantilevers with lesser cross-section sizes provide better sensitivity to picoNewton forces. Cantilevers of the AFM have been uniformly coated with gold and platinum thin films. These films are nanostructured, with thickness between 18 nm and 73 nm.

Microcantilevers consisting of a silicon or silicon nitride substrate (relatively passive side), coated with a layer of special film (high affinity to the targeted analyte), deflect because of heat transfer or chemical sorption. They have been used, for example, to distinguish between oligonucleotide (DNA) molecules of different base sequences, to detect an individual biomolecule, to measure pH changes, and to measure the surface stress associated with molecular adsorption or absorption. In many of these applications, the deflection of the cantilever is driven by the stress buildup in the layer of the metal film. Microcantilever sensors used for recognition of chemical or biological environment components work similarly to AFM cantilevers in quasistatic and dynamic mode. In the dynamic mode, they work in the resonant mode. The main difference follows from the assumed ability of some chemicals (e.g., NaCl solution) to change the cantilever's bending spring constant k_{fl} even at a small oscillation amplitude. Chen et al. (1995) showed the relationship between the changes of the bending spring constant Δk_{fl} and the change of the surface stress on the top $\Delta\sigma_{\text{top}}$ and bottom $\Delta\sigma_{\text{bottom}}$ of the microcantilever, as follows:

$$\Delta k_{\text{fl}} = \frac{\pi^2}{4} n^* \left(\Delta\sigma_{\text{top}} + \Delta\sigma_{\text{bottom}} \right),$$

where n^* is the effective corrected mass factor that depends on the cantilever geometry.

In quasistatic mode, they work based on the cantilever's deformation that is detected by modulation of optical, electrical, or magnetic fields, and modulation of mechanical stress, that is, different surface stresses (see expression (2.11) with the relationship between the surface stress and strain).

By measuring the cantilever deflection and knowing its geometrical parameters, the difference in surface stress per unit length $\Delta\sigma_{\text{top/bottom}} = \sigma_{\text{top}} - \sigma_{\text{bottom}}$ can be determined using Stoney's equation (Stoney, 1909):

$$\Delta\sigma_{\text{top/bottom}} = \frac{Eh^2}{3(1-\nu_{\text{p}})L^2} s_{\text{d}}, \tag{2.11}$$

where s_{d} is the deflection of the free-end of the cantilever. The applicability of the classical curvature–stress relation for thin film was discussed with positive conclusions by Preissig (1989).

Nanocantilevers. Nanowire, carbon nanotube, single crystal, and nanobelt technology allow us to manufacture cantilevers that are ~35–1800 times smaller than conventional microcantilevers of the previously mentioned micrometer sizes. Their high frequency options are shown in Chapter 3.

2.5.6 AFM Dynamics Parameters

Techniques. Let us recall that AFM microcantilevers may work in the quasistatic regime and dynamic oscillation regime. The latter is more sensitive. Four techniques for atomic-resolution force microscopy in vacuum are in use today: (i) the classic frequency modulation technique with large amplitudes (30 nm to 50 nm) and soft cantilevers (with a spring constant in the tenths of Newton per meter) in which both parasitic attractive and effective repulsive action forces are possible; (ii) frequency modulation with small amplitude (less than 1 nm) for sensing only repulsive forces and stiff cantilevers (spring constant in the order of 10 N/m to several kiloNewton per meter); (iii) the off-resonance technique; and (iv) the amplitude modulation method. To avoid cycle distortion at small amplitudes, the product of the spring constant and the amplitude, $k_{\text{s}}A_{\text{ca}}$, has to be larger than 100 nN in order to provide a sufficiently strong withdrawing force. The mode may be noncontact and tapping, with fundamental and higher eigenmodes. The Kelvin probe microscopy method was developed to compensate for different contact potentials of tip-sample materials at noncontact step height measurement. Rectangular, triangular, and step profile silicon microcantilevers are used in biosensors (Ansari and Cho, 2009). Souayeh and Kacem (2014) developed computational models for large amplitude nonlinear vibrations of electrostatically actuated carbon nanotube-based mass sensors.

Parameters. Let us recall that the more advanced frequency modulated "noncontact" AFM has five important operating parameters (for more detail see Morita et al., 2009): (i) the spring constant of the cantilever k_s; (ii) the eigenfrequency (natural frequency) of the cantilever f_0 (tens of kHz to GHz); (iii) the quality factor value of the cantilever Q_c (up to tens of thousands); (iv) the oscillation amplitude A_{caf}; and (v) the frequency shift of the cantilever oscillations $\Delta f = -\frac{f_0}{k_s A_{ca}^2} < F_{ts} A_{caf} \cos(2\pi f_0 t) >$ from 4 Hz to 100 Hz. The first three parameters (i–iii) are determined by the type of cantilever that is used, while the last two parameters can be adjusted. In the fifth parameter, F_{ts} is the force acting between the cantilever's tip and the monitoring sample, and t is the time of signal averaging. The averaging is denoted by the angular brackets in the formula shown earlier. The long-range attraction forces between the tip of cantilever and the monitored sample have to be canceled by immersing the tip and sample in liquid, or partly pulling off at the cantilever after its jump-to-contact occurred.

Let us recall that the cantilever itself dissipates energy (internal dissipation):

$$\Delta E_{cL} = 2\pi\, E_c / Q_c,$$

where $E_c = k_s\, A_{caf}^2 / 2$.

When the tip oscillates close to the sample (the tip-sample contact stiffness k_{ts}), the oscillation amplitude changes, according to the expression $A_{caf} = \frac{A_{ca0}}{1 + k_{ts} / k_s}$, additional damping occurs, and the driving signal should be increased in order to keep the oscillation amplitude equal A_{ca0}.

Stick-slip effect. The possible slip jumps δS_k in the system, with moving mass under the action of friction, and the elastic drive force is equal to: $\delta S_k = 2\delta F_{fr}/j_{dr}$, where δF_{fr} is the difference between the static and kinetic friction and j_{dr} is the spring constant of the drive. If we accept that, at low speed v_{dr} of the drive, the stick time is enough to cause significant quasistatic friction F_s in comparison with much smaller kinetic friction, then we can assume that $\delta F_{fr} \approx F_s$ and $j_{dr}\delta S_k = 2F_s$. One can see that after multiplying both sides of this equation by δS_k, its left side will represent the potential energy in the drive's spring, accumulated during the stick stop. Then, on the basis of Hamilton's principle with equating potential and kinetic energy, we obtain the relationship between the velocity v_{dr} of the drive motion and the probable slip jump of the driven probe with mass m_g as follows:

$$\frac{1}{2} j_{dr} (\delta S_k)^2 = F_s \delta S_k = \frac{1}{2} m_g v_{dr}^2.$$

Therefore, the critical drive speed for stick-slip motion equals $v_{dr.c} = C_F \sqrt{F_s \delta S_k / m_g}$, where the constant $C_F \approx 1$ is the factor of physical nature of the sliding friction. See these problems, as well, in (Tseytlin 2006, Sections 1–3), Amosov and Tseytlin (1969), and Luan and Robbins (2004).

The friction between the movement with the nanocantilever probe and the sample of measurand can be characterized by a nanofriction coefficient that, for contact force range $0 < F_c < 6$ μN, is equal (in accordance with Salvadori et al., 2008) to 0.047 ± 0.002 for Si on Si, 0.173 ± 0.0009 for Si on DLC (diamondlike carbon), and 0.0080 ± 0.0005 for DLC on DLC. For contact force range $9 < F_c < 13$ μN, the DLC on DLC coefficient of friction increases to 0.051 ± 0.003.

2.5.7 Contact Dynamics of Tapping Mode

In the tapping mode AFM dynamics, the tip mass, nonlinear contact forces, damping, and intermittent contact can all excite higher modes harmonics into the system motion through the mode coupling mechanism. The multimodal analysis should be taken as it offers a more general and accurate approach of studying the AFM intermittent contact dynamics. The presence of adhesion is an important factor that may change the system response significantly to a "tip-stuck-to-sample" motion. Zhang et al. (2012) considered two contact situations: (i) Hertz deformation without adhesion and (ii) DMT contact deformation with adhesion. However, the adhesion situation can be more sophisticated (Tseytlin, 2006). One of the first rigorous theoretical treatments of the adhesion of elastic spheres with their attraction is due to the research work of Johnson et al. (1971). Their results are called JKR theory. The JKR theory is applicable to soft compliant materials with high surface energy and large radii of indenter. Another DMT (Derjaguin et al., 1975) theory applies to harder, less compliant materials with low surface energy and small radii of indenter. This theory assumes that the contact deformation profile remains Hertzian. The transition solution MD (Maugis, 1992) is for materials between JKR ($\lambda_{MD} > 5$) and DMT ($\lambda_{MD} < 0.1$) approach, where $\lambda_{MD} = 2\sigma_0/(\pi\gamma_A j_K{}^2/R)^{1/3}$, σ_0 is a constant maximum adhesive traction, γ_A is the work of adhesion, R is the effective radius of contacting spheres, and j_K is the effective elastic modulus. We can show, for example, that in the DMT case, exponent $\chi = \dfrac{2}{3}\dfrac{\ln(F_Q + F_{adh})}{\ln F_Q}$ because, in these conditions, the indentation depth may be presented as $w_c = \mu_\varepsilon{}^{-2/3}R^{-1/3}(F_{adh} + F_Q)^{2/3} = A_\varepsilon F_Q{}^\chi$, where F_{adh} is the acting adhesive force and R is the radius of indenter. Therefore, $\chi = 2/3$ if F_{adh} is equal to zero, as in the Hertzian solution.

A phase shift between the oscillatory motion and drive motion of an AFM cantilever used for tapping mode AFM imaging can be related to the adhesive and elastic properties of polymeric surface layers. Thormann et al. (2010) evaluated how the optimal contrast between hard and soft surface layers can be achieved while minimizing the surface damage. This was investigated by performing classical force-distance measurements while driving the cantilever as in tapping mode imaging. The amplitude and phase response as a function of the average tip-surface separation was recorded. Five different cantilevers with a wide range of spring constants and four different tapping amplitudes were investigated and compared. Based on these experiments, it was concluded that a too stiff cantilever, a high free tapping amplitude and a low amplitude set point value often lead to surface damage, while a too low spring constant and a low free tapping amplitude result in poor phase image contrast. Intermediate values, where little surface damage and significant image contrast are obtained, were identified. In all cases, it was observed that the best image contrast was obtained when the amplitude set point was chosen, such that the amplitude during imaging was reduced to approximately 50% of the free amplitude of the cantilever vibration. If the measurand surface in liquid has nanobubbles, these must be measured with a sharp hydrophilic tip and with the cantilever having a very low spring constant (Walczyk and Schönherr, 2014).

The study by Zhang et al. (2012) on tapping modes shows that the modal amplitude of the second mode, initiated by short contact, is larger than that of the first regular one when the tip is in contact with the sample. When the tip is not in contact with the sample, the first mode amplitude dominates. The dynamics of tapping mode AFM in liquids also shows the presence of second mode vibration (Section 3.4.1). Parlak et al. (2014) studied AFM liquid contact resonance with analytical models and experiments. Abdi et al. (2015) studied dynamics of the nanoneedle probe in trolling mode AFM.

2.5.8 Nanocontact Mechanics in Manipulation

Korayem et al. (2012) provided sensitivity analysis of nanocontact mechanics in the manipulation of biological cells (DNA) on the basis of the Sobol method. They found that variation of contact deformation of nano-objects at a constant contact force is more sensitive to adhesion and modulus of elasticity, than to the surface radius of curvature and Poisson's ratio. This analysis uses Hertzian (without adhesion), JKR, DMT, and PT (Piétrement and Troyon, 2000) relations. The latter (PT) relations are more comprehensive, but more complex. The members of the PT general dimensionless equation describing elastic indentation depth, $\bar{w}_c = w_c\left(\frac{j_k}{\pi\gamma_A R^2}\right)^{1/3}$ versus load $\bar{F}_Q = F_Q/(\pi\gamma_A R)$ in the PT case, are given in a fourth-degree polynomial form of the parameter $\alpha = \left(1 - e^{-\lambda_{MD}/0.924}\right)/1.02$ for the dimensionless contact radius $\bar{r}_{a0(\alpha)} = r_{a0}\left(\frac{j_k}{\pi\gamma_A R^2}\right)^{1/3}$ at zero load, and exponent parameter $\beta_{a(\alpha)}$, and the second degree for the dimensionless adhesive force $\bar{F}_{adh(\alpha)} = F_{adh}/(\pi\gamma_A R)$.

It can be shown that, with an uncertainty of 3% or less, these adhesion parameters may be represented by lower polynomials of second and first degree, respectively, if we substitute Chebyshev's polynomials of fourth, third, and second degree. As a result, we obtain (Tseytlin, 2007) the following simplified expressions for the parameters, and the general elastic indentation equation:

$$\bar{r}_{a0(\alpha)} = -0.09606\alpha^2 + 0.62996\alpha + 1.264;\ \beta_{a(\varepsilon)} = 0.8954\alpha^2 + 0.39093\alpha;$$

$$\bar{F}_{adh(\alpha)} = -0.5458\alpha + 2.000, \text{ and}$$

$$\bar{w}_c = \bar{r}^2_{a0(\alpha)} \left[\left(\frac{\alpha + RT\left(\bar{F}_Q / \bar{F}_{adh(\alpha)}\right)}{1+\alpha} \right)^{4/3} - S_{(\alpha)} \left(\frac{\alpha + RT\left(\bar{F}_Q / \bar{F}_{adh(\varepsilon)}\right)}{1+\alpha} \right)^{\frac{2}{3}\beta_{a(\alpha)}} \right]$$

where

$$RT\left(\bar{F}_Q / \bar{F}_{adh(\alpha)}\right) = \sqrt{1 + \bar{F}_Q / (-0.5458\alpha + 2.000)}$$

and

$$S_{(\alpha)} = -2.160\alpha^{0.019}\left(1 - 1.27458\alpha^{0.045} - 0.03379\alpha^{1.9}\right)$$

Contact deformation and surface roughness. A serious practical limitation of the JKR, DMT, MD, PT and Hertz theories is that they assume an interaction of perfectly smooth surfaces. However, most real surfaces are rough. Even asperities of tiny particles as small as 1–2 nm can lower surface adhesion significantly, and change the contact deformation. Surface roughness, material-dependent phase shift of reflected rays, and other nonuniform features, can have a large influence on the uncertainty in noncontact measurements, as well.

Let us recall that surface roughness in accordance with ISO and American standards (see ISO 4287-1997, ASME B46.1-2002) is specified by kurtosis (R_{ku}), skewness (R_{sk}), arithmetic average (R_a), root mean square value (R_q), and average maximum height of the profile (R_z), the latter of which previously (see ISO 4287-1984) was estimated as an average of 10 points' heights (R_{z10}). The roughness of optical surfaces is usually specified by the RMS parameter (R_q) and the roughness of mechanical surfaces is usually specified by the arithmetic average parameter R_a. There are no simple statistically supported mathematical ratios between these two parameters in the mentioned standards.

Therefore, we provide the following estimates (Tseytlin, 2007).

In accordance with the standards, R_a is given by

$$R_a = (1/L)\int_0^L |Z(x)|\,dx.$$

However, it is easy to show that this expression corresponds to the first initial moment of statistical data for roughness irregularities $Z(x)$ because

$$\int_0^L |Z(x)|\,dx = \int_0^L \{[Z^+(x)|_{L'} - \bar{Z}] + \text{abs}[Z^-(x)|_{L''} - \bar{Z}]\}dx = \int_0^L Z_{\text{init}}(x)\,dx,$$

where $L' \approx L'' = L/2$, $Z_{init} \geq 0$ are statistical data from initial level, and $\bar{Z}$ is a mean line position.

A surface roughness profile here is imaginably transformed with the mirror reflection of the negative part (Fig. 2.8). Panel a shows the original surface

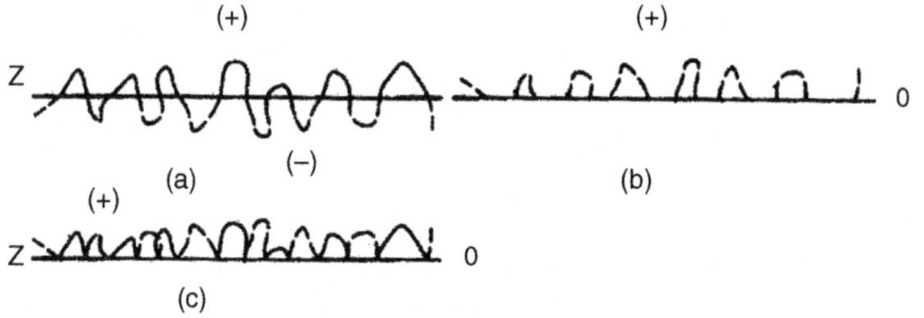

FIGURE 2.8 Schematic transformation of surface roughness profile. (a) Original surface profile; (b) mirror reflection of the negative part to the positive field; and (c) combined profile in the positive field. *(After Tseytlin, 2007, reprinted with permission.)*

roughness profile, panel b shows the mirror reflection of the negative part of the profile to the positive field, and panel c shows images of the positive and negative part of the profile that are combined in the positive field. It is clear that the first initial moment of the combined profile is equal to R_a of the original roughness profile, and the RMS is the same for the combined and original profiles by definition.

We know that true RMS is equal to the square root from the sum of the squares of the first initial moment, and the standard deviation (variance) for any variables distribution. Therefore, we particularly derived on this basis the following formula for the R_q/R_a ratio estimation:

$$R_q/R_a = \sqrt{1+\left(\mu/\aleph_{ip}\right)^2}, \tag{2.12}$$

where $2\mu = R_z/R_a$ and $\aleph_{ip}$ is the coverage factor for the acceptable maximum level of confidence p (fraction of distribution), at a certain type of irregularities distribution.

We yielded, thus, a simple theoretical proportion between true RMS (R_q) and mean arithmetic (R_a) roughness parameters, at different statistical distributions of roughness irregularities with the fit by the Student's $(R_q/R_a) = 1.11^*$ at $k = 9$ degree of freedom, $p = 99.73\%$, $\aleph_{ip} = 4.09$; normal $(R_q/R_a) = 1.2^*$ at $R_{ku} = 3$, $p = 99.73\%$, $\aleph_{ip} = 3$; uniform $(R_q/R_a) = 1.53^*$ at $R_{ku} = 1.8$, $\aleph_{ip} = 1.73$; Pearson's $(R_q/R_a) = 1.274^*$ at $R_{sk} = -0.249$, $R_{ku} = 5.547$, $p = 99.5\%$, $\aleph_{ip} = 2.53$; and Weibull's $(R_q/R_a) = 1.281^*$ at $R_{sk} = -0.6$, $R_{ku} = 3.5$; $p = 99\%$, $\aleph_{ip} = 2.5$ models, and compared them to the experimental data. The calculated * data well correspond to practical experimental estimations.

If the distribution of variables is complex and contaminated ($3 < R_{ku} < 6$) with suspected but not cancelled outliers, then the $L_{p=1.5}$ estimator may be more effective (less variance) than a mean and median estimator. The ratio of $L_{p=1.5}$ and R_a can be calculated with the Excel® "Solver" tool at discrete data Y_i presentation.

The influence of indenter-surface misalignment on the results of instrumented indentation tests of polymers is presented in research by Pelletiera et al. (2007).

A simpler practical way to estimate the contact deformation w_c at significant surface roughness is to use the method of two loads (Tseytlin, 1994) with

$$w_c = \delta w F_Q^{\chi} / \left(F_{Q2}^{\chi} - F_{Q1}^{\chi}\right) = A_\varepsilon F_Q^{\chi},$$

where $\delta w = w_c(F_{Q2}) - w_c(F_{Q1})$.

Thus, in order to evaluate the local deformation under the force F_Q, it suffices to determine the difference δw between local deformations under two values of the measuring force $F_{Q1,2}$. The measuring force can be changed either by changing the setting of the measuring device's loading mechanism, or by using an additional load attached to the body of the measuring tip. In order to find the more accurate value of coefficients A_ε and χ, we evaluate their values with the aid of the least square method applied to N pairs of experimental data w_{ci}, F_{Qi} (Tseytlin, 1994) and the maximization of the correlation coefficient ρ_χ

$$\chi = \frac{N\sum_i \left(\ln F_{Qi} \cdot \ln w_i\right) - \sum_i \ln F_{Qi} \sum_i \ln w_i}{N\sum_i \left(\ln F_{Qi}\right)^2 - \left(\sum_i \ln F_{Qi}\right)^2}; \tag{2.12a}$$

$$\max \rho_\chi = \frac{\chi\sqrt{\left[N\sum_i \left(\ln F_{Qi}\right)^2 - \left(\sum_i \ln F_{Qi}\right)^2\right]}}{\sqrt{N\sum_i \left(\ln F_{Qi}^{\chi}/\Omega_{ij}\right)^2 - \left(\sum_i \ln F_{Qi}^{\chi}/\Omega_{ij}\right)^2}}, \tag{2.12b}$$

where $\Omega_{ij} = (F_{Qi}^{\chi} - F_{Qj}^{\chi})/\delta w_{ij}$.

We can see that the correlation coefficient ρ_χ increases as the denominator in this expression decreases. However, the minimum of the denominator as a second central moment corresponds to centralizing it within the most probable value of the contact deformations and the exponent χ. The coefficient A_ε can be found according to a well-known algorithm $\ln A_\varepsilon = (1/N)\sum_i \ln w_i = (1/N)\sum_i \chi \ln F_{Qi}$.

This method does not require knowledge of a real size of the measuring tip, and may still be effective for the estimation of contact deformations at the measurement *in situ*, despite the fact that the accurate ultra-microindentation (like UMIS-2000, CSIRO-Australia) and oscillatory indentation (Cheng et al., 2006) systems have now been developed. However, knowledge of the size of the measuring tip may be useful in some cases (Maragliano et al., 2015).

A dynamic contact of two colliding spheres of radius R_1 and R_2, having relative velocity V before central impact, follows (Kachanov et al., 2003) with the maximal relative approach of the spheres δ_0 and time duration t_0 of the collision that are equal to:

$$\delta_0 = \left(\frac{\mu}{h}\right)^{2/5} V^{4/5} \quad \text{and} \quad t_0 = 2.94\left(\frac{\mu^2}{h^2 V}\right)^{1/5},$$

where $\mu = \dfrac{m_1 m_2}{m_1 + m_2}, h = \dfrac{32}{15K}\sqrt{\dfrac{R_1 R_2}{R_1 + R_2}}, K = \dfrac{2(1-\nu_{p1}^2)}{E_1} + \dfrac{2(1-\nu_{p2}^2)}{E_2}$, and m_i, E_i, ν_{pi} are the mass, Young's modulus, and Poisson's ratio of ith sphere (i = 1,2).

2.6 CONCAVE NOTCH HINGES

Coordinate measuring machines (CMM) and other precision measuring systems with compliant micromechanisms have flexible supports of different kind. Concave (a wider term than used by some authors, the conic type) notch hinges with cylindrical, elliptic (Fig. 2.9), parabolic, and hyperbolic contours, segmented and V-shaped notch hinges are useful for the purposes of precision measuring, and for robotic systems' miniaturization. Many publications in the scientific

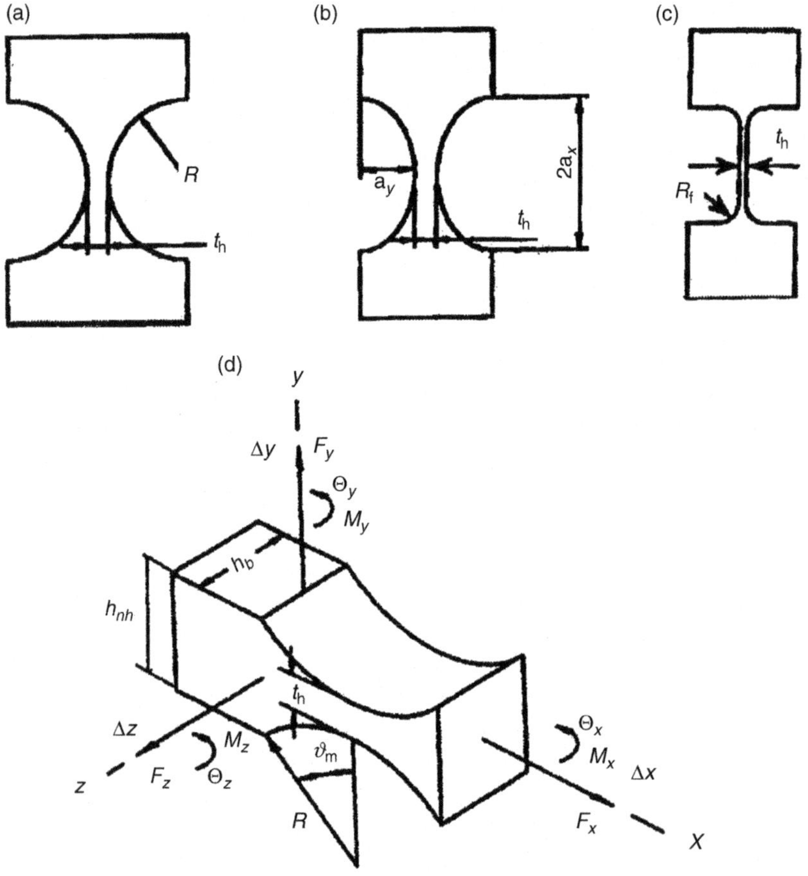

FIGURE 2.9 Notch hinges with different geometric parameters and shapes. (a) Cylindrical, (b) elliptical, (c) corner filleted, and (d) 3D image. *(After Tseytlin, 2006, reprinted with permission.)*

and engineering literature have been devoted to the problems of such hinges' accuracy, calibration, and design. However, serious attention to these miniature flexure hinges as kinematic elements has been developed only recently with the definition of their virtual instantaneous center of rotation (ICR). Let us recall that these elements do not have real guiding surfaces.

Therefore, their design and rotational compliance accurate calculation are not effective without the evaluation of the ICR position, as for the rotational joint with the certain stable axis. This evaluation has been found in the author's research on the basis of inverse conformal mapping (Tseytlin, 2002, 2011, 2012) of approximating contours (Fig. 2.10). An approximating contour in this case is presented by circles R_c shifted l_x from the hinge geometrical center to contiguity, with minimum possible deviation from a real concave hinge contour estimated by using Chebyshev's polynomials for uniform or relative approximation. Relative approximation works better for segmented and V-shaped hinges.

Elastic mechanisms, transducers, flexures, and cantilevers in various forms have found application in a variety of high precision instruments. These systems provide smooth, external friction-free, and wear-free precision motion. Notch flexure hinges are more effective than traditional elastic hinges with flexible strips because they have smaller dimensions, provide stabler positions of the rotational axis, and are simpler in assembly. However, their stress state is more complicated, and requires special calculation methods.

The calculation of elastic bodies stress and motion is based on the strain and stress tensors relations, matrices of ordinary and partial differential equations with boundary-value conditions, potential energy variational methods, linear and nonlinear, local and nonlocal elasticity principles. Some solutions may be very complicated (unwieldy) in mathematical terms. However, we should remember that any successful solution of these mathematical problems contains

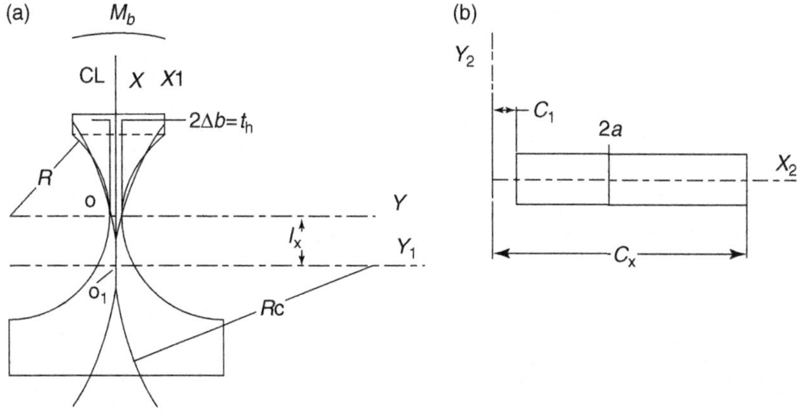

FIGURE 2.10 Effective concave approximating contour (a) and its conformal inversion (b). *(After Tseytlin, 2002, 2006, reprinted with permission.)*

elastic and dimensional parameters of the body under consideration, whose values are known only with some degree of accuracy. Data for Young's modulus, E; rigidity (shear) modulus, G; Poisson's ratio, ν_p; and the temperature expansion coefficients presented in the technical literature differ for the same materials up to 15% or more. These properties are different on the near-surface layers and for thin films. In crystals, they depend on the direction of lattice planes. Modified Young's modulus $E^* = E/(1 - \nu_p^2)$ is used for calculations of wide strips and plates bending.

Therefore, in practice, it is better to derive approximate, yet simple and tractable solutions with satisfactory accuracy, instead of deriving complicated and unwieldy "exact" formulas. This is especially true if the accuracy of a simple tractable solution is practically the same as that of a more complicated "exact" solution, in comparison with the finite element method (FEM) and experimental (Ex) results.

Application of a different theoretical approach based on the inverse conformal mapping of approximating contour allowed us to derive more tractable equations for the rotational compliance of notch hinges with circular, elliptic, conic, and even wider range of smooth concave profiles.

Inverse conformal mapping of a concave flexure hinge is based on the approximation of real contours with radius-vector R by two circles contiguous in the point O_1 with conditional radius R_c (Fig. 2.10a), whose centers are shifted from the middle of the hinge O a distance l_x.

There have been many attempts to derive satisfactory theoretical equations for the compliance, stiffness, and stress predictions of notch hinges, including traditional methods with integration of linear differential equations over the actual hinge contour.

Exact solution. The full theoretical equation (Paros and Weisbord, 1965) for the rotational compliance around the Z-axis (Fig. 2.9a,d) of notch hinges with a circular profile is:

$$Clrpw\gamma = \theta_z / M_z = 3\left(2Eh_b R^2\right)^{-1}\left(2\beta+\beta^2\right)^{-1}\left\{\left[(1+\beta)/\gamma^2 + \frac{3+2\beta+\beta^2}{\gamma\left(2\beta+\beta^2\right)}\right]\right.$$
$$\left.\left[\sqrt{1-(1+\beta-\gamma)^2}\right] + \left[6(1+\beta)\left(2\beta+\beta^2\right)^{-3/2}\right]\times\left[\tan^{-1}\left(\sqrt{\frac{2+\beta}{\beta}}\frac{(\gamma-\beta)}{\sqrt{1-(1+\beta-\gamma)^2}}\right)\right]\right\} \quad (2.13)$$

where t_h is the thickness and $\beta = t_h/2R$ is the relative thickness of the hinge crosspiece; h_{nh} (Fig. 2.9d) is the width and $\gamma = h_{nh}/2R$ is the relative width of the hinge; h_b is the depth of the hinge; and R is the radius of the hinge contour. This "exact" equation is "unwieldy," as noted by its authors Paros and Weisbord (1965). For a full semicircular notch (right circle) the dimensionless parameter

γ is equal to $\gamma = 1 + \beta$ and the expression (2.13) for hinge compliance slightly reduces, but still remains an "unwieldy" one:

$$Clrpw = \theta_z / M_z = 3\left(2Eh_bR^2\right)^{-1}\left(2\beta+\beta^2\right)^{-1}\left\{\left[\left(3+4\beta+2\beta^2\right)(1+\beta)^{-1}\left(2\beta+\beta^2\right)^{-1}\right]\right.$$
$$\left.+\left[6(1+\beta)\left(2\beta+\beta^2\right)^{-3/2}\right]\tan^{-1}\left(\sqrt{(2+\beta)/\beta}\right)\right\} = 3\left(2Eh_bR^2\right)^{-1} f(\beta). \tag{2.14}$$

Derived later, Slocum's integral formula (Slocum, 1992) for semicircular notches is also a complicated one, as it is clear from the following slightly reduced presentation:

$$Clrsl = \theta_z / M_z$$
$$= 3R\left(2Eh_b\right)^{-1}\rho_{n2}^{-1}\left\{\rho_{n1}^{-1} + \left[\frac{2R^2+\rho_{n1}^2}{\rho_{n1}} + \frac{3R\rho_{n1}\left(\frac{\pi}{2} - \tan^{-1}\left(-R/\sqrt{\rho_{n2}}\right)\right)}{\sqrt{\rho_{n2}}}\right]\rho_{n2}^{-1}\right\}, \tag{2.15}$$

where $\rho_{n1} = 0.5t + R; \rho_{n2} = \rho_{n1}^2 - R^2$.

Paros and Weisbord derived a more concise approximate equation for circular thin notch hinge rotation compliance:

$$Clr_{\mathrm{SMP}} = \theta_z / M_z = 9\pi R^{1/2} / \left(2Eh_b t_h^{5/2}\right). \tag{2.16}$$

This equation corresponds to the full theoretical formula (2.13) within 1% for circular hinges with relative thickness $\beta = t_h/2R$ in the range from 0.01 to 0.05, and within 5–12% for thicker circular hinges with β in the range from 0.1 to 0.3. However, comparison of the calculations obtained from these theoretical equations with experimental data and finite element analysis results shows a much larger difference of up to 25% or more (Smith, 2000).

A direct finite element analysis method is more reliable and corresponds better to experimental data, but it is not convenient for a synthesis of new hinge systems. For that purpose, Smith has derived the following empirical equation (Smith and Chetwynd, 1992) from the finite element analysis results for circular hinges:

$$Clr_{\mathrm{SEM}} = \theta_z / M_z = 2K_{\mathrm{tR}}R / EI_{zz};\ K_{\mathrm{tR}} = 0.166 + 0.565t/R, I_{zz} = (1/12)h_{\mathrm{bt}}^3 \tag{2.17}$$

This equation is concise and adequate, but only for thick circular hinges with relative thickness β in the range from 0.1 to 0.5. Notch hinges in real structures have β in the range from 0.01 to 0.3, and may have different actual profiles. As such, it is clear that some different kind of theoretical equations, specifically based on the continuum mechanics for the whole range of the relative thickness and profiles, are necessary. In the paper by Paros and Weisbord (1965), as well as in Slocum's book, the initial differential equation is presented as $y''_x = \theta' = M_z(x)dx/EJ(x)$ that is typical for small bending of straight elastic beams. Moreover, the compliance calculation results by Slocum's formula (2.15), and Paros and Weisbord's formula (2.14) are identical. If this is the case, we suppose that all sources are based on the same assumption of straight beam bending that they claim to be exact. But it can be shown that the behavior of the circular notch hinge, especially the thick one, is rather similar to the superposition of two bending imaginary parallel and opposite directed curved beams, for which the initial differential equation is different than for a straight beam. Note that the section flexural rigidity constant for a curved beam is equal to:

$$J_{zz} = \int_{-h/2}^{h/2} y^2 h_b \, dy/(1 + y/R).$$

When h/R is less than 0.1 or $\beta < 0.05$, the value of J_{zz} is very close to the section moment of inertia, that is

$$J_z = \int_{-h/2}^{h/2} y^2 h_b \, dy.$$

However, when h/R approaches value 1, the difference between J_{zz} and J_z becomes significant. Therefore, it is possible that the exact solution of the previously used differential equation is not the exact solution of the problem. With this in mind, we use here a different theoretical approach to the problem (Tseytlin, 2002, 2006).

2.6.1 Inverse Conformal Mapping of Approximating Contour

Contour maps. The contour maps of circular notch hinges consist of two separated by $2\Delta b = t_h$ noncontiguous circular arcs with radius R that correspond to concave hinge surfaces. It is evident that each concave contour may be approximated by two contiguous circles with radius $R_c = l_x^2/2\Delta b$ shifted a distance, l_x (Fig. 2.10a). This approximation is necessary for the possible application of inverse conformal mapping with the transformation of the circular contours to a rectilinear one, and excluding an infinite point from the calculation field.

Without such approximation, inverse conformal mapping would not be effective, unlike that developed for sharp arc sectors, as a result of the shift of real profiles relative to the center of inversion. In this case, circles would be transformed again into circles, and that is not helpful. If the origin O_1 of new coordinates is positioned at the point of contiguity, the resulting coordinates will be equal to: $X_1 = X + l_x$ and $Y_1 = Y$. The approximating contour of the hinge will be converted with inversion to the rectangular band defined by X_2 and Y_2 (Fig. 2.10b) coordinates with dimensions

$$a = 1/2R_c, C_x = 1/l_x, C_1 = 1/(l_x + X_{max}). \qquad (2.18)$$

Now, the elastic deformation of the circular hinge can be effectively analyzed by using an inversion of the approximating contour. The two-dimensional mode displacements in the plane U_x and V_y along axes X_1 and Y_1 are equal to:

$$U_x = (1/2G)\left[(1-\nu_p)S(X_1,Y_1) - \partial F_{sm}/\partial X_1\right] + U_o - \vartheta_o Y_1, \qquad (2.19a)$$

$$V_y = (1/2G)\left[(1-\nu_p)T(X_1,Y_1) - \partial F_{sm}/Y_1\right] + V_o + \vartheta_o X_1, \qquad (2.19b)$$

where G is the shear modulus and ν_p is Poisson's ratio.

Under the loading moment M, Airy's stress-functions (Tseytlin, 2006) are

$$F_s = -MY_2^3/(4a^3), \qquad (2.20)$$

and

$$F_{sm} = -MY_1^3/\left[(4a^3)(X_1^2 + Y_1^2)^2\right] \qquad (2.21)$$

in inversion and the hinge fields, respectively (Table 2.5).

In addition,

$$S(X_1,Y_1) + iT(X_1,Y_1) = \int_z f(Z)dZ, \qquad (2.22)$$

and

$$f(Z) = s_\nabla + it_\nabla. \qquad (2.23)$$

After yielding the Laplacian, $s_\nabla = \nabla^2 F_{sm}$, and conjugate the harmonic function, t_∇, we have found by integration the expressions for the functions U_x, V_y,

TABLE 2.5 Airy's Stress-Function in the Hinge and Inverse Fields

Type of load	Inverse stress-function, F_s	*Hinges' stress-function, F_{sm}
Extension force, P_F; $\Omega_K = P_F/2a$	$(\Omega_K/2)Y_2^2$	$(\Omega_K/2)Y_1^2/r_1^2$
Moment, M; $\Omega_K = 3M/2a^3$	$(\Omega_K/6)Y_2^3$	$(\Omega_K/6)Y_1^3/r_1^4$
Transverse force, Q_F; $\Omega_K = 3Q_F/4a^3$	$\Omega_K(a^2Y_2 - Y_2^3/3)(C_x - C_1 - X_2)$	–

*In polar coordinates r_1, θ; $r_1 = (X_1^2+Y_1^2)^{1/2}$; 2a is the inverse strip height (Fig. 2.10a,b).
Source: After Tseytlin (2006), reprinted with permission.

$\partial F_{sm}/\partial X_1$, and $\partial F_{sm}/\partial Y_1$ at $U_o \approx V_o \approx \vartheta_o \approx 0$ in accordance with the boundary-value conditions:

$$U_x = \frac{M_z x_1 y_1}{4Ga^2 w\Delta b\left(x_1^2 + y_1^2\right)^2}\left[2\left(1-\nu_p\right) - \frac{y_1^2}{x_1^2 + y_1^2}\right]; \tag{2.24}$$

$$V_y = \frac{M_z}{8Ga^2 w\Delta b\left(x_1^2 + y_1^2\right)^2}\left[\frac{3}{2}\left(1-\nu_p\right)\left(x_1^2 - y_1^2\right) - \frac{y_1^2\left(y_1^2 - 3x_1^2\right)}{x_1^2 + y_1^2}\right]. \tag{2.25}$$

Let us recall that concentrated moments at inversion are expressed as a single-valued unambiguous function:

$$M_b / h_b \Delta b = M / a \text{ and } M = M_b a / (h_b \Delta b), \tag{2.26}$$

where M_b is the bending moment in the hinge field. Stress σ_x in the hinge body under bending moment is equal to the second partial derivative of the stress function F_{sm}. Therefore, at the side surface center point with $Y_1 = \Delta b$, $X_1 = l_x$, we have

$$\sigma_x = \partial F_{sm}^2 / \partial Y_1^2 \Big|_{\Delta b,\, lx} \approx 1.5\, M_b / \left(\Delta b^2 h_b\right). \tag{2.27}$$

It is known that the boundary in the initial system that is free of distributed tension before the inversion is transformed into a boundary, subject to normal distributed tension in the inverse system. This tension rate has the same value at all points of the transformed boundary, and its effect on the hinge's kinematics is negligible. The stress concentration factor will be discussed later in this section. The tilt angle of the hinge center line may be derived from (2.24) and (2.25) such that

$$\theta = \partial V_y / \partial X_1 \Big|_{Y_1=0} = 3M_b\left(1-\nu_p\right) / \left(8Ga^2 h_b \Delta b^3 X_1\right). \tag{2.28}$$

The rigid body rotation angle θ_z on the hinge equals to two times the center line tilt at the center hinge point ($X_1 = l_x$, $Y_1 = 0$) and therefore

$$\theta_z = 3M_b\left(1-\nu_p\right) l_x / \left[4G(\Delta b)^3 h_b\right]. \tag{2.29}$$

Hence, the rotational compliance of the hinge is consequently equal to:

$$Clr = \theta_z / M_b. \tag{2.30}$$

2.6.2 Circular Notch Hinge

Contour approximation. The circles' effective shift l_x and effective radius R_c of the contour map may be found by using the least maximum absolute error approximation with Chebyshev's polynomial of the second order $R_2(Z)$ that has the smallest deflection $\tilde{\gamma}(Z)$ from zero in the range of $Z \subseteq [0,1]$. If the ratio δ_b

of the effective circle deflection from the real profile to the crosspiece thickness of the hinge is assumed to be uniformly small, then the following equation with this condition can be used:

$$\delta_b = (1/\Delta b)\left\{\left[X^2/\left(RA_\xi \xi^2\right)\right] + \Delta b - \left(X + l_x\right)^2/\left(R_c A_\xi \xi^2\right)\right\} \subseteq \{\gamma(X)\}, \quad (2.31)$$

where $\{\gamma(X)\}$ is the allowed relative error of the contour approximation.

The values of the constants A_ξ and ξ depend on the range of the hinge's active contour polar angle (Fig. 2.9d),

$$\vartheta_m = \sin^{-1}\left(X_{max}/R\right), \quad (2.32)$$

on the hinge body at the profile under consideration, as well as its presentation in accordance with the corresponding trigonometric function expansion by a series with one variable member (minimal polynomial). Note that the overall width of the hinge body is equal to: $W = h_{nh} = 2R(1 - \cos\vartheta_m) + t_h$. Here we have the following three options: (i) the thin notch hinges where the most significant part of deformation occurs within the boundaries of the hinge body corresponding to the polar angle $\vartheta_m \subseteq \pm 0.8$ rad; (ii) the thick notch hinges where the boundaries of approximation pertain to the whole profile at $X/R = 1.0$ and $\vartheta_m \subseteq \pm(\pi/2)$ rad; the width of the hinge body in this case is $W = 2R + t_h$ and $W/2R = 1 + \beta$; and (iii) the intermediate notch hinges where the polar angle of the hinge body is considered within the limits $\vartheta_m \subseteq \pm 1.2$ rad.

(i) For the thin notch hinge with small ratio, $2\beta = t_h/R$ in the range $2\beta \leq 0.07$, the best approximation of the functions $\sin\vartheta_m$ and $(1 - \cos\vartheta_m)$ by a series with one variable member corresponds to: $A_\xi = (1/0.478) = 2.092$ and $\xi = 0.922$; (ii) for the thick notch hinge with the ratio $2\beta = t_h/R$ in the range $0.20 < 2\beta \leq 1.0$, the best possible expansion of the functions $\sin\vartheta_m$ and $(1 - \cos\vartheta_m)$ by a series with one variable member corresponds to $A_\xi = (1/0.421) = 2.375$ and $\xi = 0.72$; and (iii) for the intermediate notch hinges with the ratio $2\beta = t_h/R$ within $0.07 < 2\beta \leq 0.20$, the effective expansion of the functions $\sin\vartheta_m$ and $(1 - \cos\vartheta_m)$ by a series with one variable member corresponds to $A_\xi = 2.375$ and $\xi = 0.82$.

These approximations give us the option to derive concise formulae with comparatively moderate uncertainty. The maximum uncertainty of approximation for the angle $\vartheta_m \subseteq \pm 0.8$ in the sin function is within ±0.02, and in the $(1 - \cos\vartheta)$ function is within ±0.003. We see their effectiveness in comparison with the "exact" but "unwieldy" formulae of Slocum, Paros, and Weisbord, their simplified representation for thin circular notch hinges, and the empirical Smith's formula derived from the finite element analysis data.

Substitution of $X = X^\circ X_{max}$ into (2.31) yields an equation that may be approximated by Chebyshev's polynomial $R_2(X^\circ)$ of second order from which a system of following equations is derived:

$$R_c = l_x^2/Q_b, \quad (2.33)$$

$$l_x = \left\{ R + \left[R^2 + (0.8284)^2 RX_{\max}^2 / Q_b \right]^{1/2} \right\} / \left(0.8284\, X_{\max} / Q_b \right), \quad (2.34)$$

where $Q_b = A_\xi \xi^2 \Delta b$.

General equation. Substitution of l_x expression from (2.34) into (2.29) yields the general equation for the angle θ_z of the circular notch hinge rotation under bending moment M_b as

$$\theta_z = \frac{3M_b\left(1-v_p^2\right)\left\{1+\left[1+0.6862\left(X_{\max}/R\right)^2 (R/\Delta b)/\left(A_\xi \xi^2\right)\right]^{1/2}\right\}}{2(0.8284)Eh_b(\Delta b)^2\left(X_{\max}/R\right)/\left(A_\xi \xi^2\right)}. \quad (2.35)$$

Let us recall that Young's modulus $E = 2G\,(1 + v_p)$. Substituting Poisson's ratio by $v_p = 1/3$ (typical for many metals), we finally obtain

$$\theta_z = \frac{4M_b\left\{1+\left[1+0.6862\left(X_{\max}/R\right)^2 (R/\Delta b)/\left(A_\xi \xi^2\right)\right]^{1/2}\right\}}{3(0.8284)Eh_b(\Delta b)^2\left(X_{\max}/R\right)/\left(A_\xi \xi^2\right)}. \quad (2.36)$$

Therefore, for thin circular notch hinges after the substitution of the corresponding A_ξ and ξ into (2.36), the rotational compliance in the thin hinge case (i) is equal to:

$$Clr\,\mathrm{i} = 4\left\{1+[1+0.1986(R/\Delta b)]^{1/2}\right\}/\left[Eh_b(\Delta b)^2\right]. \quad (2.37)$$

The coefficient 0.1986 may be changed to 0.215 if we assume that angle $\vartheta_m \subseteq \pm\, 0.9$ rad.

For thick notch hinges (ii), the rotational compliance is

$$Clr\,\mathrm{ii} = 4\left\{1+[1+0.5573(R/\Delta b)]^{1/2}\right\}/\left[2Eh_b(\Delta b)^2\right]. \quad (2.38)$$

If Poisson's ratio $v_p \neq 0.333$, multiply Clr ii obtained from (2.38) by factor $(1-v_p^2)/0.889$.

For intermediate notch hinges (iii), we have

$$Clr\,\mathrm{iii} = 4\left\{1+[1+0.373(R/\Delta b)]^{1/2}\right\}/\left[1.45Eh_b(\Delta b)^2\right]. \quad (2.39)$$

2.6.3 Elliptical Notch Hinge

Contour approximation. In Cartesian coordinates, the elliptical profile may be specified by two semiaxes, a_x and a_y (Fig. 2.9b) along X and Y coordinates at elliptical ratio, $\varepsilon = a_x/a_y$. It is easy to show that for elliptical notch hinge's compliance (stiffness), the radius R in (2.35)–(2.39) should be replaced by the ratio $a_x^2/a_y = \varepsilon\, a_x = R_e$ because for an ellipse $Y_e = a_y \cos\vartheta$, $X_e = a_x \sin\vartheta$, $\vartheta = X/(a_x\xi)$, and $Y = a_y - a_y \cos\vartheta = a_y(1 - \cos\vartheta) = a_y X^2/a_x{}^2 A_\xi \xi^2$. Hence, the relative deflection for the elliptical profile is

$$\delta_{be} = (1/\Delta b)\left\{\left[a_y X^2/\left(a_x^2 A_\xi \xi^2\right)\right] + \Delta b - \left(X + l_x\right)^2/\left(R_e A_\xi \xi^2\right)\right\} \subseteq \{\gamma(X)\}, \quad (2.40)$$

and the general and thickness-dependent equations for elliptical hinge compliance remain the same as those for the circular hinges, if we accept that $a_x^2/a_y = R_e$ and radius R is substituted by a conditional radius, $R_e = \varepsilon\, a_x$. In this case, we have

$$\theta_{ze} = \frac{3M_b\left(1-\nu_p^2\right)\left\{1+\left[1+0.6862\left(X_{max}/R_e\right)^2\left(R_e/\Delta b\right)/\left(A_\xi \xi^2\right)\right]^{1/2}\right\}}{2(0.8284)Eh_b(\Delta b)^2\left(X_{max}/R_e\right)/\left(A_\xi \xi^2\right)}. \quad (2.41)$$

At $\nu_p = 1/3$, compliance for thin elliptical hinges will be predicted by equation $(i)_e$

$$Clr\,\mathrm{ie} = 4\left\{1+\left[1+0.1986\left(R_e/\Delta b\right)\right]^{1/2}\right\}/\left[Eh_b(\Delta b)^2\right]. \quad (2.42)$$

For thick elliptical hinges, compliance will be predicted by equation $(ii)_e$

$$Clr\,\mathrm{iie} = 4\left\{1+\left[1+0.5573\left(R_e/\Delta b\right)\right]^{1/2}\right\}/\left[2Eh_b(\Delta b)^2\right]. \quad (2.43)$$

If Poisson's ratio $\nu_p \neq 0.333$, *Clriie* should be multiplied by the factor $\left(1-\nu_p^2\right)/0.889$.

For intermediate elliptical hinges, compliance will be predicted by equation $(iii)_e$

$$Clr\,\mathrm{iiie} = 4\left\{1+\left[1+0.373\left(R_e/\Delta b\right)\right]^{1/2}\right\}/\left[1.45Eh_b(\Delta b)^2\right]. \quad (2.44)$$

Relative thickness ($\beta = t_h/2R$) for elliptical hinges should be replaced by $\beta_e = t_h/2R_e$. The thin elliptical hinge equation $(i)_e$ is recommended at $\beta_e < 0.007$ and elliptical ratio $\varepsilon > 5$; the thick elliptical hinge equation $(ii)_e$ is for hinges with $0.1 < \beta_e < 0.3$; the intermediate elliptical hinge equation $(iii)_e$ is effective for hinges with $0.007 < \beta_e \leq 0.1$. If $\varepsilon = 1$ and $a_x = R$, any of these equations (2.41–2.44), will converge to similar equations for circular hinges (2.36–2.39).

Our solutions for cylindrical and elliptical notch flexure hinges (Eqs 2.36–2.44) were used by many researchers and developers as convenient, and sufficiently accurate options. Lin et al. (2013), for example, concluded: "Tseytlin (2002) used the inverse conformal mapping of circular approximating contour to deduce closed-form equations of the circular flexure hinges and the elliptical flexure hinges. Predictions of the developed models are likely to be much closer to FEA and experimental data than other theoretical equations." Now we can show that the solution with the inverse conformal mapping may be applied to many other forms of concave flexure hinges.

2.6.4 Computerized Approximation of Concave Contours

Using Eqs (2.29) and (2.30), we yield a quasiuniversal formula that is working for rotational compliance (or rotational stiffness) estimation of hinges with many smooth profiles, including cylindrical, elliptical, parabolic, hyperbolic, segmented, and V-shaped forms. It is presented by the following equation with only one functional variable shift l_x of approximating circles:

$$\alpha_z/M_z = 3(1-v_p)l_x/[4G(\Delta b)^3 w], \qquad (2.45)$$

where α_z is the angle of hinge rotation under the action of moment M_z.

We see how simple and tractable is this equation in comparison with the "precision" complicated, unwieldy equations for conic flexures (Chen et al., 2008, 2009, 2010) developed by regular integration method for Euler–Bernoulli beam.

The general Eq. (2.45) is common for all types of hinges with a possible approximation of their profiles by two shifted contiguous circles, and includes material parameters with Young's (or shear) modulus and Poisson's ratio. The latter is totally omitted in the known theoretical solutions by other authors. New concise equations are derived from the original general equation on the basis of trigonometric functions' simplified series expansion in certain ranges of hinge crosspiece relative thickness, and should be used accordingly: (i) at $\beta <= 0.035$; (ii) at $0.1 < \beta \le 0.3$; (iii) at $0.035 < \beta \le 0.1$ for the thin, thick, and intermediate circular hinges, respectively; $(\text{i})_e$ at $\beta_e = t_h/2R_e = (t_h a_y/2a_x^2) \le 0.007$, $\varepsilon > 5$; $(\text{ii})_e$ at $0.1 < \beta_e \le 0.3$; $(\text{iii})_e$ at $0.007 < \beta_e \le 0.1$ for the thin, thick, and intermediate elliptical hinges, respectively, with elliptical ratio ε of up to 10.

In these cases, the difference between theoretical rotational stiffness, properly built finite element analysis, and experimental models should not be more than 10%.

Let us now compare the results of rotational compliance α_z/M_z estimation for the elliptical arc and circular hinges (Table 2.6) with the "precision" Chen et al. (2008) noted as (C), Tseytlin $(\text{Ts})_i$ approximate formula (2.45), FEM (Fi), and experimental (Ex) methods data in Table 2.7.

$(\text{Ts})_i$ size-dependent circular and elliptical formula in the publication by Tseytlin (2011) is as follows:

$$\alpha_z/M_z = 4\{1+[1+q_\alpha(R_e/\Delta b)]^{1/2}\}/[\mu_\alpha Ew(\Delta b)^2]$$

with $R_e = a_x^2/b_x = \varepsilon a_x$ for ellipse and for circular arc $R_e = R$.

TABLE 2.6 Design Samples (Chen et al., 2008) of Flexure Hinges ($E = 2.07 \times 10^2$ GPa, $G = 0.81 \times 10^2$ GPa, $v_p = (E/2G) - 1 = 0.278$, depth $w = 10$ mm, thickness $t_h = 2\Delta b = 1$ mm, and $X_{max} = c = 5$ mm)

Sample	Type	ϕ_m (°)	a (mm)	b (mm)	R_e (mm)	q_α	μ_α	ε
1	Elliptic arc	45	7.071	4	12.5	0.1986*	1	1.77
2	Elliptic arc	60	5.774	4	8.33	0.373	1.45	1.44
3	Elliptical	90	5	4	6.25	0.5573	2	1.25
4	Circular	60	5.774	5.774	5.774	0.373	1.45	1
5	Right-circular	90	5	5	5	0.5573	2	1

Here $q_\alpha = [2e/(1+e)] + 1/s'$, $s' = p/t_h$; p is the focus distance parameter, e is the eccentricity, X_{max} is the maximum active contour coordinate under consideration; $\phi_m = \vartheta_m$ (Fig. 2.9); $\varepsilon = a/b$, μ_α is a thickness factor (1 – thin, 1.45 – intermediate, 2 – thick), and a and b are the semiaxis of an ellipse.
*May be changed to $q_\alpha = 0.215$ at $\phi_m = \vartheta_m = 52°$.
Source: After Tseytlin (2011), reprinted with permission.

(C) formula in the publication by Chen et al. (2008) is presented as follows:

$$\left(12a/Ewb^3\right)\frac{12s^4(2s+1)}{(4s+1)^{5/2}}\arctan\left(\sqrt{4s+1}\tan\frac{\varphi_m}{2}\right)$$
$$+\frac{2s^3(2s+1)(6s^2+4s+1)\sin\varphi_m}{(4s+1)^2(1+2s-2s\cos\varphi_m)^2}-\frac{2s^4(12s^2+4s+1)\sin\varphi_m\cos\varphi_m}{(4s+1)^2(1+2s-2s\cos\varphi_m)^2},$$

where $s = b/t_h$.

The uncertainties |δ(Fi)| and |δ(Ex)| show the difference between those calculated with certain equations and corresponding FEM and experimental Ex data. The average uncertainty for C calculations is equal to 3.75% = (2.9%Fi + 4.6%Ex)/2 in comparison with FEM and experimental Ex data and average uncertainty for $(Ts)_i$ calculations in comparison with FEM and experimental Ex data is 3.85% = (5.5%Fi + 2.2%Ex)/2. Therefore, we see that the accuracy of calculations with a correctly applied approximation method by inverse conformal mapping at the more tractable equation $(Ts)_i$ with only one functional variable l_x in the first power, and the more complex so-called "precision" equation of generalized integration method (C) are similar. Note that differences between FEM|δ(Fi)|and experimental |δ(Ex)| results in the publication by Chen et al. (2008) are up to 9.5%.

Data for approximating circles displacement (shift) distance l_x for $(Ts)_i$ function in Table 2.7 are found by calculation with the formula for circular and elliptical arc contours:

$$l_x = \left\{R_e + \left[R_e^2 + (0.8284)^2 R_e X_{max}^2/Q_b\right]^{1/2}\right\} / \left(0.8284\, X_{max}/Q_b\right), \tag{2.46}$$

where $Q_b = A_\xi \xi^2 \Delta b$ with $A_\xi \xi^2 = 1.2312$ for thick hinges case.

TABLE 2.7 **Comparison Between FEM (Denoted Fi), Experimental (Ex), and Calculated Results (C) for Chen et al. (2008), and $(Ts)_i$ for Tseytlin (2002, 2006, 2011) of Rotational Compliance α_z/M_z**

Sample (method)	α_z/M_z (rad/N·m)	$\lvert\delta(Fi)\rvert$ (%)	$\lvert\delta(Ex)\rvert$ (%)	l_x (mm)
1 (C)	2.311×10^{-2}	4.5	5.2	
1 (Fi)	2.421×10^{-2}	–	0.7	
1 (Ex)	2.437×10^{-2}	0.7	–	
1 $(Ts)_i$*	2.565×10^{-2}	5.9	5.2	5.1964
2 (C)	1.904×10^{-2}	3.3	2.3	
2 (Fi)	1.970×10^{-2}	–	5.8	
2 (Ex)	1.861×10^{-2}			
2 $(Ts)_i$*	1.885×10^{-2}	4.8	1.9	3.8189
3 (C)	1.652×10^{-2}	1.9	7.4	
3 (Fi)	1.684×10^{-2}	–	9.49	
3 (Ex)	1.538×10^{-2}	8.85	–	
3 $(Ts)_i$*	1.530×10^{-2}	9.1	0.5	3.099
4 (C)	1.603×10^{-2}	1.7	4.8	
4 (Fi)	1.631×10^{-2}	–	6.67	
4 (Ex)	1.529×10^{-2}	6.25	–	
4 $(Ts)_i$*	1.546×10^{-2}	5.2	1.1	3.1313
5 (C)	1.488×10^{-2}	2.98	3.1	
5 (Fi)	1.445×10^{-2}	–	0.14	
5 (Ex)	1.443×10^{-2}	0.14	–	
5 $(Ts)_i$*	1.401×10^{-2}	3.0	2.9	2.8398

Data for (Fi) and (Ex) are from Chen et al. (2008), α_z is the angle of the hinge rotation at action of applied moment M_z.
*With modified $E^* = E/(1 - \nu_p^2) = (E/0.923)$ and $G^* = G/(1 - \nu_p^2) = (G/0.923)$ moduli.
Source: After Tseytlin (2011), reprinted with permission.

Note that Eq. (2.46) is similar to Eq. (2.34). In these equations, the shift l_x has an analytical expression. However, l_x can be easily found with the help of AutoCAD by indicating only three corresponding points (3P method) of hinge profile, especially for a thin notch hinge with any smooth concave contour. Figure 2.11 shows these points for the arcs with parameters from Table 2.8: ellipses 1 and 2 and circles 3–5. Point A in each curve corresponds to coordinate $x = 0.8284X_{max}$ ($x = 3.3136$ mm for curves 1–4 and $x = 2.4852$ mm for curve 5), point O_p corresponds to the position of the hinge contour center with $x = 0$,

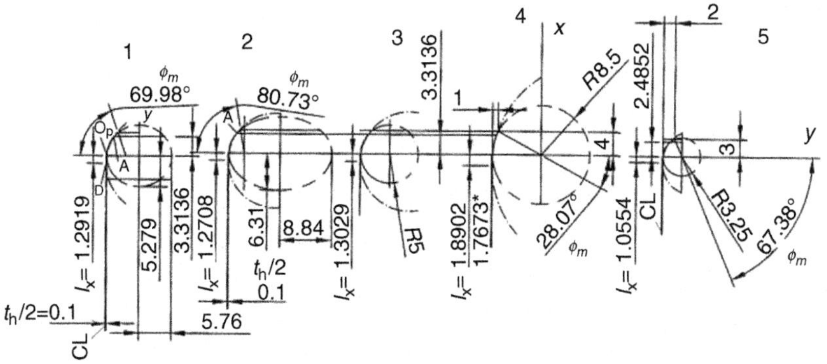

FIGURE 2.11 AutoCAD semiautomatic approximation of elliptical and circular profiles 1–5 (solid lines with their imaginary continuation in dashed lines) at specified arc angles with approximating circles shifted by l_x (dashed-dotted lines) contiguous to the hinge center line CL. *(After Tseytlin, 2011, reprinted with permission.)*

$y = 0$, and point D is the tangent point between corresponding approximating circles (dash-dot lines) and the center straight line (CL) of the hinge in initial position with $y = -t_h/2$. AutoCAD finds the latter point automatically by snap. Note also that the factor 0.8284 corresponds to the root of Chebyshev's second degree polynomial. Use the smaller of two l_x values in the case of estimation by analytical and AutoCAD based approximating methods for the same notch flexure hinge sample.

TABLE 2.8 Design Samples of Conic Flexure Hinges ($w = 10$ mm, $t_h = 0.2$ mm, $E = 2.07 \times 10^2$ GPa, $G = 0.81 \times 10^2$ GPa, $v_p = (E/2G) - 1 = 0.278$)

Sample	Figures	Type	a (mm)	b (mm)	ϕ_m (°)	X_{max} (mm)	l_x (mm)
1	2.11	Elliptic	5.76	5.279	69.98	4	1.2919
2	2.11	Elliptic	8.84	6.31	80.73	4	1.2708
3	2.11	Circular	5.00	5.00	53.13	4	1.3029
4	2.11	Circular	8.5	8.5	28.07	4	1.7673*
5	2.11	Circular	3.25	3.25	67.38	3	1.0554
6	2.12a	Parabolic	$p = 4$ mm; $e = 1$		90	4	1.2442
7	2.12b	Parabolic	$p = 5$ mm; $e = 1$		43.6	2	1.3549
8	2.12c	Hyperbolic	$P = 0.73$ mm; $e = 2.2$		115.99	4	2.2981

Calculated with Eq. (2.46) at $A_\xi \xi^2 = 1.778$ for thin notch hinge.
Source: After Tseytlin (2011), reprinted with permission.

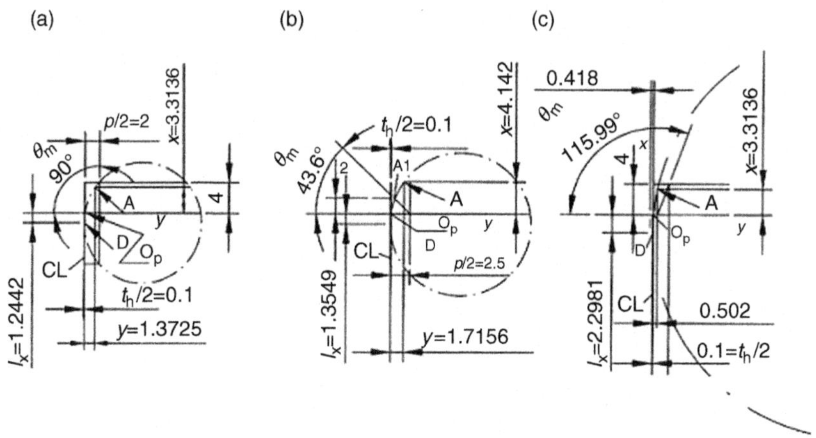

FIGURE 2.12 AutoCAD semiautomatic approximation of circles shifted by l_x and contiguous to the hinge CL for parabolic contours, with specified focus parameter. (a) p = 4 mm and arc angle θ_m = 90°; (b) p = 5 mm and θ_m = 43.6°; and (c) for hyperbolic contour (Table 2.8, sample 8), β = (major semiaxis/p) = 0.418/0.73 = 0.57. *(After Tseytlin, 2011, reprinted with permission.)*

We estimate in Figure 2.12a,b the approximating circles (dash-dot lines) shift distance l_x for the parabolic curves, with the help of the same AutoCAD 3P circle method, even without presentation of these curves' contour images. Point A of an approximating circle in the case of parabola 6 (Table 2.8, and Fig. 2.12a) has coordinates $x = 0.8284p = 3.3136$ mm, and $y = x^2/2p = 1.3725$. The latter value follows from a parabola common equation. Point A of an approximating circle for parabola 7 (Table 2.8) has coordinates $x = 0.8284p = 4.142$ mm, and $y = x^2/2p = 1.7156$ mm (Fig. 2.12b), which are estimated by the same method as for parabola 6. The deflection of the specified end point A_1 from the approximating circle in this case is only 0.13 mm.

Estimation of the approximating circles shift to contiguity l_x for a hyperbolic curve may be executed with the same AutoCAD 3P circle method, as for the parabolic one, by using the regular hyperbola equation. Chen et al. (2009), meanwhile, evaluates the "extended" nonsymmetric hyperbolic profile that requires the consideration of the hyperbola's major semiaxis as a thickness factor.

The example of such approximation (major semiaxis 0.418 mm, $\beta = 0.57$) is shown in Figure 2.12c, and Tables 2.8 and 2.9 (sample 8).

2.6.5 Instantaneous Center of Rotation

An elastic hinge body doesn't have actual guiding surfaces, and its center of rotation is a virtual instantaneous one. The position of this center in the hinge body can be estimated by an experimental study of its motion (Tseytlin, 2002). The hinge rotation virtual center also can be estimated by calculations because

TABLE 2.9 Comparison (Tseytlin, 2011) of Rotational Compliance α_z/M_z and Shift of Instantaneous Rotation Center d_{ea} Estimated by (2.45) and (2.51a) at Evaluated l_x Values with the Corresponding Values in Chen et al. (2009)

Sample no.	1	2	3	4	5	6	7	8
α_z/M_z, (l_x),*rad/N·m Tseytlin (2011)	0.863	0.849	0.871	1.09**	0.706	0.767**	0.836**	0.210
α_z/M_z,*rad/N·m Chen et al. (2009)	0.846	0.822	0.859	1.109†	0.682†	0.775	0.863 Ex	0.216
$d_{ea} \times 10^4$, m,* Tseytlin (2011)	2.06	2.02	2.07	2.81	1.68	1.98	2.16	$l_x/3 = 0.77$
$R_{ae} \times 10^4$, m,* Chen et al. (2009)	2.05	1.99	2.09	≈2.75	1.65	1.91	2.18†	≈0.80

The difference of corresponding values α_z/M_z are not more than 5% and for d_{ea} and R_{ae} are not more than 4%.
**Calculated with modified shear modulus $G^*=G/(1-\nu_p^2)$.
†Calculated with Chen et al. (2008, 2009) formulas because calculated data for these samples are not provided in Chen et al. (2008).
Source: After Tseytlin (2011), reprinted with permission.

its position corresponds to the first kinematic pole of the rigid body connected to the hinge. Its possible displacement from a circle should be less for thinner hinges with a smaller profile radius. Additionally, the elliptical hinge with the same length as the circular one, $2a_x = 2R$, has a lager conditional radius, $R_e = \left(a_x^2/a_y\right)$. Therefore, this profile may have a smaller concentration stress and a larger limit of possible elastic rotation, but probably does not provide the same level of ICR stability in the center of the hinge body. The elliptical hinge represents some kind of intermediate solution between the circular hinge and leaf spring. Additionally, the leaf spring hinge does not provide a stable position for the center of rotation that depends on the applied load and spring deflection from its initial position. Lobontiu et al. (2001) showed that, as a consequence of being more compliant than the right circular flexure hinges, the corner-filleted flexures are also less precise in keeping the center of rotation with minimum offset. The shift of the rotation center caused by a bending effect only can be 10–15 times larger in a corner-filleted flexure than the shift experienced by a similar right circular flexure hinge for the same amount of loading. For small- or microscale applications, where two links connected by a flexure must replicate pure rotation as close as possible, this aspect might pose serious limitations. Lobontiu (2002) has studied the precision of the rotation for notch flexure hinges with a conical section (cylinder, ellipse, parabola, and hyperbola) shape by evaluation of the hinge's center point displacement on the basis of the Castigliano's second theorem. However, this evaluation does not directly reflect the definition of the ICR position for the rotation of a kinematic body on the hinge under consideration.

Estimated values of the shift l_x allow us to find not only the rotational compliance α_z/M_z of the corresponding flexure notch hinges, but also the position of the instantaneous axis of rotation for a moving assembly 2 clamped to a hinge (Fig. 2.13a). This position for thin notch hinges ($A_\xi \xi^2 = 1.778$) corresponds to the first kinematic pole of a moving rigid body with coordinates

$$X_{p1} = X_o - dV_{yo}/d\alpha_z; Y_{p1} = Y_o + dU_{xo}/d\alpha_z, \tag{2.47}$$

where

$$dV_{yo}/d\alpha_z = \left(dV_{yo}/dM_z\right):\left(d\alpha_z/dM_z\right), \tag{2.47a}$$

Note that the detection of the approximating circles and their shift l_x to contiguity allows us to find the position (2.47) of the instantaneous center of the hinge rotation with necessary accuracy, without the influence of material elastic properties. The rotational compliance (stiffness) and the position of the instantaneous center of its rotation in the notch hinge plain are the main important properties of a notch hinge, requiring accurate estimation. Our tractable equations with the same one functional variable in the first power totally satisfy this requirement.

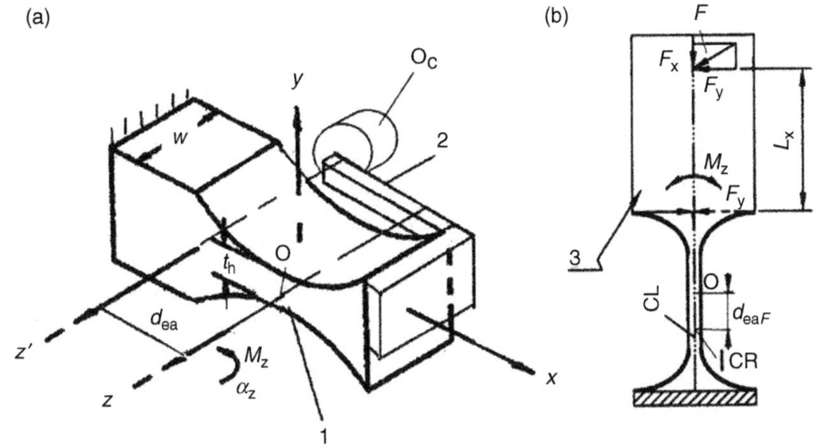

FIGURE 2.13 Position (a) of instantaneous rotation center O_c with axis z' in assembly 2 clamped to notch hinge 1, and (b) shift d_{eaF} of ICR from geometric center O on the segmented notch hinge with rigid link 3. *(After Tseytlin, 2012, reprinted with permission.)*

Using the expressions for displacements U_x and V_y in Eqs (2.24) and (2.25), we found the derivatives dU_{xo}/dM_z and dV_{yo}/dM_z which at $x_1 = l_x$ and $y_1 = 0$ in the geometrical center of notch hinges are equal to the following relations:

$$dU_{xo}/dM_z = 0,\quad dV_{yo}/dM_z = \frac{3(1-\nu_p)}{32Ga^2 l_x^2 w\Delta b}, \tag{2.48}$$

where $a = \frac{1}{2}R_c^{-1}$ and $R_c = \frac{l_x^2}{A_\xi \xi^2 \Delta b}$.

Substitution of the latter expressions into (2.48) and considering $A_\xi \xi^2 = 1.778$ as for thin notch hinges with $\beta < 0.035$ yields the following equation for the derivative dV_{yo}/dM_z:

$$dV_{yo}/dM_z = \frac{3(1-\nu_p)l_x^2}{25.29Gw(\Delta b)^3}. \tag{2.49}$$

And finally, dividing (2.49) by (2.45), we have the following sought expression:

$$dV_{yo}/d\alpha_z = d_{ea} = \frac{l_x}{6.32} \approx \frac{l_x}{2\pi}, \tag{2.50}$$

where $\pi = 3.14$.

Therefore, the best location of the instantaneous rotation center O_c and the axis z' in the assembly 2 (Fig. 2.13a) corresponds to their shift from the geometrical center O of the thin notch hinge 1 on axis z in the initial state by $d_{ea} = \frac{l_x}{2\pi}$.

For a thick hinge,

$$A_\xi \xi^2 = 1.2312 \text{ and } d_{ea} = l_x/3. \tag{2.51}$$

For intermediate hinge $A_\xi \xi^2 = 1.59695$ and $d_{ea} = l_x/5.1$. The average denominator for thick and intermediate hinges in d_{ea} formula (2.51a) is equal to 4.065, which corresponds to

$$d_{ea} = l_x/4.065. \tag{2.51a}$$

$X_o = 0$ and $Y_o = 0$ are the coordinates of the hinge center point O on the moving body clamped to it, $V_{yo} = \varepsilon_y$ and $U_{xo} \approx 0$ are the translational displacements of point O, and α_z is the body's 2 (clamped to the hinge) rotational angle. Therefore, the best location of instantaneous rotation center O_c and axis z' in the assembly 2 corresponds to their shift from the geometrical center O of the notch hinge 1 on axis z in the initial state by d_{ea}.

This shift value is not an error, and it indeed helps to minimize rotation error. However, it can cause the errors of the assembly 2 rotation if its specified axis does not coincide with the recommended position of instantaneous center. We can find a value similar to d_{ea} in Chen et al. (2009), where it is expressed as a "relative error":

$$\frac{\varepsilon_y/M_z}{\alpha_z/M_z} = R_{ae}.$$

Table 2.9 shows the comparison of rotational compliance α_z/M_z, instantaneous axis of rotation shift d_{ea} and "relative error" R_{ae} values based on our calculations with Eqs (2.45) and (2.51a) and Chen et al. (2009) results for the flexure notch hinge samples specified in Table 2.8.

The position of the ICR changes if the notch hinge is loaded by the bending force (vector $\vec{F} = \vec{F_y} + \vec{F_x}$) applied to the rigid link 3 (Fig. 2.13b) on the distance L_x from the hinge. In this case, the equivalent bending moment $M_z = F_y L_x$, constituent force $F_y = M_z/L_x$, and the constituent force F_x are applied to the end of the hinge.

We estimate that $dV_y/dF_y = Cl_{\delta_y F_y}$ and $d\theta_z/dF_y = Cl_{\theta_z F_y}$. As a result, ICR shift from the action of force F_y is $d_{ea\,F} = \frac{dV_y}{dF_y}\frac{dF_y}{d\theta_z} = \frac{Cl_{\delta_y F_y}}{Cl_{\theta_z F_y}}$. Therefore, the shift of ICR along the axis X from the geometric center point O of the flexure hinge (Fig. 2.13b) can be equal to $d_{eay} = d_{ea} \pm d_{eaF}$.

We can show that $d_{ea\,F} = \frac{dV_y}{dF_y}\frac{dF_y}{d\theta_z} \approx \frac{l^2}{3(l/2+L_x)}$ and $d_{eaF} \ll d_{ea}$ at $L_x \gg l$, where $l = X_{max}$ is a maximum distance of flexure hinge contour along X coordinate from its center.

We evaluate the shift of ICR from the geometric center point O of the segmented flexure hinge at the following relations of the load and compliances Cl_{ij}. The central part of the hinge (Fig. 2.15a) has length $l_c = 25.4 - 12 = 13.4$ mm, and $\Delta_{yF}/\alpha_z - l_c/2 = (2/3 - 1/2)l_c \approx 0.17 l_c$.

However, the influence of a lateral force F_y on the hinge kinematics will diminish with the increase of the distance L_x of this force application because the ratio of maximum bending deflections Δ_{yF} for the hinge in the direction Y by the force F_y and bending moment Δ_{yM} will diminish as well: $\Delta_{yF}/\Delta_{yM} = \dfrac{2l_c}{3L_x}$ at $L_x > l_c$.

Kinematics. As a reminder, the locus of ICRs in kinematics is called the immovable centrode C_e, and the locus of instantaneous center of velocities ICVs is called the movable centrode C_{e1}. It is clear that when the plane body moves, the previously mentioned centrodes roll one upon another without sliding, and their point of contiguity corresponds to ICR and ICV. The curvature of the trajectory of a body point's motion depends on the difference between the centrode curvatures ($K - K_1$), but not on the curvature of each of the centrodes. If the difference between the curvatures is constant, the centrodes may be replaced by a circle C_0 (Fig. 2.14) with diameter $2d_r$ and the straight line C_1 at the same curvature difference:

$$K - K_1 = K_0 - K_{10} = K_0 = 1/d_r, \tag{2.52}$$

where $K_0 = 1/d_r$ is a curvature of the circle with the radius d_r, and the curvature of straight line $K_{10} = 0$. The equation for estimating radius d_r is as follows:

$$d_r = \sqrt{\left[y_0'(\theta) + x_0''(\theta)\right]^2 + \left[y_0''(\theta) - x_0'(\theta)\right]^2}. \tag{2.53}$$

Circle C_0 with diameter $2d_r$ is called a circle of rolling, and circle C_n with diameter d_r is called a circle of rotation, or rotational circle. The straight line C_1 is the pole tangent and the perpendicular to it at point P_1 (ICR) straight line is

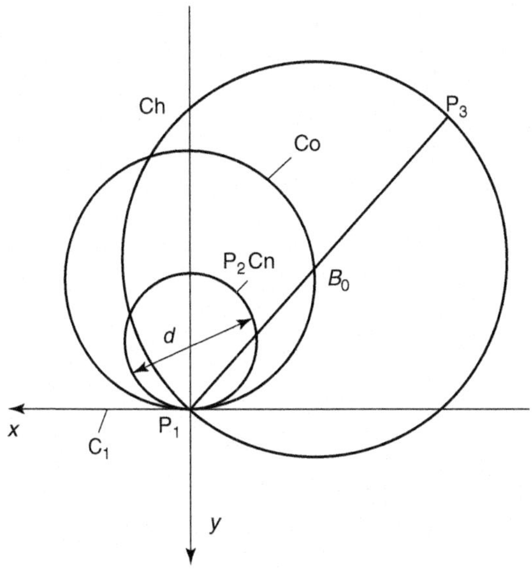

FIGURE 2.14 Rolling and rotation kinematic circles.

the pole normal. The trajectory of the points of the circle rolling on a straight line is known as a trochoid.

The point B_0 on the circle of rolling, for which the curvature of the trajectory $K = 0$ and its derivative $K' = 0$, is called the Ball's point. The trajectory of Ball's point in the limited path is close to the straight tangent line with the order that is not less than three. Chebyshev's point (Ch) has a trajectory even closer to a rectilinear one, whose contiguity to the tangent line is not less than the fifth order. The trajectory of the displacement of point P_c of the moving plane that corresponds at the first moment to the ICR (P_1) may be described by the following equations in x and y coordinates with the origin at P_{10};

$$x_{P_c} = \frac{2K_1 - K}{(K_1 - K)^2}\frac{\theta^3}{3!} + \frac{4K_1K' - KK' - 3K_1K'}{(K_1 - K)^3}\frac{\theta^4}{4!}; \tag{2.54}$$

$$\begin{aligned} y_{P_c} &= -\frac{1}{K_1 - K}\frac{\theta^2}{2!} + \frac{K_1' - K'}{K_1 - K}\frac{\theta^3}{3!} \\ &+ \frac{(K_1 - K)(K_1'' - K'') - 2(K_1' - K')^2 + 3K_1^2 - 3KK_1 + K^2}{(K_1 - K)^3}\frac{\theta^4}{4!}. \end{aligned} \tag{2.55}$$

At a small rotation angle θ of the circle rolling on the straight line, we have

$$x_{P_c} = (\theta - \sin\theta)/(K_0 - K_{10}) \approx -\theta^3(K_{10} - K_0)/\left[3!(K_{10} - K_0)^2\right]; \tag{2.56}$$

$$y_{P_c} = -(1 - \cos\theta)/(K_{10} - K_0) \approx -\theta^2/\left[2!(K_{10} - K_0)\right]. \tag{2.57}$$

Hence, at a small rotation angle θ of the circle rolling on the straight line, the cycloid trajectory of point P_c is close to the semicubic parabola. Therefore, the small shift of ICR at a finite angle of rotation corresponds to a cycloid that is close to semicubic parabola. At a finite angle of rotation within $\alpha_z = 0.1$ rad (5.7°), we have a small $X_{ICR} = \alpha_z^3 d_{ea}/6 = 0.00017 d_{ea}$ and $Y_{ICR} = \alpha_z^2 d_{ea}/2 = 0.005 d_{ea}$ shift of ICR coordinates from their initial $(-d_{ea}, 0)$ position, corresponding to the rolling without sliding for circle of rolling on the pole tangent (Tseytlin, 2006, 2012). This may be considered in many cases as a practically negligible shift.

2.6.6 Segmented Hinges

Segmented flexure notch hinges with changeable thickness and straight constant thickness parts can be also studied by estimation of approximating circles shift to contiguity l_x. It is better in this case to use Chebyshev's polynomial with minimum relative deflection from zero: $P_4^0(X) = XP_3(X) = X^4 - (3/4)X^2$ with X in the range $[-1, +1]$.

Figure 2.15a,b shows such an example of a corner-filleted flexure notch hinge rotational compliance calculation under the action of a bending moment M_z. The hinge has an elliptic part with semiaxis 6 and 8 mm, and a straight part

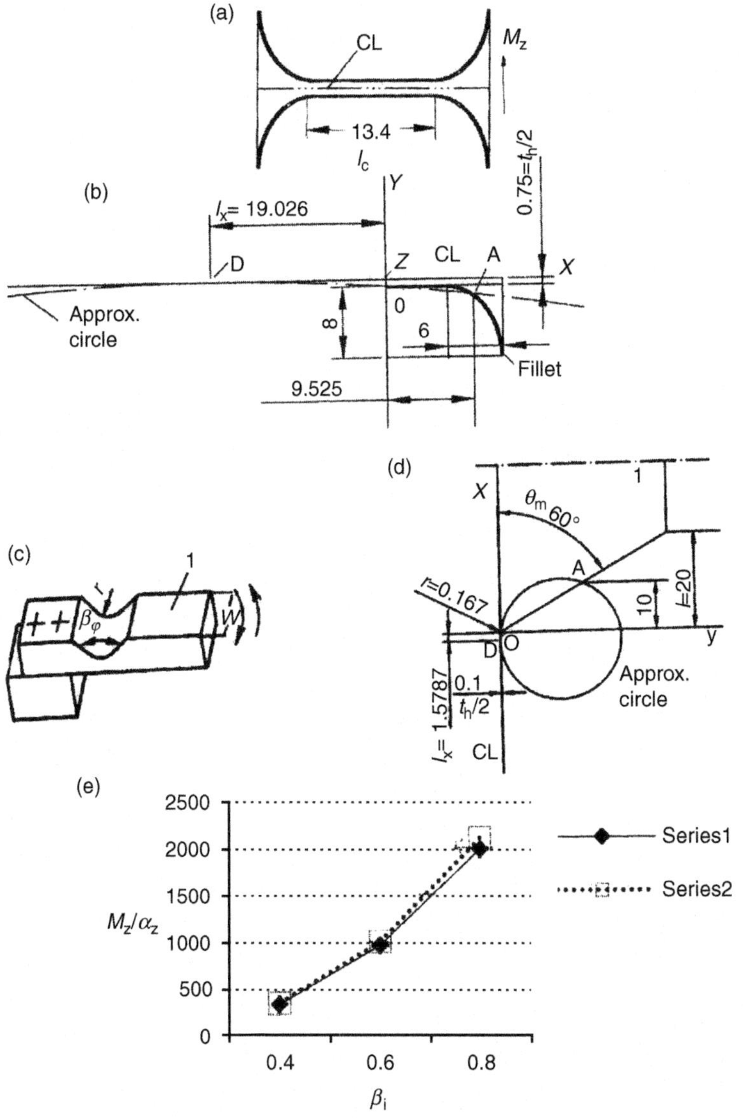

FIGURE 2.15 Segmented flexure notch hinge (a), the schematic of its approximating circles estimate (b); V-shaped flexure hinge (c), and schematic of its approximating circles estimate (d); and charts for (1) calculated and (2) FEM-supported data of the stiffness M_z/α_z in a V-shaped hinge (e). *(After Tseytlin, 2012, reprinted with permission.)*

with length $l_c = 13.4$ mm at a width of 1.5 mm and depth of the hinge $w_h = 6$ mm. Using the 3P method in AutoCAD circle identification program (Fig. 2.15b), we can find the necessary approximating circle that passes through point A on the elliptic contour with coordinate $X = 0.75(6 + l_c/2) = 0.75(6 + 6.7) = 9.525$, point O in the center of the straight segment contour, and automatically snaps to a tangential position on the center line (CL with dash-three dot line in Fig. 2.15a) of the hinge at point D on Figure 2.15b. As a result, we find the shift to contiguity of the approximating circles as $l_x = 19.026$ mm. Using Eq. (2.45), we estimate the rotational compliance of the segmented transverse symmetric hinge $\alpha_z/M_z = 0.145$ rad/N·m that deviates from corresponding experimental value $\alpha_z/M_z = 0.137$ rad/N·m (Lobontiu et al., 2011) by $|\delta|\% = 5.8\%$ at $G = 24.9$ GPa and $\nu_p = 0.32$ for aluminum (Table 2.10).

Figure 2.15c–e present an example of V-shaped notch flexure hinges computerized approximation with the help of AutoCAD circle 3P option and estimation on this basis of the shift to contiguity l_x for rotational compliance (or stiffness M_z/α_z) evaluation with the Eq. (2.45). For the sake of a comparison with Tian et al. (2010) results, we choose the following parameters for the V-shaped hinge: $\theta_m = 60°$, $\beta_1 = t_h/2r = 0.6$, $r = 0.167$ mm at $t_h = 0.2$ mm, $l = X_{max} = 20$ mm, and depth of the beam $w_h = 10$ mm; Poisson's ratio of the material $\nu_p = 0.28$, Young's modulus $E = 210$ GPa, and shift modulus $G = E/2(1 + \nu_p) = 82$ GPa for spring steel. Using Chebyshev's "relative" polynomial $P_3^0 = XP_2(X) = X(X^2 - 1/2)$ and providing the approximating circle through 3 points (A at 0, 5 X_{max}, O in the hinge contour center, and D as the snap tangential to the hinge center line CL),

TABLE 2.10 Comparison of Calculated (Tseytlin, 2012) and Experimental (FEM) Data for Segmented and V-Shaped Cantilevers

Type of flex hinge	3-Segment corner-filleted Fig. 2.15a,b, Al*		V-shaped, Fig. 2.15c–e Spring Steel		
Rotational compliance (stiffness)	Calculated α_z/M_z 0.145 rad/N·m $\lvert\delta\rvert\% = 5.8$	Experiment α_z/M_z 0.137 rad/N·m (Lobontiu et al., 2011)	**Calculated M_z/α_z 961.9 N·m/rad for β_1 1980 for β_2 343 for β_3	FEM M_z/α_z 1000.0 N·m/rad for β_1 (Tian et al., 2010) 2100 for β_2 357 for β_3	$\lvert\delta\rvert\%$ 3.8 5.7 3.9
$d_{ea} = dV_{y0}/d\alpha_z = \Delta y/\alpha_z$ shift of ICR, mm Fig. 2.13a,b	$l_x/2\pi$ 3.03	Experiment, d_{ea} 2.8 ± 0.5 (Tseytlin, 2002)	$l_x/4.065$ 0.312...$\beta_3 = 0.4$ 0.388...$\beta_1 = 0.6$ 0.452...$\beta_2 = 0.8$	$\Delta y/\alpha_z$, FEM (Tian et al., 2010) 0.324...$\beta_3 = 0.4$ 0.381...$\beta_1 = 0.6$ 0.466...$\beta_2 = 0.8$	$\lvert\delta\rvert\%$ 3.7 1.8 3.0

*Material is aluminum (Al).
**With scale factor of $\times 10^3$.
Source: After Tseytlin (2012), reprinted with permission.

we have $l_x = 1.5787$ mm and find $M_z/\alpha_z = 961.9$ N·m/rad from Eq. (2.45) with dimensional scale factor 10^3. This value differs from the result $M_z/\alpha_z = 1000$ N·m/rad for the same parameters on Figure 5a of Tian et al. (2010) that was supported by FEM, by only $|\delta|\% = 3.8\%$. We have $l_{x2} = 1.8399$ for β_2 case and find $M_z/\alpha_z = 1980$ N·m/rad with scale factor 10^3. This value differs by $|\delta|\% = 5.7\%$ from the FEM supported result $M_z/\alpha_z = 2100$ N·m/rad for the V-shaped hinge with the same parameters shown in Figure 5a of Tian et al. (2010). Figure 2.15e shows $l_{x3} = 1.2725$ mm for $\beta_3 = 0.4$ case that corresponds to $M_z/\alpha_z = 343$ N·m/rad from Eq. (2.45) with scale factor 10^3. This value differs by $|\delta|\% = 3.9\%$ from FEM supported result $M_z/\alpha_z = 357$ N·m/rad for a V-shaped hinge with the same parameters shown in Figure 5a of Tian et al. (2010). These data are presented in Figure 2.15e where charts series 1 and 2 correspond to our calculations and the FEM supported data, respectively.

Estimated values of the shift l_x allows us to find not only the rotational compliance α_z/M_z of the corresponding flexure notch hinges, but also the position of the instantaneous axis of rotation for a moving assembly clamped to the hinge.

The stress concentration factor. A rectangular beam with V-shaped notch has a stress concentration factor that depends on angle β_φ and radius r.

Eq. (2.27) defines the normal stress on the edges of a circular hinge crosspiece at its bending. However, the real stress in the beam's notch at its deformation increases by the so-called stress concentration factor K_t and is equal to:

$$\sigma_x = 6M_b K_t / t_h^2 h_b. \tag{2.58}$$

The problem with the stress concentration factor is indeed a complex one because this factor depends on the shape and parameters of notches in relation to the type of applied loads (Tseytlin, 2006; Chen et al., 2014). The simplest expression for the circular notch hinge stress concentration factor is presented by Smith (2000), where $K_t = (1 + \beta)^{9/20}$. If we use this expression for K_t and derive the expression for moment $M_b = M_z$ from Eq. (2.16), then σ_x can be presented by (2.59), after K_t and M_b substitution and simple cancellations, as follows:

$$\sigma_x = 4\left(\sqrt{2}\right)E\theta\left(1+\beta\right)^{9/20}\beta^{1/2}/3\pi. \tag{2.59}$$

It is clear from (2.59) that stress (at a specified angle of rotation θ) is smaller for relatively thin notch hinges with smaller values of β. Furthermore, the stress concentration factor is not to exceed 1.12 for a typical range of notch hinge's relative thickness $\beta = 0.01 - 0.3$. If we use $K_t = \sqrt{(1+\beta)}$ instead of $K_t = (1 + \beta)^{9/20}$, the error in the specified range of β will be less than 2%. In this case, it is easy to derive from (2.16), (2.58), and (2.59) the expression for the maximum allowable thickness t_{hmax} of the circular hinge crosspiece that is equal to

$$t_{hmax} = -r + \sqrt{r^2 + \left(\sigma_e/\theta_{max}\right)^2 9\pi^2 r/(16E^2)}, \tag{2.60}$$

where maximum angle θ_{max} of the hinge's rotation, yield stress σ_e, Young's modulus E of the material, and radius r of the hinge's profile should be specified. But for typical alternating stress in the notch hinge at work conditions, one should use endurance stress limit σ_{en} instead of yield stress σ_e value in the calculation of the maximum allowable thickness of the crosspiece. Therefore,

$$t_{hmax} = -r + \sqrt{r^2 + (\sigma_{en}/\theta_{max})^2 9\pi^2 r/(16E^2)}. \quad (2.61)$$

REFERENCES

Abdi, A., Pishkenari, H.N., Keramati, R., Minary-Jolandan, M., 2015. Dynamics of the nanoneedle probe in trolling mode AFM. Nanotechnology 26 (20), 205702.

Allemand, J.F., Bensimon, D., Croquette, V., 2003. Stretching DNA and RNA to probe their interactions with proteins. Curr. Opin. Struct. Biol. 13 (3), 266–274.

Amosov, I.S., Tseytlin, Y.M., 1969. Technique for evaluating the measurement effort. Meas. Tech. 12, 1459–1461.

Ansari, M.Z., Cho, C., 2009. Deflection, frequency, and stress characteristics of rectangular, triangular, and step profile microcantilevers for biosensors. Sensors 9, 6046–6057.

Arutyunov, P.A., Tolstikhina, A.L., 2002. Atomic force microscopy and universal means of measuring physical quantities in the mesoscopic length range. Meas. Tech. 45, 714–723.

Baumann, C.G., Smith, S.B., Bloomfield, V.A., Bustamante, C., 1997. Ionic effects on the elasticity of single DNA molecules. Proc. Natl. Acad. Sci. USA 94, 6185.

Bodensiek, K., Li, W., Sanchez, P., Nawaz, S., Schaap, I.A.T., 2013. A high-speed vertical optical trap for the mechanical testing of living cells at piconewton forces. Rev. Sci. Instrum. 84, 113707.

Braginsky, V.B., Manukin, A.B., 1977. Measurements of Weak Forces in Physics Experiments. Chicago University Press, Chicago.

Bustamante, C., Rivetti, C., 1996. Visualizing protein-nucleic acid interactions on a large scale with the scanning force microscope. Annu. Rev. Biophys. Biomol. Struct. 25, 395–429.

Butt, H.-J., Siedle, P., Seirt, K., Fendler, K., Seeger, T., et al.,1993. Scan speed limit in atomic force microscopy. J. Microsc. 169, 75–84.

Carr, D.W., Evoy, S., Sekaric, L., Graighead, H.G., Parpia, J.M., 1999. Measurement of mechanical resonance and losses in nanometer scale silicon wire. Appl. Phys. Lett. 75, 920–922.

Chen, G.Y., Warmack, R.J., Thundat, T., Allison, D.P., Huang, A., 1994. Resonance response of scanning force microscopy cantilever. Rev. Sci. Instrum. 65, 2532–2537.

Chen, G.Y., Thundat, T., Wachter, E.A., Warmack, R.J., 1995. Adsorption-induced surface stress and its effects on resonance frequency of microcantilevers. J. Appl. Phys. 77, 3618–3622.

Chen, G.M., Shao, X.D., Huang, X.B., 2008. A new generalized model for elliptical arc flexure hinge. Rev. Sci. Instrum. 79, 095103.

Chen, G.M., Liu, X., Gao, X.H., Jia, J., 2009. A generalized model for flexure hinges. Rev. Sci. Instrum. 80, 055106.

Chen, G.M., Du, Y., Liu, X., 2010. Note: supplements and corrections to the generalized conic flexure hinge model. Rev. Sci. Instrum. 81, 076101.

Chen, G., Wang, J., Liu, X., 2014. Generalized equation for estimating stress concentration factors of various notch flexure hinges. J. Mech. Design 136 (3), 031009.

Chen, L., Offenhäusser, A., Krause, H.-J., 2015. Magnetic tweezers with high permeability electromagnets for fast actuation of magnetic beads. Rev. Sci. Instrum. 86 (1–9), 044701.

Cheng, Y.T., Ni, W., Cheng, C.M., 2006. Nonlinear analysis of oscillatory indentation in elastic and viscoelastic solids. Phys. Rev. Lett. 97 (7), 075506.

Chumak, A.A., Milloni, P.W., Berman, G.P., 2004. Effects of electrostatic and Casimir force on cantilever vibration. Phys. Rev. B 70, 085407.

Clausen-Schaumann, H., Rief, M., Tolksdorf, C., Gaub, H.E., 2000. Mechanical stability of single DNA molecules. Biophys. J. 78, 1997–2007.

Derjaguin, B.V., Muller, V.M., Toporov, Y.P., 1975. Effect of contact deformations on the adhesion of particles. J. Colloid Interface Sci. 53, 314–326.

Dolitskii, I.N., Fedorenko, T.A., 1972. Calculating the maximum attainable natural frequencies of elastic elements in electromechanical pressure transducers. Meas. Tech. 15, 1040–1042.

Efimov, A.A., Ivanov, V.V., Volkov, I.A., Lizunova, A.A., Lisovskii, S.V., Ermakova, M.A., 2014. The determination of the effective radius of the tip of the probe on an Atomic Force Microscope using monodispersed silicon oxide nanoparticles. Meas. Tech. 56 (12), 1343–1346.

Essevaz-Roulet, B., Bocklemann, U., Hesiot, F., 1997. Mechanical separation of the complementary strands of DNA. Proc. Natl. Acad. Sci. USA 94, 10114–11935.

Fan, L., Potter, D., Sulchek, T., 2012. Constant tip-surface distance with atomic force microscopy via quality factor feedback. Rev. Sci. Instrum. 83, 023706.

Gecevičius, M., Drevinskas, R., Beresna, M., Kazansky, P., 2014. Single beam optical vortex tweezers with tunable orbital momentum. Appl. Phys. Lett. 104, 231110.

Gosse, C., Croquette, V., 2002. Magnetic tweezers: micromanipulation and force measurement at the molecular level. Biophys. J. 82, 3314–3329.

Husain, A., Hone, J., Postma, H.W.Ch., Huang, X.M.H., Drake, T., Barbic, M., Scherer, A., Roukes, M.L., 2003. Nanowire-based very-high-frequency electromechanical resonator. Appl. Phys. Lett. 83, 1240–1242.

Ilic, B., Krylov, S., Craighead, H.G., 2010. Young's modulus and density measurements of thin atomic layer deposited films using resonant nanomechanics. J. Appl. Phys. 108, 044317.

Jing, G.V., Ma, J., Yu, D.P., 2007. Calibration of the spring constant of AFM cantilever. J. Electron Microsc. 56 (1), 21.

Johnson, K.L., Kendall, K., Roberts, A.D., 1971. Surface energy and the contact of elastic solids. Proc. R. Soc. London Ser. A 324 (1558), 301–313.

Jones, R.V., 1988. Instruments and Experience. Wiley, London.

Joshi, C.H., 2000. Energen, Inc. Compact magnetostrictive actuators and linear motors. In: Actuator Conference, Bremen, Germany, June.

Kachanov, M., Shafiro, B., Tsurkov, I., 2003. Handbook of Elasticity Solutions. Kluwer Academic Publishers, Dordrecht; Boston, USA.

Korayem, M.H., Rastegar, Z., Taberi, M., 2012. Sensitivity analysis of nano-contact mechanics models in manipulation of biological cell. Nanosci. Nanotechnol. 2 (3), 49–56.

Landau, L.D., Lifshitz, E.M., 1969. Statistical Physics (J.B. Sykes, M.J. Kearsley, Trans. from Russian). Pergamon Press, Oxford, UK.

Lange, D., Brand, O., Batles, H., 2002. CMOS Cantilever Sensor Systems: Atomic Force Microscopy and Gas Sensing Applications. Springer, Berlin.

Lansdorp, B.M., Tabrizi, S.J., Dittmore, A., Saleh, O.A., 2013. A high-speed magnetic tweezers beyond 10,000 frames per second. Rev. Sci. Instrum. 84, 044301.

Lin, R., Zhang, X., Long, X., Fatikow, S., 2013. Hybrid flexure hinges. Rev. Sci. Instrum. 84, 085004.

Lobontiu, N., 2002. Compliant Mechanisms: Design of Flexure Hinges. CRC, Boca Raton, FL.

Lobontiu, N., Paine, J.S.N., Garcia, E., Goldfarb, M., 2001. Corner-filleted flexure hinges. Trans. ASME 123, 346–352.

Lobontiu, N., Cullin, M., Ali, M., Brock, J.M., 2011. A generalized analytical compliance model for transversely symmetric three-segment flexure hinges. Rev. Sci. Instrum. 82 (10), 105116.

Love, A.E.H., 1944. A Treatise on the Mathematical Theory of Elasticity. Dover Publication, New York, NY.

Luan, B., Robbins, M.O., 2004. Effect of inertia and elasticity on stick-slip motion. Phys. Rev. Lett. 93, 036105.

Magrab, E.B., 1979. Vibration of Elastic Members. Sijthoff & Noordhoff, Alphen aan den Rijn.

Maragliano, C., Glia, A., Stefancich, M., Chiesa, M., 2015. Effective AFM cantilever tip size: methods for *in-situ* determination. Meas. Sci. Technol. 28, 015002.

Mathew-Fenn, R.S., Das, R., Silverman, J.A., Walker, P.A., Harbury, P.A.B., 2008. A molecular ruler for measuring quantitative distance distribution. PLoS One 3, e3229, Supplementary materials section.

Maugis, D., 1992. Adhesion of spheres: the JKR-DMT transition using a Dugdale model. J. Colloid Interface Sci. 150 (1), 243–269.

Morita, S., Giessibl, F.J., Wiesendanger, R. (Eds.), 2009. Noncontact Atomic Force Microscopy, vol. 2, Springer, Berlin.

Nguyen, T.-H., Lee, S.-M., Na, K., Yang, S., Kim, J., Yoon, E.-S., 2010. An improved measurement of dsDNA elasticity using AFM. Nanotechnology 21 (7), 075101.

Noeth, N., Keller, S.S., Boisen, A., 2014. Integrated cantilever-based flow sensors with tunable sensitivity for in-line monitoring of flow fluctuations in microfluidic systems. Sensors 14, 229–244.

Panovko, Y.G., 1968. Natural and forced vibrations of beams and beam systems. In: Birger, I.A., Panovko, Y.G. (Eds.), Strength, Stability, Vibrations, Handbook, vol. 3. Mashinostroenie, Moscow, pp. 285–346 (in Russian).

Park, K., et al., 2007. Topography imaging with a heated atomic force microscope cantilever in tapping mode. Rev. Sci. Instrum. 78, 043709.

Parlak, Z., Tu, Q., Zausher, S., 2014. Liquid contact resonance AFM: analytical models, experiments, and limitations. Nanotechnology 25, 445703.

Paros, J.M., Weisbord, L., 1965. How to design flexure hinges. Machine Design 37, 151–156.

Pelletiera, C.G.N., Dekkersb, E.C.A., Govaerta, L.E., den Toondera, J.M.J., Meijera, H.E.H., 2007. The influence of indenter-surface misalignment on the results of instrumented indentation tests. Polym. Test. 26, 949–959.

Perkins, T.T., Smith, D.E., Larson, R.G., Chu, S., 1995. Stretching of a single tethered polymer in a uniform flow. Science 268, 83–87.

Perkins, T., Smith, D., Chu, S., 1997. Single polymer dynamics in an elongation flow. Science 276, 2016–2021.

Piétrement, O., Troyon, M., 2000. General equations describing elastic indentation depth and normal contact stiffness versus load. J. Colloid Interface Sci. 226, 166–171.

Preissig, F.J. von, 1989. Applicability of the classical curvature-stress relation for thin films on plate substrates. J. Appl. Phys. 66, 4262–4268.

Sader, J.E., 2003. Susceptibility of atomic force microscope cantilever to lateral forces. Rev. Sci. Instrum. 74, 2438–2443.

Sader, J.E., Green, C.P., 2004. In-plain deformation of cantilever plates with applications to lateral force microscopy. Rev. Sci. Instrum. 75, 878–883.

Sader, J.E., Sanelli, J.A., Adamson, B.D., Monty, J.P., et al., 2012. Spring constant calibration of atomic force microscope cantilevers of arbitrary shape. Rev. Sci. Instrum. 83, 103705.

Sader, J.E., Lu, J., Mulvaney, P., 2014. Effect of cantilever geometry on the optical lever sensitivities and thermal noise method of the atomic force microscope. Rev. Sci. Instrum. 85, 113702-1-6.

Salvadori, M.C., Brown, I.G., Vaz, A.R., Melo, L.L., Cattani, M., 2003. Measurement of the elastic modulus of nanostructured gold and platinum thin films. Phys. Rev. B 67 (4), 153404.

Salvadori, M.C., Lisboa, F.S., Fernandes, F.M., Brown, L.G., 2008. Novel method for measuring nanofriction by atomic force microscope. J. Vac. Sci. Technol. B 26 (2), 643.

Sarid, D., 1997. Exploring Scanning Probe Microscopy with Mathematica. Wiley, New York, NY.

Sheard, B.S., Gray, M.B., Mow-Lowry, C.M., McClelland, D.E., 2004. Observation and characterization of an optical spring. Phys. Rev. A 69, 051801.

Simmons, R.M., Finer, J.T., Chu, S., Spudlich, J.A., 1996. Quantitative measurements of force and displacement using an optical trap. Biophys. J. 70, 1813–1822.

Slocum, A.H., 1992. Precision Machine Design. Prentice Hall, New Jersey.

Smith, S.T., 2000. Flexure Elements of Elastic Mechanisms. Gordon and Breach, Amsterdam.

Smith, S.T., Chetwynd, D.G., 1992. Foundations of Ultraprecision Mechanism Design. Gordon and Breach, Amsterdam.

Smith, S.B., Finzi, L., Bustamante, C., 1992. Direct mechanical measurements of the elasticity of single DNA molecules by using magnetic beads. Science 258 (5085), 1122–1126.

Souayeh, S., Kacem, N., 2014. Computational models for large amplitude nonlinear vibrations of electrostatically actuated carbon nanotube-based mass sensors. Sens. Actuat. A 208, 10–20.

Stoney, G.G., 1909. The tension of metallic films deposited by electrolysis. Proc. R. Soc. London Ser. A 82, 172–175.

Sugimoto, Y., Pou, P., Abe, M., Jelinek, P., Pérez, R., Morita, S., Custance, Ó., 2007. Chemical identification of individual surface atoms by atomic force microscopy. Nature 446, 64–67.

Thormann, E., Pettersson, T., Kettle, J., Claesson, P.M., 2010. Probing material properties of polymeric surface layers with tapping mode AFM: which cantilever spring constant, tapping amplitude and amplitude set point gives good image contrast and minimal surface damage? Ultramicroscopy 110 (4), 313.

Tian, Y., Shirinzadeh, B., Zhang, D., 2010. Closed-form compliance equations of filleted V-shaped flexure hinges for compliant mechanism design. Precis. Eng. 34, 408.

Timoshenko, S.P., 1955. Strength of Materials. Van Nostrand, New York, NY.

Timoshenko, S.P., Young, D.H., 1964. Vibration Problems in Engineering, third ed. Van Nostrand, Princeton, NJ.

Tseytlin, Y.M., 1994. Special measures for thin coatings measurement. In: Proceedings of the Fortieth International IIS, ISA, Baltimore, USA, pp. 181–189.

Tseytlin, Y.M., 2002. Notch flexure hinges: an effective theory. Rev. Sci. Instrum. 73, 3363–3368.

Tseytlin, Y.M., 2006. Structural Synthesis in Precision Elasticity. Springer, New York, NY.

Tseytlin, Y.M., 2007. Nanostep and film-coating thickness traceability. In: Proceedings of Fifty-Third IIS ISA, Tulsa, OK, USA, 470, Papertp007iis006.

Tseytlin, Y.M., 2011. Tractable model for concave flexure hinges. Rev. Sci. Instrum. 82 (1), 015106.

Tseytlin, Y., 2012a. Note: rotational compliance and instantaneous center of rotation in segmented and V-shaped notch hinges. Rev. Sci. Instrum. 83, 026102:1–126102.

Tseytlin, Y.M., 2012b. Flexible helicoids, atomic force microscopy (AFM) cantilevers in high mode vibration, and concave notch hinges in precision measurements and research. Micromachines 3, 480.

Velthuis, A.J.W. te, Kerssemakers, J.W.J., Lipfert, J., Dekker, N.H., 2010. Quantitative guidelines for force calibration through spectral analysis of magnetic tweezers data. Biophys. J. 99 (4), 1292–1302.

Walczyk, W., Schönherr, H., 2014. Characterization of the interaction between AFM tips and surface nanobubbles. Langmuir 30 (24), 7112–7120.

Wang, M.D., Yin, H., Landick, R., Gelies, J., Block, S.M., 1997. Stretching DNA with optical tweezers. Biophys. J. 72, 1335–1346.

Weaver, W., Timoshenko, S.P., Young, D.H., 1990. Vibration Problems in Engineering, fifth ed. Wiley, New York, NY.

Yamakawa, H., 1997. Helical Wormlike Chains in Polymer Solutions. Springer, Berlin.

Yu, Z., Dulin, D., Cnossen, J., et al., 2014. A force calibration standard for magnetic tweezers. Rev. Sci. Instrum. 85, 123114.

Zhang, Y., Zhao, H., Zuo, L., 2012. Contact dynamics of tapping mode atomic force microscopy. J. Sound Vibration 331, 5141–5152.

Chapter 3

AFM with Higher Mode Oscillations and Higher Sensitivity

3.1 EFFECTS OF THE RESONANCE MODES. KINETOSTATIC METHOD

Higher order resonance modes enhance the sensitivity and resolution of atomic force microscopes operating in a dynamical regime. Higher modes may also improve the determination of the force constant of a microcantilever with less uncertainty. However, higher modes are also expected to demonstrate large spring constants, and greater sensitivity to change in experimental conditions.

The kinetostatic method (Tseytlin, 2005, 2006, 2008, 2010) is based on the representation of a cantilever beam vibration at higher eigenmodes, as a simply supported beam at the nodal points, with virtual forces to prevent their lateral displacement. To implement the kinetostatic method, one must first evaluate the position of the nodal points that depends on the frequency of the vibration, and the vibration form of the elastic beam for given boundary conditions. We show that this method is applicable not only to rectangular cantilevers, but to V-shaped and other cantilevers, as well.

3.1.1 Vibrating Beams

3.1.1.1 Free End Cantilever

The resonance frequency of each normal mode of vibration for an undamped rectangular cantilever (Fig. 3.1) with one end free can be expressed by $f_n = \dfrac{h}{6.928\pi}\dfrac{(\alpha L)_n^2}{L^2}\sqrt{\dfrac{E}{\rho_\gamma}}$, where factor $(\alpha L)_n$ = 1.8751, 4.6941, 7.8548, 10.996 for mode n_f = 1, 2, 3, 4, respectively, and for higher values of mode n $(\alpha L)_n = (2\,n-1)\pi/2$. In addition,

$$f_n / f_1 = [(\alpha L)_n / (\alpha L_1)]^2 = \sqrt{Q_{\text{fac}(n)} / Q_{\text{fac}(1)}},$$

where Q_{fac} represents corresponding Q-factor at certain mode n of vibration.

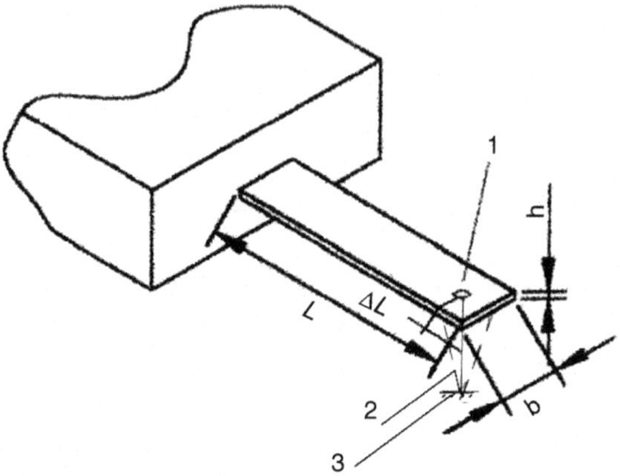

FIGURE 3.1 Rectangular cantilever, 1; probe, 2; and measurand object, 3. *(After Tseytlin, 2005, 2008, reprinted with permission.)*

The higher modes of the cantilever at lateral vibration have more nodal points with zero displacements (Fig. 3.2a,b).

The position of these points (x_{ij}) for the free end rectangular cantilever can be found from the equation of oscillation amplitudes $X(x)$ that may be written using Krylov's functions (Table 3.1) as follows:

$$X(x) = 4[T(\alpha L)\,U(\alpha x) - S(\alpha L)\,V(\alpha x)]. \tag{3.1}$$

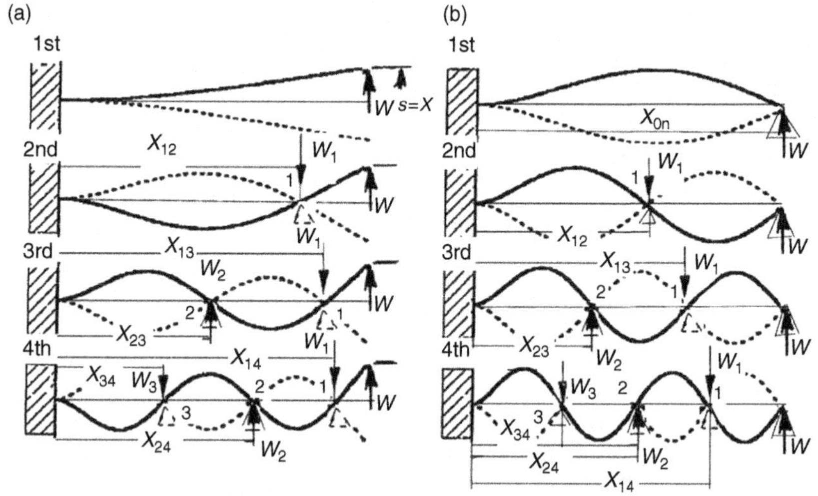

FIGURE 3.2 Vibration amplitude form and the nodal points coordinates at the first, second, third, and fourth modes of the rectangular cantilever with free moving end (a) and supported-contact end (b). *(After Tseytlin, 2005, 2008, reprinted with permission.)*

TABLE 3.1 Krylov's Beam Functions

Type	Formula	Type	Formula
$S(\alpha x)$	½ (coshαx + cosαx)	$T(\alpha x)$	½ (sinhαx + sinαx)
$U(\alpha x)$	½ (coshαx − cosαx)	$V(\alpha x)$	½ (sinhαx-sinαx)
$Y_1(m_w\xi)$	$S^2(\alpha x) - U^2(\alpha x) = \cosh m_w\xi \cos m_w\xi$;	$Y_2(m_w\xi)$	$S(\alpha x)T(\alpha x) - U(\alpha x)V(\alpha x) = ½ [\cosh m_w\xi \sin m_w\xi + \sinh m_w\xi \cos m_w\xi]$;
$Y_3(m_w\xi)$	$[T^2(\alpha x) - V^2(\alpha x)]/2 = ½ \sinh m_w\xi \sin m_w\xi$;	$Y_4(m_w\xi)$	$U(\alpha x)V(\alpha x) + U(\alpha x)T(\alpha x) = ¼ [\cosh m_w\xi \sin m_w\xi - \sinh m_w\xi \cos m_w\xi]$;

Note: here $\alpha x = m_w\xi$; Krylov's functions (see Krylov, 1936) may be presented through exponential functions as well because $\sinh(z) = ½ (e^z - e^{-z})$, $\cosh(z) = ½ (e^z + e^{-z})$, $\sin(z) = (1/2i)(e^{iz} - e^{-iz})$, and $\cos(z) = ½ (e^{iz} + e^{-iz})$.
Source: After Tseytlin (2006), reprinted with permission.

The nodal point positions correspond to $X(x) = 0$ and the expression

$$U(\alpha x)\left[\frac{T(\alpha L)}{S(\alpha L)}\right] - V(\alpha x) = 0, \tag{3.2}$$

where x is the nodal point length coordinate on the cantilever 1 with length L (Fig. 3.1).

The numerical solution of Eq. (3.2) for the second mode of the cantilever vibration with $(\alpha L)_2 = 4.6941$ and $T(\alpha L)_2/S(\alpha L)_2 = 0.98187$ yields the nodal point position coordinate at $\alpha x_{12} = 3.678$, or $x_{12} = 0.7836\,L$ from the clamped end. The ratios $T(\alpha L)_1/S(\alpha L)_1 = 1.36221$ for the first mode, and $T(\alpha L)_i/S(\alpha L)_i$ have the values 1.00078 and 0.999966 for the third and fourth modes, respectively, whose nodal points positions are $\alpha x_{13} = 6.8144$, $x_{13} = 0.8675\,L$; $\alpha x_{23} = 3.9553$, $x_{23} = 0.5036\,L$ (for the third mode) and $\alpha x_{14} = 9.95444$, $x_{14} = 0.90528L$; $\alpha x_{24} = 7.0823$, $x_{24} = 0.64408\,L$; $\alpha x_{34} = 3.9401$, $x_{34} = 0.35832L$ (for the fourth mode).

3.1.1.2 Both Sides Clamped Beam

The higher roots of frequency equation

$$\cos(\alpha L)\cosh(\alpha L) = 1 \tag{3.3}$$

for a beam clamped on both sides (doubly clamped) are equal to 7.859, 10.996, 14.137, and $\pi(2n+1)/2$ at $n = 2, 3, 4, \ldots$, respectively. The nodal point positions in this case correspond to the expression

$$S(\alpha x)\,[T(\alpha L)/U(\alpha L)] - V(\alpha x) = 0. \tag{3.4}$$

3.1.1.3 Clamped-Supported Beam

The higher roots of the frequency equation

$$\tan(\alpha L) = \tanh(\alpha L) \tag{3.5}$$

are applicable to a beam clamped on one side and simply supported on the other. It is very important in these applications to isolate and discard the static component of a detector signal in order to enhance the resolution and sensitivity of a sensor.

Boundary conditions and frequency equations for these cantilevers are shown in Table 3.2. For given boundary conditions, the eigenvalues α (Table 3.3) allow us to find nodal point coordinates x_{in}. The corresponding values of nodal point coordinates x_{in} are shown in Table 3.4 and Figure 3.2a,b. Note that the first index in the expression x_{in} designates the nodal point number, and the second index designates the mode number.

Let us recall that the coordinates x_{in} are expressed as a fraction of the cantilever's length L. Calculation with Eq. (3.1) shows that the amplitude of rectangular cantilever vibrations, estimated for the second through the fourth modes, are approximately 1.3 times larger for the cantilever's moving end, than in other antinodes.

We have found (Krylov, 1936) that the same Eq. (3.1) is applicable to the clamped-supported cantilever. Therefore, the nodal point position (Fig. 3.2b) expression for this case remains the same as Eq. (3.2). However, the boundary conditions and the frequency equation for the case of the clamped-supported cantilever are different (Tables 3.2 and 3.3). The numerical solutions of Eq. (3.2), using an Excel spreadsheet for this case, are as follows:

1. for the second mode of the clamped-supported cantilever vibration, $(\alpha L)_2 = 7.06886$, $T(\alpha L)_2/S(\alpha L)_2 = 0.9999986$, and $(\alpha x)_{12} = 3.9405$;
2. for the third mode, $(\alpha L)_3 = 10.2102$, $T(\alpha L)_3/S(\alpha L)_3 = 0.9999999$, $(\alpha\ x)_{13} = 7.06799$, and $(\alpha\ x)_{23} = 3.9407$;
3. for the fourth mode, $(\alpha L)_4 = 13.3518$, $T(\alpha L)_4/S(\alpha L)_4 = 1.0000$, $(\alpha\ x)_{14} = 10.2102$, $(\alpha\ x)_{24} = 7.06799$, and $(\alpha\ x)_{34} = 3.94073$. $(\alpha L)_n = \pi\ (4\ n + 1)/4$ at the higher modes n.

The corresponding values of the nodal points coordinates x_{in} are shown in Table 3.4 for a free end, and supported (in hard and flexible contact) rectangular cantilever, with different masses on the moving end. Note that $x_{0n} = 1.0$ for all modes n of vibrating clamped-supported through hard contact cantilevers.

3.1.1.4 End Concentrated Mass

The boundary conditions and frequency equation for the clamped-free cantilever with mass M^* on the moving end are shown in Table 3.2. The oscillation amplitude $X(x)$ for this case is as follows:

$$X(x) = C_1 S(\alpha x) + C_2 T(\alpha x) + C_3 U(\alpha x) + C_4 V(\alpha x), \tag{3.6}$$

TABLE 3.2 Boundary Conditions and Frequency Equations for Rectangular Cantilevers

Type of clamping	Clamped free	Clamped supported	Clamped-elastic support	Clamped-free with mass M^*
Boundary conditions	$X(0)=0, X'(0)=0$ $X''(L)=0, X'''(0)=0$	$X(0)=0, X'(0)=0$ $X(L)=0, X''(L)=0$	$X(0)=0, X'(0)=0$ $X''(L)=0, X'''(L)=jX(L)/EJ$	$X(0)=0, X'(0)=0$ $X''(L)=0,$ $M^*\omega^2 X(L)=EJX'''(L)$
Frequency equation	$\cos\alpha L \cosh\alpha L=-1$	$\tan\alpha L=\tanh\alpha L$	$0=1+\cosh\alpha L\cos\alpha L-$ $\frac{jL^3}{EJ(\alpha L)^3}(\sinh\alpha L\times\cos\alpha L-\cosh\alpha L\sin\alpha L)$	$M^*/m_c(\alpha L)_n\times$ $[\tanh(\alpha L)_n$ $-\tan(\alpha L)_n]+1$ $+\sec(\alpha L)_n \sec h(\alpha L)_n$ $=0$

Source: After Tseytlin (2008), reprinted with permission.

TABLE 3.3 Natural Frequency Eigenvalues α for Rectangular Cantilevers

End motion	Clamped-free				Elastic support $M/m_c = 0$			Support contact
N Mode	$M/m_c = 0$	$M/m_c = 0.001$	$M/m_c = 0.01$	$M/m_c = 0.1$	$M/m_c = 0.2$	$jL^3/EJ = 12$	100	$M/m_c = 0$
1	1.875 1	1.873 233	1.856 787	1.722 74	1.616 399 66	2.721 27	3.640 543	3.9266
2	4.694 1	4.689 424	4.649 726 5	4.399 52	4.267 061 57	4.814 26	5.615 996	7.0686
3	7.854 76	7.846 972	7.782 671 2	7.451 06	7.318 372 67	7.879 89	8.074 843	10.2102
4	10.996	10.984 67	10.897 588	10.521 78	10.401 56	11.004 63	11.078 484	13.3518

Source: After Tseytlin (2008), reprinted with permission.

TABLE 3.4 Spring Constants' Ratio at Higher Eigenmodes of a Rectangular Cantilever

Free end	Mode	2	3	4	2	3	4
	End mass	$M^*/m_c = 0.1$			$M^*/m_c = 0.2$		
	Nodes coordinates	$x_{12} = 0.841096$	$x_{13} = 0.92071$ $x_{23} = 0.53041$		$x_{12} = 0.874396$	$x_{13} = 0.943434$ $x_{23} = 0.539645$	$x_{14} = 0.96839$ $x_{24} = 0.68017$ $x_{34} = 0.378834$
1	Calculated (Tseytlin, 2008)	$K_{21} = 75.39$ $q = 0.985$ $W_1 = 1.23029$	$K_{31} = 999.19$ $q = 0.985$ $W_1 = -0.1583$ $W_2 = 0.1930$		$K_{21} = 81.07$ $q = 1$	$K_{31} = 762.195$ $q = 1$ $W_1 = -1.17505$ $W_2 = 0.21429$	$K_{41} = 3268.68$ $q = 1$
2	Experiment (Melcher et al., 2007)	74.9	950				
	$M^*/m_c = 0$; clamped-free end				$M^*/m_c = 0$; clamp-supported hard Contact end		
3	Nodes coordinates	$x_{12} = 0.7836$	$x_{13} = 0.8675$ $x_{23} = 0.5036$	$x_{14} = 0.90528$ $x_{24} = 0.64408$ $x_{34} = 0.35832$	$X_{12} = 0.5575$ $X_{02} = 1.0$	$x_{13} = 0.6922$ $x_{23} = 0.3859$ $x_{03} = 1.0$	$x_{14} = 0.7647$ $x_{24} = 0.5294$ $x_{34} = 0.2951$ $x_{04} = 1.0$

(Continued)

TABLE 3.4 Spring Constants' Ratio at Higher Eigenmodes of a Rectangular Cantilever (*cont.*)

4	Calculated (Tseytlin, 2008)	$K_{21} = 39.3$ $q = 0.98$ $W_1 = 1.33844$ $\theta_{A2}^{-1} = 7.2$ $q_* = q$	$K_{31} = 265.5$ $q = 0.98$ $W_1 = -1.318$ $W_2 = 0.3858$ $\theta_{A3}^{-1} = 9.43$ $q_* = 1$	$K_{41} = 1089.3$ $q = 0.98$ $W_1 = -1.2647$ $W_2 = 0.32638$ $W_3 = -0.07623$ $\theta_{A4}^{-1} = 20.2$	$K_{21} = 7.6\_q = 0.972$ $5.93\_q = 1$ $W_1 = 2.19058$ $\theta_{A2}^{-1} = 3.13$	$K_{31} = 26.9$ $\_q = 0.972$ $W_1 = -2.0366$ $W_2 = 1.289$ $K_{31} = 18.3\_q = 1$ $W_1 = -2.2636$ $W_2 = 1.57186$ $\theta_{A3}^{-1} = 6.68$	$K_{41} = 76.7$ $\_q = 0.972$ $42.3\_q = 1$ $W_1 = -2.28233$ $W_2 = 1.64237$ $W_3 = -0.45068$ $\theta_{A4}^{-1} = 11.56$
	Thermal noise (Butt and Jaschke, 1995)	$K_{21} = 39.56$ $Z^*_{2/1} = 3.25?$	$K_{31} = 308.69$ $Z^*_{3/1} = 9.42$	$K_{41} = 1181.4$ $Z^*_{4/1} = 18.54$	$Z^*_{2/1} = 3.45$	$Z^*_{3/1} = 7.16$	$Z^*_{4/1} = 12.2$
5	Calculated (Sahin et al., 2004)	$K_{21} = 40.43$	$K_{31} = 256$	$K_{41} = 1218.1$	$M^*/m_c = 0$; clamped-elastic support $jL^3/EJ = 100$		
					$x_{12} = 0.6729$	$x_{13} = 0.84697$ $x_{23} = 0.48918$	$x_{14} = 0.89939$ $x_{24} = 0.63946$ $x_{34} = 0.355577$
6	Calculated (Tseytlin, 2005)	$K_{21} = 47.96$ $q = 0.972$	$K_{31} = 409.84$ $q = 0.972$	$K_{41} = 2204$ $q = 0.972$	$K_{21} = 11.237\_q = 1$ $W_1 = 1.7292$ $\theta_{A2}^{-1} = 4.61$	$K_{31} = 118.35$ $\_q = 1$ $W_1 = -1.4502$ $W_2 = 0.55504$ $\theta_{A3}^{-1} = 10.81$	$K_{41} = 301.31q = 1$ $W_1 = -1.48635$ $W_2 = 0.600813$ $W_3 = -0.142592$ $\theta_{A4}^{-1} = 39.49$

Source: After Tseytlin (2008), reprinted with permission.

where C_i are constants that can be estimated through the solution of the simultaneous system of four boundary conditions' equations (Table 3.2) as

$$X(0) = C_1 S(0) + C_2 T(0) + C_3 U(0) + C_4 V(0), \tag{3.7}$$

$$X' = \alpha[C_1 V(0) + C_2 S(0) + C_3 T(0) + C_4 U(0)] = 0, \tag{3.8}$$

$$X''(L) = \alpha^2 [C_3 S(\alpha L) + C_4 T(\alpha L)] = 0, \tag{3.9}$$

$$X'''(L) = \frac{M^* \omega^2 X(L)}{EJ} = \alpha^3 [C_3 V(\alpha L) + C_4 S(\alpha L)], \tag{3.10}$$

where ω is the angular frequency of the cantilever vibrations, E is Young's modulus of elasticity for the material, J is the moment of a cross-section inertia, $\alpha = \sqrt[4]{m_l \omega^2 / EJ}$, $m_l = \rho_\gamma A_c$ is a length unit mass of the cantilever, ρ_γ is the density of the cantilever material, and A_c is its cross-section area. Note that the angular frequency of the vibration is proportional to α^2.

Let us recall that $S(0) = 1$, $T(0) = U(0) = V(0) = 0$ that yields $C_1 = 0$, and $C_2 = 0$. Now, we can find the following expressions for the constants C_3 and C_4 from the last two boundary conditions' equations:

$$C_3 = -\frac{M^* \alpha X(L)}{EJ \rho_\gamma A_c} [S^2(\alpha L) - T(\alpha L) V(\alpha L)]^{-1} T(\alpha L),$$

and

$$C_4 = \frac{M^* \alpha X(L) S(\alpha L)}{EJ \rho_\gamma A_c} [S^2(\alpha L) - T(\alpha L) V(\alpha L)]^{-1}.$$

Substitution of these values of C_i into Eq. (3.6) yields the following amplitudes equation for a vibrating cantilever, with the mass M^* on its moving end:

$$\begin{aligned} X(x)_n = &-\frac{M^* \alpha X(L)}{EJ \rho A_c} [S^2(\alpha L)_n - T(\alpha L)_n V(\alpha L)_n]^{-1} T(\alpha L)_n U(\alpha x) \\ &+ \frac{M^* \alpha\, X(L) S(\alpha L)_n}{EJ \rho A_c} [S^2(\alpha L)_n - T(\alpha L)_n V(\alpha L)_n]^{-1} V(\alpha x). \end{aligned} \tag{3.11}$$

And the nodal point equation with $X(x) = 0$ is as follows:

$$X(x)_n = U(\alpha x)[T(\alpha L)_n / S(\alpha L)_n] - V(\alpha x) = 0. \tag{3.12}$$

We see, however, that Eq. (3.12) is exactly the same as the nodal point position expression (3.2) for the vibrating cantilever without a mass attached to its end.

The numerical solutions of Eq. (3.12) using an Excel spreadsheet for this case at $M^*/m_c = 0.1$ are as follows:

1. for the second mode of the cantilever vibration, $(\alpha L)_2 = 4.39952$, $T(\alpha L)_2/S(\alpha L)_2 = 0.98317$, $(\alpha x)_{12} = 3.7004$;
2. for the third mode, $(\alpha L)_3 = 7.45106$, $T(\alpha L)_3/S(\alpha L)_3 = 1.0006$, $(\alpha x)_{13} = 6.8603$, $(\alpha x)_{23} = 3.95214$.

   The numerical solutions for Eq. (3.12) in the case with $M^*/m_c = 0.2$ are as follows: (1) for the second mode of cantilever vibration, $(\alpha L)_2 = 4.26706$, $T(\alpha L)_2/S(\alpha L)_2 = 0.98621$, and $(\alpha x)_{12} = 3.7311$; (2) for the third mode, $(\alpha L)_3 = 7.31837$, $T(\alpha L)_3/S(\alpha L)_3 = 1.0046$, $(\alpha x)_{13} = 6.9044$, and $(\alpha x)_{23} = 3.94932$; (3) for the fourth mode, $(\alpha L)_4 = 10.40156$, $T(\alpha L)_4/S(\alpha L)_4 = 0.99998$, $(\alpha x)_{14} = 10.0728$, $(\alpha x)_{24} = 7.0748$, and $(\alpha x)_{34} = 3.94046$, the corresponding values of nodal point coordinates x_{in} as a fraction of L are shown in Table 3.4 for the rectangular cantilever of mass m_c and with a mass M^* on the moving end at the mass ratios $M^*/m_c = 0.1$ and $M^*/m_c = 0.2$, respectively.

The same numerical solutions can be found for nodal point positions at other mass ratios with known frequency eigenvalues (Table 3.3).

3.2 EFFECTIVE SPRING CONSTANTS RATIO

Let us recall that straight rectangular prismatic cantilever has a static spring constant at its bending that is given by

$$K_{1\,\text{eff}} = W / s = 3EJ / L^3,$$

where W is the load on the cantilever's end, and s is the displacement of the cantilever end under this load.

This spring constant can be used for the first oscillation mode calculation, as well (some authors show an increase of this value for the first oscillation mode by 3–8%). We assume further that the application of a load W and an additional mass to the cantilever has practically offset from the cantilever's very end by some distance ΔL. Therefore, we use a modified value of the cantilever working length $L^* = L - \Delta L = qL$, and modified expressions for the coordinates of the nodal points in higher oscillation modes n as $x^* = (x_{in}/q)$, where the factor q for the position of the probe or additional mass is $q \leq 1$. This assumption, however, does not change the position of the nodal points on the vibrating cantilever. We also assume, in accordance with the kinetostatic method, that in the second, third, fourth, and other higher oscillation modes, the cantilever beam may be represented as a simply supported one in the nodal points, in order to prevent their lateral displacement. As a result, we have indeterminate beams to the degree that is equal to the number of such supports (nodal points). The support on the end of the clamped-supported beam, on the contrary, is a real one (triangles with the solid lines in Fig. 3.2b) that represents the contact of cantilever's probe with the hard surface of the measurand object, or a reference cantilever. In this case, $x_{01} = 1 = \text{const}$.

We eliminate the indeterminacy by replacing the virtual supports with the unknown redundant forces W_1, W_2, W_3... in the corresponding nodal points 1, 2, 3..., respectively. Then it is easy to estimate the values of these forces in relation to the quasi-end load W by the so-called force superposing method. This method was applied before (Tseytlin, 2005, 2008) where it is shown in detail. On this basis, we obtained the following results for the spring constants in the second, third, and fourth modes:

1. *The second mode's spring constant* ($K_{2\text{eff}}$) ratio to the first one ($K_{1\text{eff}}$) is equal to

$$K_{21} = K_{2\text{eff}} / K_{1\text{eff}} = A_{q2}^{-1},$$

and

$$A_{q2} = q^3 - \frac{(W_1 / W)x_{12}^2}{2q^2}\left[2\frac{x_{12}}{q} + \left(q - \frac{x_{12}}{q}\right)\right], W_1 / W = \frac{(3q^2 - x_{12})}{2x_{12}} \quad (3.13)$$

2. *The third mode's spring constant* ($K_{3\text{eff}}$).

In this case, we use the beam that is indeterminate to the second degree, and can be represented by the matrix superposition equation as

$$[s_{ij}][W_j] + [s_{iw}]W = [0], \quad (3.14)$$

where s_{ij}, s_{iw} are the flexibility coefficients, W_j are the virtual redundant forces in the nodal points, the first index $i = 1, 2$ corresponds to the nodal point number, and the second index $j = 1, 2, w$ corresponds to the acting virtual force.

The solution of this second rank matrix equation, with respect to the redundant forces, can be obtained with the inverse matrix $[s_{ij}]^{-1}$ as follows:

$$[W_j] = [s_{ij}]^{-1}[s_{iw}]W, \quad (3.15)$$

where $s_{ii} = \dfrac{(x_{i3})^3}{q^3}$, $s_{ij} = s_{ji}$, $s_{ij} = \dfrac{(x_{i3})^2}{2q}(3 - x_{i3} / q^2)$.

Now, we estimate the ratio between the effective spring constants at the third and first modes as

$$K_{31} = K_{3\text{eff}} / K_{1\text{eff}} = A_{q3}^{-1}, \quad (3.16)$$

where

$$A_{q3} = \left| q^3 + \sum_{i=1}^{2} W_i s_{iw} \right|.$$

3. *The fourth mode's spring constant* ($K_{4\text{eff}}$).
The matrix superposition equation for this mode is

$$[d_{ij}][W_j]+[d_{iw}]W=[0], \tag{3.17}$$

where d_{ij}, d_{iw} are the flexibility coefficients, W_j are the virtual redundant forces in the nodal points, and W is the quasi-end force. The first index $i = 1$–3 corresponds to the nodal point number, and the second index $j = 1$–3, w corresponds to the acting virtual force. The solution of this third rank matrix equation with respect to the redundant forces may be obtained with the inverse matrix $[d_{ij}]^{-1}$ as follows:

$$[W_j]=-[d_{ij}]^{-1}[d_{iw}]W, \tag{3.18}$$

where

$$d_{ii}=\frac{(x_{i4})^3}{q^3},\ d_{ij}=d_{ji};\ d_{ij}=\frac{(x_{j4})^2}{2q^3}(3x_{i4}-x_{j4})\text{ at } i<j \text{ and}$$
$$d_{iw}=\frac{(x_{i4})}{2q}(3-x_{i4}/q^2). \tag{3.18a}$$

Hence, the ratio between the spring constants, at the fourth and first modes, is equal to

$$K_{41}=K_{4\text{eff}}/K_{1\text{eff}}=A_{q4}^{-1}, \tag{3.19}$$

where

$$A_{q4}=\left|q^3+\sum_{i=1}^{3}W_i d_{iw}\right|.$$

The numerical solutions of Eqs (3.18) and (3.19) can be easily found with the help of Excel's "minverse" and "mmult" functions.

3.3 CANTILEVER END INCLINATION SPRING CONSTANTS

In atomic force microscopy with optical ray deflection, the inclination θ at the end of the cantilever is measured, rather than the deflection itself. We recall that the inclination in the end of the cantilever by the force W_i that is applied to the point i with coordinate x_{in} is equal to $\theta_{\text{iend}}=W_i x_{in}^2/2EJ$, where W_i are the same virtual forces, found in Sec. 3.2, acting in the nodal points. Therefore, the resulting inclination, on the end of the cantilever, by all acting forces is given by

$$\theta_{\Sigma n}=\frac{WL^2}{2EJ}\left(q_*^2+\sum_{n-1}W_i x_{in}^2\right)=\frac{WL^2}{2EJ}\theta_{\text{An}}, \tag{3.20a}$$

and the inclination spring constants' ratio for modes higher than that in the first mode, $\theta_1^{-1} = \dfrac{2EJ}{L^2}$, equals

$$\theta_{\text{A}n}^{-1} = \left(q_*^2 + \sum_{n-1} W_i x_{in}^2 \right)^{-1}, \tag{3.20b}$$

where n is the mode's number, $q_* = q$ or 1, and the forces W_i are expressed as a fraction of the end force W.

Table 3.4 and Figure 3.3 (Tseytlin, 2008) show calculation results for all of the above cases of rectangular cantilever application.

The use in these calculations of the corresponding q factor for the position of the probe, or additional mass on the cantilever, is presented as well. These

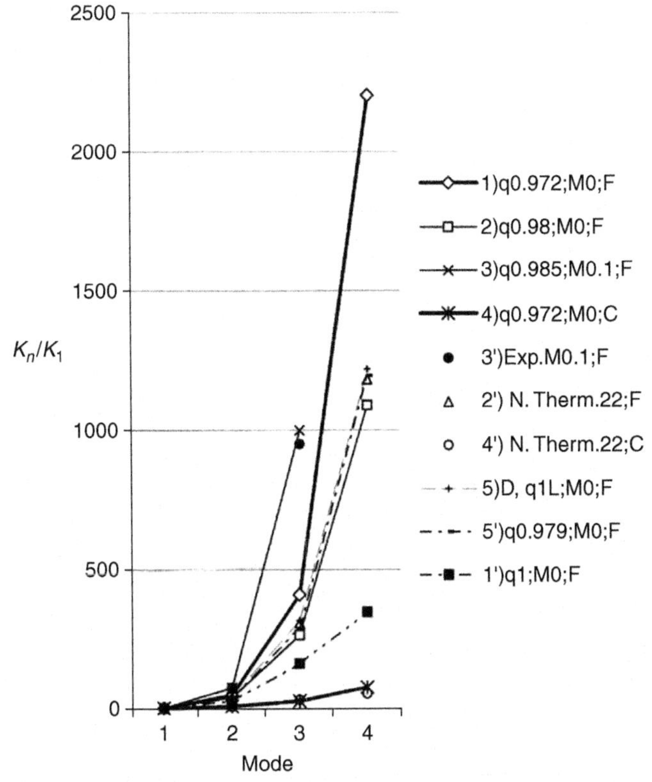

FIGURE 3.3 Charts (1)–(5) and adequate (1′)–(5′) of the spring constants' ratios at the specified modes, positioning factor q, and mass (M^*/m_c) ratio for the rectangular cantilever with the free end (F), supported contact end (C), thermal noise (Butt and Jaschke, 1995) evaluation (N. Therm.), experimental (Melcher et al., 2007, Exp.), and dynamic (Rast et al., 2000) evaluation (D). *(After Tseytlin, 2008, reprinted with permission.)*

calculations do not require knowledge of the cantilever's real dimensions. The results of the calculations are compared with the data presented by other researchers. It is evident that all our results are satisfactory, and within a few percent, coincide with the data of other authors. However, the inclination spring ratio $\theta_{A2}^{-1} = 7.2$ for free end cantilever without clamped mass ($M^*/m = 0$) shows significant difference than the same data presented by Butt and Jaschke (1995) for the equivalent ratio $Z_{n/1}^{*} = Z_1^{*2} / Z_2^{*2} = (5.77 / 3.2) = 3.25$.

There is apparently a misprint in that publication, with the correct value $Z_2^{*2} = 2.3$. The spring constants ratios at different (M^*/m_c) masses on the cantilever's end are found with the help of the frequency eigenvalues presented in Table 3.3. We show in Table 3.4 the numerical solutions for the mass ratio $M^*/m_c = 0.1$ that we compare with the experimental estimates (Melcher et al., 2007).

If we recall that angular and linear eigenfrequencies are proportional to the factor α^2, we would not recommend using this approximate (Sahin et al., 2004) proportionality between spring constant ratio and the square of the corresponding frequency ratios without taking certain precautions, even for a clamped-free cantilever. The position factor q has a large influence on these relationships.

One can see, however, the following approximate relationship for the supported-contact cantilever, in the case of the angular inclination spring ratio, as proportional to the first power of the corresponding frequencies ratio, because of $q = 1$, and

$$\theta_{An}^{-1} = (\alpha_n / \alpha_1)^2 = f_n / f_1.$$

The summary results of our evaluations for spring constants ratios for rectangular cantilever are shown in Table 3.5.

3.4 QUALITY FACTOR INFLUENCE

All of the calculations shown earlier were done without considering the ambient and internal cantilever material resistance to its vibration. This resistance is usually expressed through the Q-factor value.

It is known that the Q-factor value changes in a wide range at the cantilever's vibration in vacuum, air, and gases, or liquid, where the Q-factor is significantly lower. However, the change of the eigenfrequencies in the transition of the cantilever from vacuum to air is insignificant. The changes of the frequency eigenvalues are more significant in liquids, but the oscillation amplitude remains the same as in Eq. (3.1). Therefore, we assume that in the latter case the same nodal point position equation, Eq. (3.2), can be used, but with a different set of vibration frequency eigenvalues.

TABLE 3.5 Relations of Virtual Forces and Spring Constants' Ratios for a Rectangular Cantilever

	Mode n		
Parameter	Second with one node	Third with two node	Fourth with three node
Virtual forces W_i with respect to end load W	$W_i/W=(3q^2-x_{12})/2x_{12}$	$[W_i]=-[s_{ij}]^{-1}[s_{iw}]W$	$[W_i]=-[d_{ij}]^{-1}[d_{iw}]W$
Spring constants' ratio	$K_{21}=K_{2\mathrm{eff}}/K_{1\mathrm{eff}}=A_{q2}^{-1}$	$K_{31}=K_{3\mathrm{eff}}/K_{1\mathrm{eff}}=A_{q3}^{-1}$	$K_{41}=K_{4\mathrm{eff}}/K_{1\mathrm{eff}}=A_{q4}^{-1}$
K_{n1} at mode n to mode 1	$A_{q2}=q^3-\frac{(W_1/W)}{2q^2}\times\left[2\frac{x_{12}}{q}+3\left(q-\frac{x_{12}}{q}\right)\right]$	$A_{q3}=\left\lvert q^3+\sum_{i=1}^{2}W_is_{iw}\right\rvert$	$A_{q4}=\left\lvert q^3+\sum_{i=1}^{3}W_id_{iw}\right\rvert$
Inclination spring constant ration θ_{An}^{-1}	$\theta_{An}^{-1}=\left(q_*^2+\sum_{n-1}W_ix_{in}^2\right)^{-1}$		

s_{ij}, s_{iw}, d_{ij}, and d_{iw} are the flexibility coefficients.
Source: After Tseytlin (2008, 2010), reprinted with permission.

On the same basis, the sensitivity of flexural and torsional vibration modes of AFM cantilevers to normal and lateral stiffness ($j_{\mathrm{sn}}, j_{\mathrm{sl}}$) variations (see Turner and Wiehn, 2001) can be studied. The dimensionless form of the flexural sensitivity for a cantilever with a uniform section is given by Turner and Wiehn as

$$\sigma_r=\frac{df/d\beta_n}{(1/2\pi)L^2\sqrt{EI/(\rho_\gamma A_c)}},\ \text{and}\ \frac{\partial f}{\partial \beta_n}=\frac{\partial f}{\partial(\alpha L)_n}\frac{d(\alpha L)_n}{d\beta_n},$$

where A_c is the cross-sectional area of the cantilever, and $\beta_n=j_{\mathrm{sn}}/(EI/L^3)$. Similar estimates are given for the torsional sensitivity of the uniform section beam, as well as the flexural and torsional sensitivity of the V-shaped cantilever with "nonuniform" cross section. Studies of the sensitivity for the first four flexural modes of an AFM cantilever with the sidewall probe are provided on the same theoretical basis by Chang et al. (2008).

The spring constants j_i of rectangular cantilevers in AFM systems in permanent and tapping modes with repulsive (contact) and attractive (noncontact) regime change, with the change of the cantilever's frequency mode $n = i$ as $j_2/j_1 = 49.76$, $j_3/j_1 = 409.84$, and $j_4/j_1 = 2204$ (Fig. 3.4; Tseytlin, 2007).

If a cantilever with 29–40 N/m spring constant is oscillating at an amplitude 2 nm, and it experiences contact repulsion during one quarter of its

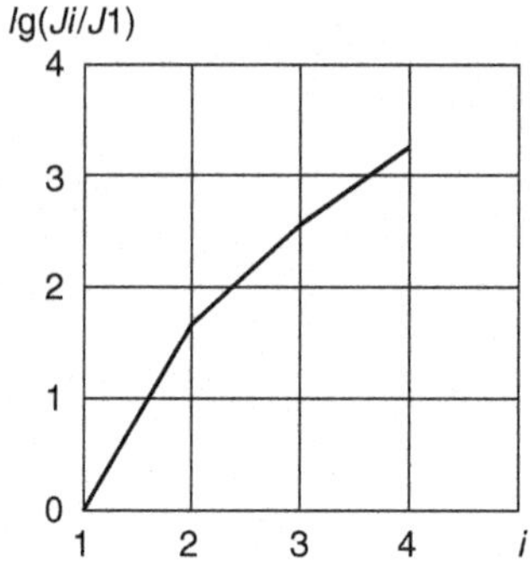

FIGURE 3.4 Change of spring constants j_i of rectangular cantilever with cantilever frequency mode i. *(After Tseytlin, 2007, reprinted with permission.)*

journey (0.5 nm), then the generated contact force is 20–25 nN (Anczykowski et al., 1996), This regime may have the attractive and a repulsive contact forces.

3.4.1 AFM Tapping Mode Dynamics in Liquid

Basak and Raman (2007) showed that the two-mode model for AFM tapping dynamics in liquid corresponds better with experimental data than the point-mass model. They presented two corresponding equations:

$$\frac{q_{i_{tt}}}{\omega_i^2} + \frac{1}{\omega_i Q_i} q_{i_t} + q_i = \frac{F_i e^{J\omega_d t}}{K_i} + \frac{F_{ts}(Z_c - q_1 - q_2)}{K_i},$$

where $i = 1$ or 2, $F_1 = 0.39\ LF_d$, $F_2 = -0.22\ LF_d$, F_d is a magnetic driving force per unit length, L is the length of the cantilever, $q(t) = q_1(t) + q_2(t)$ is the absolute tip motion in time t, ω_1, and ω_2 are the first and second cantilever vibration frequencies, ω_d is the frequency of the driving force, Q_1 and Q_2 are the corresponding Q-factors in liquid near the sample surface, K_1 and K_2 are the effective stiffness of two modes, Z_c is the distance of the tip from the surface at the absence of any interaction forces, and F_{ts} is the tip-sample interaction force. The conventionally calibrated photodiode output corresponds to $q_1(t) + 3.47q_2(t)$ for rectangular lever, and $q_1(t) + 2.69q_2(t)$ for triangular lever.

See also Parlak et al. (2014), about liquid contact resonance AFM analytical models, experiments, and limitations.

3.5 END EXTENDED MASS (V-SHAPED CANTILEVER)

V-shaped cantilevers are widely used in AFM, despite the fact that they have a more complicated design, and some uncertain features of their lateral stability, when compared with rectangular ones; these depend on the geometric parameters and the plane of lateral force application. V-shaped cantilevers can find even more use at higher vibration modes in the atomic force microscopy, as a

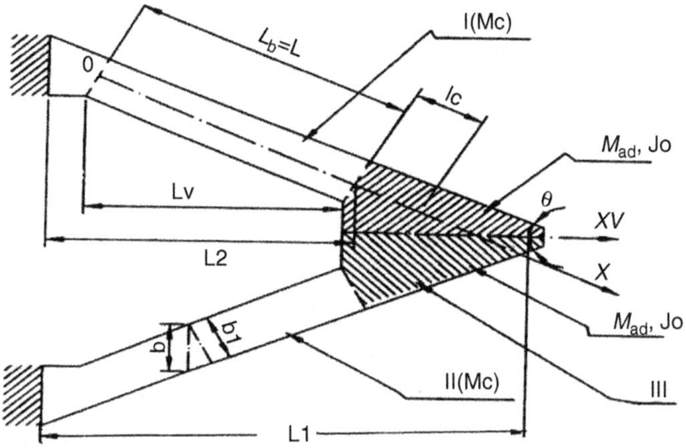

FIGURE 3.5 Schematic of V-shaped cantilever.

result of improved accuracy in the detection of their deflection by the optical lever, because the sign of the slope does not vary along their triangular (trapezoid) part in the incident optical lever area, unlike in a rectangular cantilever at higher flexural modes, where the nodal points become increasingly closer to each other and to the moving end of the rectangular cantilever, with the increase of its higher flexural eigenmodes.

A typical V-shaped cantilever has two (I and II, Fig. 3.5) parallel working rectangular cantilevers ("legs"), the moving ends of which are connected to a common triangular or trapezoid part III with the some functional tip. Therefore, many authors consider a model of a V-shaped cantilever using the parallel beam approximation (PBA). However, the spring constants of a free V-shaped cantilever for the first and higher eigenmodes of vibration have a more complicated connection with the V-shaped cantilever parameters than for the rectangular one. This is caused by the extension in the axial direction of the constructional triangular (trapezoid) part of mass M_{ad} with a moment of rotational inertia J_0 and by the mutual skew angle θ of the rectangular cantilever parts. Let us first estimate the features of a rectangular cantilever with an extended end mass because in previous research (Chen et al., 1994) the effective mass has been considered as a concentrated in the end of the cantilever.

3.5.1 Outline of the Theory. *Boundary Conditions*

An extended end mass usually has a moment of rotational inertia and offset of the center of gravity, beyond the moving end of the rectangular cantilever. The latter feature is unlikely for the end concentrated mass. Let us recall that the general expression for the vibration form of an elastic beam in Krylov's functions is as follows:

$$X(X) = C_1 S(\varepsilon x) + C_2 T(\alpha x) + C_3 U(\alpha x) + C_4 V(\alpha x), \tag{3.21}$$

where C_i are constants that can be estimated through the solution of the following system of four simultaneous boundary conditions' equations:

$$X(0) = C_1S(0) + C_2T(0) + C_3U(0) + C_4V(0), \tag{3.22}$$

$$X'(0) = \alpha[C_1V(0) + C_2S(0) + C_3T(0) + C_4U(0)], \tag{3.23}$$

$$EJX''(L) = [(I_0 + M_{ad}l_c^2)\omega^2 X'(L) + M_{ad}l_c\omega^2 X(L)], \tag{3.24}$$

$$EJX'''(L) = [-M_{ad}\omega^2 X(L) - M_{ad}l_c\omega^2 X'(L)], \tag{3.25}$$

where M_{ad} is the end extended mass with a moment of rotational inertia J_0, l_c is the offset of the mass M_{ad} center of gravity (Fig. 3.5), $\omega^2 = \alpha^4 EJ / m$, ω is the angular frequency of the cantilever vibration, E is the Young's modulus of the material, J is the moment of cross-section inertia, $\alpha = \sqrt[4]{m\omega^2 / EJ}$ is the frequency factor, and m is the mass per unit length of the cantilever.

We see that $C_1 = 0$ and $C_2 = 0$, as a result of the features shown earlier for the Krylov's functions at zero argument.

Therefore, the displacement $X(L)$ and its derivatives with respect to the length coordinate at $x = L$ are as follows:

$$X(L) = C_3U(\alpha L) + C_4V(\alpha L), \tag{3.26a}$$

$$X'(L) = \alpha[C_3T(\alpha L) + C_4U(\alpha L)], \tag{3.26b}$$

$$X''(L) = \alpha^2[C_3S(\alpha L) + C_4T(\alpha L)], \tag{3.26c}$$

$$X'''(L) = \alpha^3[C_3V(\alpha L) + C_4S(\alpha L)]. \tag{3.26d}$$

3.5.2 Frequency Equation

Substitution of Eq. (3.26) into Eqs (3.22–3.25) with $M_{ad}/mL = \alpha_M$; $J_0/mL^3 = J_M$, and $l_c/L = \varepsilon_m$ yields (through the second order nodal point position determinant) the following frequency equation:

$$\begin{aligned} 1 + C_F + \alpha_M A_1(R_F - P_F) - 2\alpha_M (A_1)^2 \varepsilon_m H_F - (J_M + \alpha_M \varepsilon_m^2)(A_1)^3 (R_F + P_F) \\ + J_M (A_1)^4 \alpha_M (1 - C_F) = 0, \end{aligned} \tag{3.27}$$

where $A_1 = (\alpha L)_n; C_F = \cosh A_1 \cos A_1; R_F = \sinh A_1 \cos A_1;$
$P_F = \cosh A_1 \sin A_1; H_F = \sinh A_1 \sin A_1.$

This equation is accurate, and practically identical to a similar equation presented by Bhat and Wagner (1976).

The x coordinates of the nodal points should satisfy the following requirement: $X(x) = C_3U(\alpha x) + C_4V(\alpha x) = 0$, and the values of constants C_3 and C_4 can be obtained from the boundary conditions (3.24, 3.25) and (3.26b, c, d). We multiply for that both sides of Eq. (3.25) by the offset l_c and then add the left and right members of Eq. (3.24), respectively, to the resulting equation, yielding the following expression:

$$X''(L)+X'''(L)l_c = J_0\omega^2 X'(L)/(EJ)=\alpha^4 J_0 X'(L)/m.$$

Then, after substituting into it the corresponding expressions for $X''(L)$, $X'''(L)$, and $X(L)$ from Eq. (3.26b, c, d), we find a relationship between the constants C_3 and C_4 and the desired equation for the positions of the nodal points:

$$\frac{J_M(\alpha L)^3 U(\alpha L)-T(\alpha L)-\varepsilon_m(\alpha L)S(\alpha L)}{S(\alpha L)+(\alpha L)\varepsilon_m V(\alpha L)-J_M(\alpha L)^3 T(\alpha L)}U(\alpha x)+V(\alpha x)=0. \quad (3.28)$$

One can see that this equation will be the same as the equation previously derived for the position of the nodal points for the vibrating clamped-free cantilever, with or without a concentrated mass on its moving end, if $J_M = 0$ and $\varepsilon_m = 0$.

3.5.3 Ratios of Spring Constants

The formulas for the ratios of the spring constants' deflection K_{n1} and inclination θ_{An}^{-1} in the rectangular beam vibrating at mode n and 1, respectively, in a direction perpendicular to the axis of the beam, are shown in the Tables 3.4 and 3.5.

Now, we develop the kinetostatic model of a free-end V-shaped cantilever.

3.5.3.1 *Transformation of the Constituent Beam's Flexural Rigidity*

We assume that the flexural rigidity of a constituent cantilever beam ("leg") at static deformation and natural vibration (quasistatic spring constant) can be estimated as for the PBA, in the following terms:

$K_1 = Eh^3 b/(4L_v^3)$ for bending, and $\theta_{A1}^{-1} = Eh^3 b/(6L_v^2)$ for the end inclination, where E is Young's modulus of the material, h is the thickness, b is the section width (Fig. 3.5), and L_v is the projection of the beam length, with respect to the V-shaped cantilever axis x_v. We suppose, however, that each of the beams (I and II), at higher eigenmode oscillations and increased rigidity, will vibrate with the nodal points on its own axis x, and the corresponding quasistatic flexural rigidities will be equal to $K_1^* = Eh^3 b_1/(4L^3)$ for bending, and $\theta_{A1}^{-1*} = Eh^3 b_1/(6L^2)$ for the end inclination, where b_1 and L (Fig. 3.5) are the width and length of the beam. Therefore, we express the transformation factors of these orientation changes through the cosine function of the half angle $\theta/2$ (Fig. 3.5) of the rectangular beams' mutual skew, as follows:

$$n_k = K_1/K_1^* = \frac{1/\cos(\theta/2)}{\cos^2(\theta/2)} = \cos^{-4}(\theta/2) \text{ and}$$
$$n_\theta = (\theta_{A1}^{-1*}/\theta_{A1}^{-1})(L_2/L_1) = \cos^3(\theta/2)(L_2/L_1),$$

where L_1 corresponds to the V-shaped cantilever representative tip node, and L_2 corresponds to the beam's end. Now, we can estimate the displacement K_{vn1} and the end inclination spring constants' ratios at modes n and 1, respectively, for the V-shaped cantilever, through the product of the corresponding spring constants' ratios for a rectangular cantilever K_{n1}, $\theta_{\mathrm{An}}^{-1}$ and the transformation factors as

$$K_{vn1} = n_{\mathrm{k}} K_{n1} \tag{3.29a}$$

and

$$\theta_{\mathrm{vAn}}^{-1} = n_{\theta} \theta_{\mathrm{An}}^{-1}. \tag{3.29b}$$

Our study of Eq. (3.28) has shown that mode 2 does not have any nodal points, mode 3 has only 1 nodal point, mode 4 has 2 nodal points, and mode 5 has 3 nodal points, that is, mode n has $n-2$ nodal points. These results are different from the features of a vibrating rectangular cantilever with a free end, or an end with a concentrated mass, where the number of nodal points is equal to $n-1$ for mode number n. We take this difference into account in the indexing of the nodal points' coordinates, by including a second index $(n-1)/n$ as a fraction, with the numerator being equal to $(n-1)$ at the number of nodal points $(n-2)$, and the denominator equal to the one used for frequency calculation mode n. We obtained the spring constant ratios K_{n1} and $\theta_{\mathrm{An}}^{-1}$ of the rectangular constituent beams by calculation, with the formulas presented in Table 3.5 for the modes with the same number of nodal points as those for the corresponding V-shaped cantilever model at the factor $q = 1$. The transformation of these values into the corresponding V-shaped cantilever spring constant ratios K_{vn1} and $\theta_{\mathrm{vAn}}^{-1}$ is done in Table 3.6, in accordance with Eq. (3.29a) and (3.29b), respectively.

Meanwhile, the kinetostatic model shows the same specifics of a V-shaped cantilever's behavior as the calculations with the FEA model (Stark et al., 2001), where mode 2 does not have any nodal points, mode 3 has only 1 nodal point, mode 4 has 2 nodal points, and mode 5 has 3 nodal points, that is, mode n has $n-2$ nodal points.

The calculation results for the particular cantilever, and the comparison with the thermomechanical noise data, are shown in Table 3.6. The corresponding frequency factor $(\alpha L)_{\mathrm{n}}$ for each eigenmode of the cantilever vibration is obtained by numerical calculation using Eq. (3.27). The number of corresponding nodal points and their coordinates are obtained from Eq. (3.28). The values of transformation – n_{k} and n_{θ} – are calculated for the angle $\theta = 52°$. The difference δ between the obtained values K_{vn1}, $\theta_{\mathrm{vAn}}^{-1}$ and the corresponding values for deflections $(<u_1^2>/<u_2^2>)$, and for adequate to V-shaped cantilever end rotation photodiode signals $(<s_1^2>/<s_2^2>)$, presented in Stark et al. (2001), have been estimated as an uncertainty: $\delta_{\mathrm{k}} = |\, K_{vn1}/\,(<u_1^2>/<u_2^2>) - 1|100\%$ and $\delta\theta = |\theta_{\mathrm{vAn}}^{-1}\,/\,(<s_1^2>/<s_2^2>) - 1|100\%$. The values with one and two asterisks in Table 3.6 correspond to a V-shaped cantilever with the skew angles $\theta = 48°$

TABLE 3.6 High Mode Oscillations and Spring Constants' Ratio for the V-Shaped Cantilevers

	Parameters of the vibrating constituent beam I or II (Fig. 3.5)					Free V-shaped cantilever spring constants' ratios at mode n and natural mode 1 vibration			
Mode n	Factor $(\alpha L)_n$ Eq. (3.27)	Number of nodal points Eq. (3.28)	x_{ij} Nodal points coordinates Eq. (3.28)	W_i/W Virtual point forces (Table 3.5)	K_{n1} θ_{An}^{-1} (Table 3.5)	K_{vn1} Eq. (3.29a)	θ_{vAn}^{-1} Eq. (3.29b)	[Stark et al. (2001)] Displacement $\langle u_1^2\rangle/\langle u_n^2\rangle$ $\langle S_1^2\rangle/\langle S_n^2\rangle$ photo-signal	δ, % uncertain
2	3.3582	0							
3	5.64015	1	$x_{12/3}$	$W_1 = -1.58916$	15.32	23.47		26.85	$\delta_k = 12.6$
			0.71799		5.53		2.68	2.57	$\delta_o = 4.3$
4	8.32101	2	$x_{13/4}$		133	203.76		222.9	$\delta_k = 8.6$
			0.8740776	$W_1 = -1.39702$		211**			δ_k 4.9**
			$x_{23/4}$		22.18		10.74	10.52	δ_k 2.1
			0.4727073	$W_2 = 0.50311$			10.47**		δ_0 0.5**
5	11.28284	3	$x_{14/5}$						
			0.927243	$W_1 = -1.33488$	784	1201		1089	$\delta_k = 10.3$
			$x_{24/5}$			1125.8*			$\delta_k = 3.4$*
			0.625867	$W_2 = 0.476211$	63.03		30.51	36.79	$\delta_k = 17.1$
			$x_{34/5}$				32.04*		$\delta_k = 12.9$*
			0.3493883	$W_3 = -0.18815$					

The V-shape cantilever data: $L_2/L_1 = 0.667$, $\alpha_M = 0.765$; $J_M = 0.0146$; $\varepsilon_m = 0.1$; $\theta/2 = 26°$, $n_k = 1.532$, and $n_\theta = 0.484$.
*$\theta/2 = 24°$, $n_k = 1.436$ and $n_\theta = 0.509$.
**$\theta/2 = 27°$, $n_k = 1.587$ and $n_\theta = 0.472$. *Values* $\langle u_1^2\rangle/\langle u_2^2\rangle$ *and* $\langle s_1^2\rangle/\langle s_2^2\rangle$ *are evaluated (Stark et al., 2001) by FEA.*
Source: After Tseytlin (2010), reprinted with permission.

and 54°, respectively. One can see that the estimated uncertainties are within an acceptable range.

The kinetostatic model is simpler in application than an object oriented FEA model, and represents the functionally oriented connections of the cantilever's parameters with the spring constant ratios in the higher mode oscillations.

Uncertainties between these two methods are under the requirements of the information criterion of uncertainty negligibility (Section 3.9).

3.5.3.2 *Spring Constant of Triangular (Trapezoid) Part*

The kinetostatic model of supported (in contact) V-shaped cantilever includes the spring constant of its triangular (trapezoid) part with a probe, as consequently working with the constituent rectangular beams ("legs") spring.

In the contact case, the force F is applied to the probe at the point of the cantilever central line, at a distance VL_2 from the rectangular part (Fig. 3.6). Therefore, we consider that half of this force $F/2$ and the bending moment $(F/2)VL_2$ are applied to the end of each rectangular part of the V-shaped cantilever. We can consider this case as a rectangular cantilever which is clamped-elastically supported on the moving end ($q = 1$), with the boundary conditions as follows: $X(0) = 0$, $X'(0) = 0$, $X''(L) = 0$, and $X'''(L) = -jX(L)/EJ$, where j is the spring constant of a "triangular"(trapezoid) part of the V-shaped cantilever, and EJ is the elastic property of the rectangular cantilever's part.

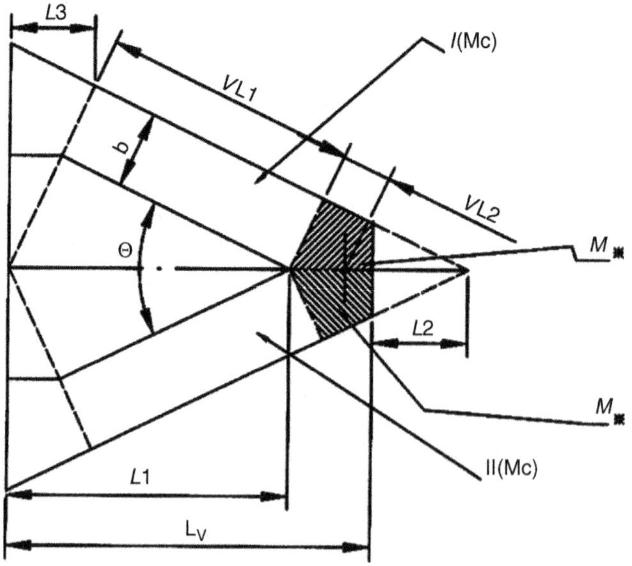

FIGURE 3.6 V-shaped cantilever. $VL_2/VL_1 = V$.

Using Eqs (3.6–3.9), we derived the following expression for these boundary conditions in terms of αL and four constants C_i: $C_1\ (0) = 0$ and $C_2(0) = 0$,

$$C_3 S(\alpha L) + C_4 T(\alpha L) = 0$$

$$C_3\left[\alpha^3 V(\alpha L) + \frac{j}{EJ}U(\alpha L)\right] + C_4\left[\alpha^3 S(\alpha L) + \frac{j}{EJ}V(\alpha L)\right] = 0. \quad (3.30a)$$

This system of simultaneous equations allows us to obtain the corresponding frequency equation from the second-order determinant:

$$\begin{vmatrix} S(\alpha L) & T(\alpha L) \\ \left[\alpha^3 V(\alpha L) + \frac{j}{EJ}U(\alpha L)\right], & \left[\alpha^3 S(\alpha L) + \frac{j}{EJ}V(\alpha L)\right] \end{vmatrix} = 0 \quad (3.30b)$$

The expansion of this determinant corresponds to the following frequency equation:

$$1 + \cosh\alpha L \cos\alpha L - \frac{jL^3}{EJ(\alpha L)^3}(\sinh\alpha L \cos\alpha L - \cosh\alpha L \sin\alpha L) = 0. \quad (3.30c)$$

One can estimate the value of j by its calculation with the known formulas for the triangular (trapezoid) cantilever, or in the process of V-shaped cantilever spring constant calibration in a static state, or a regime of the first mode vibration. If the displacement of the cantilever's probe under the applied force F equals s_1 and the displacement of each rectangular part of the cantilever with $VL_1 = L_r$ under force $F/2$ equals $s_2 = FL_r^3(1+1.5V)/6EJ$, then the value of an elastic "support" spring constant j for each rectangular part of the V-shaped cantilever is equal to $j = F/(s_1 - s_2)$. The ultimate values of this expression are as follows: (i) $j = 0$ in noncontact case, and (ii) $j = \infty$ at $s_1 \approx s_2$ that corresponds to a very rigid triangular (trapezoid) part, or a hard contact support of the cantilever's rectangular parts. Eq. (3.30c) converges (see Table 3.2) to the frequency equation for the clamped-free rectangular cantilever in case (i), and to the frequency equation for the clamped-supported rectangular cantilever with a hard contact on the moving end, in case (ii). The calculation results for the intermediate case with $jL^3/EJ = 100$ at $L = L_r$ are shown in Tables 3.3 and 3.4. Let us recall that the spring constant K_{vn} of the V-shaped cantilever and consequently working spring constant $K_{rn\,\mathrm{eff}}$ of rectangular parts, and j of the triangular part can be connected through the following expression: $K_{vn}^{-1} = (2K_{rn\,\mathrm{eff}})^{-1} + j^{-1}$. Hence, the ratio of the V-shaped cantilever spring constants at mode n to one at the first mode is equal to

$$K_{vn}/K_{v1} = (j + 2K_{r1\mathrm{eff}})K_{rn\,\mathrm{eff}}/(j + 2K_{rn\,\mathrm{eff}})K_{r1\mathrm{eff}} = \left(\frac{jL^3}{EJ} + 6\right)K_{n1}/\left(\frac{jL^3}{EJ} + 6K_{n1}\right). \quad (3.31)$$

We can show with the application of the L'Hôpital's rule that for the cases $j = \infty$ (rigid body) and $j = 0$ (free end of the rectangular part of the cantilever), Eq. (3.31) corresponds to the relation $K_{vn}/K_{v1} = K_{rn\,eff}/K_{r1eff}$. One could find from Eq. (3.31) for the intermediate case with $jL^3/EJ = 100$ that $K_{v4}/K_{v1} = 16.7$ at the fourth mode.

It is evident that the kinetostatic method for the spring constant transformation in higher eigenmodes of the AFM cantilever vibration is simple, and quasiuniversal. The frequency equation for more complicated boundary conditions can be obtained using the fourth-order determinant (Weaver et al., 1990).

Various formulae for *torsional spring constant of V-shaped* cantilever are presented by Neumeister and Ducker (1994), and also by Sader (2003) in the form that shows the complexity of this problem. The approximate solution was found by combining the torsional spring constant for the two inclined quasi-rectangular cantilevers $K_{\mathrm{tors}}^{\mathrm{q-rec}}$ (parallel beam approximation) that undergo simultaneous bending and twisting, under applied torque, on the axis of symmetry close to the vertex of a V-shaped cantilever with a small shift ΔL (Fig. 3.7a,b), and the torsional spring constant of the triangular part of V-shaped cantilever $K_{\mathrm{tors}}^{\mathrm{trian}}$.

The resulting torsional spring constant $K_{\mathrm{tors}}^{\mathrm{V}}$ for the V-shaped cantilever is equal to:

$$K_{\mathrm{tors}}^{\mathrm{V}} = \left(\frac{1}{K_{\mathrm{tors}}^{\mathrm{q-rec}}} + \frac{1}{K_{\mathrm{tors}}^{\mathrm{trian}}} \right)^{-1} \tag{3.32}$$

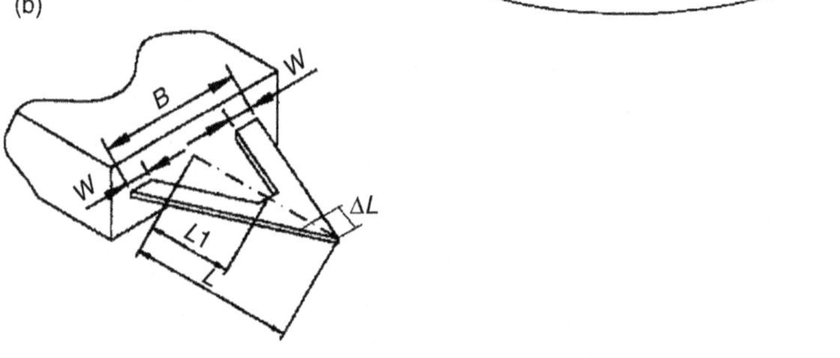

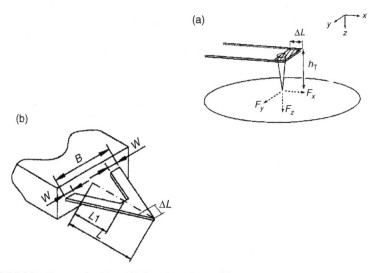

FIGURE 3.7 Rectangular (a) and V-shaped cantilevers (b) parameters.

One can find the torsional spring constant for the rectangular part of the "parallel beam approximation" by applying torque M_{tq} in the section of the V-shaped cantilever with $x = L_1$, that is, on the end of the recess. This torque produces the bending forces Q_i and twisting moments M_i in each of the inclined, by angle $\tan^{-1}(B/2L)$ strips with width w and distance between their axes $\left(B\frac{L-L_1}{L}-w\right)$.

Therefore, $Q_i = \dfrac{M_{tq} - 2M_i/\cos\tan^{-1}(B/2L)}{\left(B\frac{L-L_1}{L}-w\right)}$

and $$K_{tors}^{q-rec} = \frac{Ewh^3\left(B\frac{L-L_1}{L}-w\right)\Re(B)}{8L_1^3},$$

where

$$\Re(B) = \left(B\frac{L-L_1}{L}-w\right) + \frac{16L_1^2\cos\tan^{-1}(B/2L)}{6(1+\nu_p)\left(B\frac{L-L_1}{L}-w\right)}. \tag{3.33}$$

The torsional spring constant for the triangular part of theV-shaped cantilever is equal to:

$$K_{tors}^{trian} = \frac{Eh^3(B/2L)}{3(1+\nu_p)}\left(\log\frac{w}{\Delta LB/2L}\right)^{-1} \tag{3.34}$$

at $L_1 = L(B-2w)/B$ (triangular recess), because the torsional section rigidity I_{tors} of the thin bar with constant thickness h and varying width $b(x_1) = \left(B\frac{L-L_1}{L}\right)\left(1-\frac{x-L_1}{L-L_1}\right)$ approximately equals $I_{tors} \approx \frac{Gb(x_1)h^3}{3}$, where $x_1 = x-L_1$.

If we recall that the modulus $G = E/2(1+\nu_p)$, the angle of rotation for this triangular bar under torque M_{tr} shifted from its vertex by ΔL is

$$\theta_{tr} = \int_0^{L-L_1-\Delta L} \frac{6(1+\nu_p)M_{tr}dx_1}{Eh^3 b(x_1)}.$$

The latter integral yields a logarithm, as a result of the integrand with the variable x_1 to the first power in the denominator. The approach to the problem discussed here, as well as in the reference, is indeed a rough approximation for very small strains. However, more accurate calculations will not provide significant progress, because of our approximate knowledge of the cantilever's real tiny geometric and anisotropic elastic parameters. Meanwhile, expressions

(3.33 and 3.34) correctly show the main functional relationships for the V-shaped cantilever's constituents that are useful in its structural synthesis for precision elastic systems, for example, the large relative influence of the cantilever thickness, the effect of Poisson's ratio, etc. The necessary accuracy in this case can practically be achieved only on the basis of a traceable calibration and adjustment, including the microcantilever laser bending for high-precision curvature change.

Higher modes $n > 1$ frequencies for a tapering cantilever can be found on the basis of the following formula (after Morse, 1948):

$$f_n \approx f_{n\,\mathrm{unif}}\left[1+\frac{2c_j}{\pi^2(n+1/2)^2}\right],$$

where index c_j is the tapering factor, and $f_{n\,\mathrm{unif}}$ is the corresponding frequency for a similar prismatic uniform cantilever.

3.5.4 Sensitivity to Additional Mass at Higher Modes

As the resonance frequency mode number increases, the sensitivity of the cantilever detector to the additional mass on its moving end becomes greater. However, its sensitivity to the additional mass at the corresponding nodal points is close to zero. We can show this on the basis of the formula for the resonant frequency $f_n = \frac{1}{2\pi}\sqrt{k_{n\,\mathrm{eff}} / m_{n\,\mathrm{eff}}}$ in mode n by evaluation of the corresponding effective spring constant, $k_{n\,\mathrm{eff}}$, effective mass, $m_{n\,\mathrm{eff}} = n_n^*$ m, and the effective mass factor n_n^*. Let us assume that a cantilever vibrating at higher mode can be modeled as an indeterminate beam, to the degree that is equal to the number of nodal points (virtual removable supports). Then, the virtual redundant forces W_j (Fig. 3.2a) in the nodal points can be estimated with the force superposition method by matrix equation, such as

$$[d_{ij}][W_j] = [d_{iw}],$$

where d_{ij}, d_{iw} are the flexibility coefficients, W is the quasi-end force at the coordinate $L^* = Lq$ with $q < 1$, the first index $i = 1$ to (n−1) corresponds to the number of nodal points, and the second index $j = 1$ to (n–1), w corresponds to the acting virtual force. The solution of this matrix equation, with respect to redundant forces, may be obtained similar to Eq. (3.18, 3.18a) with the inverse matrix $[d_{ij}]^{-1}$ as follows: $[W_j] = -[d_{ij}]^{-1}[d_{iw}]W$, where $d_{ii} = \frac{(x_{in}^{**})^3}{q^3}$; $d_{ij} = d_{ji}$; $d_{ij} = \frac{(x_{jn}^{**})^2}{2q^3}(3x_{in}^{**} - x_{jn}^{**})$ at $i < j$; $d_{iw} = \frac{(x_{in}^{**})^2}{2q}\left(3 - \frac{x_{in}^{**}}{q^2}\right)$.

Note that all coordinates marked with two asterisks are equal to the corresponding nodal points' coordinates [derived from (3.12)], divided by the length of cantilever L that makes them dimensionless. Now, one can estimate the deflection at the cantilever's quasi-end offset section caused by superposition of forces W, W_1, W_2, W_j – action that is equal to:

$$s = s_{\text{qend}}(W) - \sum_{j}^{(n-1)} s_{\text{qend}}(W_j) = A_{qn}\frac{L^3}{3EJ},$$

where $s_{W\,\text{qend}} = \dfrac{WL^3q^3}{3EJ}$; $s_{W_j\,\text{qend}} = \dfrac{W_j(x_{jn}^{**})^2L^3}{6EJq}(3 - x_{jn}^{**}/q^2)$.

Substituting these expressions, we obtain the total deflection in the offset moving end section, and the effective static spring constant for the cantilever in the nth mode of resonant oscillations, as $K_{n\,\text{eff}} = \dfrac{3EJ}{A_{qn}L^3}$. This allows us to estimate the ratio between the effective spring constant $K_{n\,\text{eff}}$ at n resonant mode, and the same parameter $K_{1\,\text{eff}} = 3EJ/(L^3)$ for the first resonant mode of rectangular cantilever that equals $K_{n\,\text{eff}}/K_{1\,\text{eff}} = A_q^{-1}$. Therefore, the ratio between the effective mass factor for the nth mode and the first mode of oscillations is

$$n_n^*/n_1^* = \frac{f_1^2}{f_n^2}K_{n\,\text{eff}}/K_{1\,\text{eff}} = \frac{(\alpha L)_1^4}{(\alpha L)_n^4}A_{qn}^{-1}. \tag{3.35}$$

Let us recall that $n_1^* \approx 0.24$ for a prismatic rectangular cantilever. Therefore, the effective mass factor for nth mode of oscillations of the same cantilever is equal to:

$$n_n^* = 0.24\frac{(\alpha L)_1^4}{(\alpha L)_n^4}A_{qn}^{-1}. \tag{3.35a}$$

Hence, the ratio between sensitivity to additional mass Δm at the nth mode of resonant oscillations and the first resonant oscillation is as follows:

$$\frac{\Delta f_n(\Delta m)}{\Delta f_1(\Delta m)} = \frac{f_1(K_{n\,\text{eff}}/K_{1\,\text{eff}})n_1^*(n_1^*m_c + \Delta m)}{f_n n_n^*(n_n^*m_c + \Delta m)} \approx \frac{f_n n_1^*}{f_1 n_n^*},$$

where m_c is the mass of the cantilever. As a result, the sensitivity to additional mass at the higher mode equals

$$\Delta f_n/\Delta m = \frac{f_n n_1^*}{f_1 n_n^*}\Delta f_1/\Delta m = \frac{(\alpha L)_n^2 n_1^*}{(\alpha L)_1^2 n_n^*}(\Delta f_1/\Delta m).$$

It is clear that at the nodal points $K_{n\,\text{eff}} = \infty$, $n_n^* = \infty$, and $\Delta f_n/\Delta m = 0$.

Resonant frequencies for an undamped cantilever may be expressed by the following equation:

$$f_n = \frac{1}{2\pi}\sqrt{\frac{K_{n\,\text{eff}}}{m_{\text{eff}\,n}}},$$

where K_{neff} is the effective spring constant of the cantilever in the nth mode of vibration, $m_{\text{eff}\,n} = n_n^* m_c$ is the effective mass, n_n^* is the effective mass factor, and m_c is the actual mass of the cantilever.

Let us recall that the effective mass and effective mass factor n_n^* depend on the cantilever's geometry.

Schematics of the first four bending modes of vibration are shown in Fig. 3.2a. It is seen that certain regions of the cantilever do not take part in the vibration displacements, and the positions of these parts are changing with the mode of vibration. Therefore, the sensitivity of a cantilever to the added mass and mass responsivity is different along the cantilever, at different vibration modes. Experimental data and simulations showed that the largest response to the additional mass was achieved when its application is close to the cantilever's moving end (Burra et al., 2012). Eq. (3.35) and (3.35a) provide a numerical estimate of the ratio between the effective mass factor for the nth mode, and the first mode of oscillation.

Hence, for the second mode:

$$n_2^* = 0.24\frac{(1.8751)^4}{(4.6941)^4}A_{q2}^{-1} = 6.11\times10^{-3}A_{q2}^{-1},$$

For the third mode of vibration:

$$n_3^* = 0.24\frac{(1.8751)^4}{(7.8548)^4}A_{q3}^{-1} = 7.79\times10^{-4}A_{q3}^{-1},$$

And for the fourth mode of vibration:

$$n_4^* = 0.24\frac{(1.8751)^4}{(10.996)^4}A_{q4}^{-1} = 2.03\times10^{-4}A_{q4}^{-1}.$$

3.6 SHIFT OF RESONANT FREQUENCY

Let us now estimate the shift of the resonant frequency for a system of a cantilever with an additional mass Δm (probe, specimen, etc.) clamped to it. In this case, we have to represent the shift $\Delta f_n = f_n(m_c) - f_n(m_c + \Delta m)$ of the resonant frequency $f_n(m_c)$ of the cantilever that has mass m_c, and the resonant frequency $f_n(m_c + \Delta m)$ of the cantilev with the additional mass Δm attached to the cantilever. It is convenient to first express the difference of the squares of these frequencies, using the bending spring constant $K_{n\,\text{eff}}$ and the corresponding masses as follows:

$$f_n^2(n_n^* m_c) - f_n^2(n_n^* m_c + \Delta m) = \left(\frac{1}{2\pi}\right)^2\left(\frac{K_{n\,\text{eff}}}{n_n^* m_c} - \frac{K_{n\,\text{eff}}}{n_n^* m_c + \Delta m}\right).$$

If $(\Delta m/m_c) \ll 1$, the following approximation is helpful:

$$f_n^2(n_n^* m_c) - f_n^2(n_n^* m_c + \Delta m) = \Delta f_n [f_n(n_n^* m_c) + f_n(n_n^* m_c + \Delta m) \approx 2 f_n \Delta f_n(\Delta m).$$

Hence, we can derive from these equations a simplified expression for the shift of frequency $\Delta f_n(\Delta m)$ that is equal to:

$$\Delta f_n(\Delta m) = \frac{K_{n\,\text{eff}}}{8\pi^2 n_n^* m_c} \frac{\Delta m}{(n_n^* m_c + \Delta m)} \approx \frac{f_n \Delta m}{(n_n^* m_c + \Delta m)}.$$

Therefore, the sensitivity to additional mass in the first fundamental mode of the cantilever vibration equals $\Delta f_1 / \Delta m = 2.08 f_1 / m_c$ as shown in Chapter 2. It is clear that the shift of the frequency in the higher modes of vibration may be evaluated by comparison with the shift in the first fundamental mode, on the basis of the following ratio:

$$\frac{\Delta f_n(\Delta m)}{\Delta f_1(\Delta m)} = \frac{f_1(K_{n\,\text{eff}} / K_{1\,\text{eff}}) n_1^* (n_1^* m_c + \Delta m)}{f_n n_n^* (n_n^* m_c + \Delta m)} \approx \frac{f_n n_1^*}{f_1 n_n^*}.$$

Hence, the sensitivity to additional mass at the higher mode is again equal to:

$$\Delta f_n / \Delta m = \frac{f_n n_1^*}{f_1 n_n^*} \Delta f_1 / \Delta m = \frac{(\alpha L)_n^2 n_1^*}{(\alpha L)_1^2 n_n^*} (\Delta f_1 / \Delta m).$$

This sensitivity increases at higher oscillation modes, as a result of the increased spring constant. The reader can see some numerical examples in Tseytlin (2005). See also Ilic et al. (2010) about attogram detection, Hähner and Parkin (2011) about adsorbed humidity by cantilever estimation at higher vibration modes, and in Bauza et al. (2005), with application of a vibration shank virtual probe.

Analysis indicates that nanomechanical resonators offer immense potential for mass sensing, with resolution at the level of individual molecules. A mass responsivity of approximately 5 fg/Hz has been observed for micrometer sized cantilevers, when operating the cantilever in the fourth mode. The largest attainable natural frequency of the triangular cantilever is higher than that of the rectangular one of the same dimensions (h,L) and of the same material.

3.7 ACTUATORS AND DETECTORS

For a cantilever probe to oscillate with sufficient amplitude (up to 100 nm), its resonant frequency should be supported by an actuator: bimorph ceramic/metal, piezotube, piezo-stacks, etc.

Cantilever oscillations are conventionally monitored optically by focusing a laser ray onto a small and curved cantilever; this reflects the ray onto a photoreceptor. The optical feedback from light scattering can be reduced with high-frequency laser current modulation. The advantage inherent in miniaturizing scanning probe microscopes has led to an AFM array probe with an integrated deflection sensor and amplifiers for signal readout, all combined on the same chip. In addition, the magnetovibrational coupling in small dielectric (such as silicon) cantilevers with a single-domain ferromagnetic film, deposited at the free end, was studied. Vogel et al. (2003) demonstrated the optically tunable mechanics of microcantilevers. A cantilever in liquid should be isolated by parylene, silicon nitride film, etc. "Minute size and mass of microfabricated cantilevers make them susceptible to thermally induced noise…Therefore, cantilever sensor can operate in the resonant mode either with or without external excitation" (Lavrik et al., 2004). The thermal frequency fluctuation of the oscillating cantilever is given by (Albrecht et al., 1991):

$$\frac{\delta f_{\mathrm{th}}}{f_0} = \sqrt{\frac{k_B T B_f}{2\pi k_s \langle A_{\mathrm{ca}}^2 \rangle f_0 \, Q_{\mathrm{fac}}}}$$

and the force gradient noise is $\delta F_{\mathrm{th}} = \frac{2k_{\mathrm{s}}}{f_0} \delta f_{\mathrm{th}}$, where k_{s} is the spring constant; B_{f} and $\langle A_{\mathrm{ca}}^2 \rangle$ are the bandwidth and mean-square amplitude of thermal oscillations, respectively, and Q_{fac} is the cantilever's Q-factor. The influence of elastic boundary conditions on the microfibers' oscillation at clamping by glue with different stiffness is discussed in Chen et al. (1994).

3.7.1 Nanotubes (Hollow Cylinders)

Even with a coherent state of a carbon nanotube resonator, where the electronic and vibrational time scales are of the same order, the smallest resolvable mass should be of the order of a thousandth of an electron mass. If two more orders of magnitude could be gained, one can reach the regime where one can weigh the relative mass change, due to the formation of a chemical bond of the absorption of a photon (Braun (2011) who used the Cramér-Rao quantum bound). Experiments reached a resolution level of zeptogram (10^{-21} g). This shows that, with existing carbon nanotube resonators, it should be possible in principle to measure a thousandth of the mass of an electron, and future improvements might allow us to reach a regime where one can measure the relativistic change of mass due to the absorption of a single photon, or the creation of a chemical bond.

Many experimental technologies of the AFM are based on the larger amplitudes (Souayeh and Kacem, 2014) and higher eigenmodes vibration (Huang and Hou, 2011), such as the atomic force acoustic microscopy, the signal inverting dynamic AFM, higher harmonic imaging, and higher eigenmode dynamic force spectroscopy. These atomic force microscope cantilevers cover at least several resonant frequency bandwidths.

A cantilever model with length L, outer diameter d_1 and inner diameter d_2 has the following frequencies:

- bending or transverse vibration (Panovko, 1968; Poncharal et al., 1999) frequency

$$f_{\mathrm{bt}} = \frac{\lambda_n^2}{8\pi L^2}\sqrt{(d_1^2 + d_2^2)}\sqrt{E / \rho_\gamma}, \text{ with } \lambda_n = \alpha_n,$$

- axial or longitudinal vibration (Panovko, 1968) frequency

$$f_{\mathrm{al}} = \frac{2n-1}{4L}\sqrt{E / \rho_\gamma},$$

- torsional vibration frequency

$$f_{\mathrm{to}} = \frac{2n-1}{4L}\sqrt{G / \rho_\gamma}.$$

Nanotubes may also be used as mechanically bistable nanoswitches (at diameter 2–6 nm), nanotweezers, nanosensors, and nanoactuators.

3.7.2 Nonlocal and Asymmetric Elasticity

3.7.2.1 Nonlocal Elasticity

We usually use the Euler-Bernoulli model for the calculation of elastic beams with regular dimensions. The Timoshenko model considers shear and rotation effects in shorter and rigid beams bending and vibration. A fundamental notion of classical continuum elasticity is that the length scale, over which deformation varies, is much larger than the discrete length of matter. However, discreteness and surface effects are significant to nanostructures due to their increased surface-to-volume ratio. In the theory of nonlocal elasticity, the stress at a reference point x is considered to be a function of the strain field at every point x in the body. Eringen (1983) shows kernel forms for nonlocal moduli H_α in (i) one-dimensional, (ii) two-dimensional, and (iii) three-dimensional problems. For example, in the case (ii)

$$H_\alpha(|x|, \tau) = (2\pi l^2 \tau^2)^{-1} K_{0\mathrm{b}}(\sqrt{x \bullet x} / l\tau),$$

where $\tau = e_0 a/l$, a is an internal characteristic length (e.g., lattice parameter, granular distance) and l is an external characteristic length, e_0 is a material constant, and $K_{0\mathrm{b}}$ is a modified Bessel function.

The experimental data for copper and silicon shows that their nonlocal length $e_0 a$ is of the order 0.3 nm (Maranganti and Sharma, 2007) and can be larger for other materials where classical elasticity breaks down. Maranganti also shows that discreteness is not the only cause of nonlocality. The long-range nature of interatomic forces is another factor. The effects of nonlocal parameters and surface tension are significant for the natural frequency of nanobeams, rotating nanocantilevers, films, and nanotubes (Pradhan and Murmu, 2010; Lee and Chang, 2010; and Zang and Liu, 2008). The governing equations of transverse vibration for the simply supported nanotube, with consideration of surface effects, can be expressed using the nonlocal Timoshenko beam model as

$$E*I*\frac{\partial^2\Phi}{\partial X^2}+\left[1-(ea_0)^2\frac{\partial^2}{\partial X^2}\right]\left[KGA\left(\frac{\partial Y}{\partial X}-\Phi\right)-\rho\frac{\partial^2\Phi}{\partial t^2}\right]$$

$$=0,\frac{\partial}{\partial X}KGA\left(\frac{\partial Y}{\partial X}-\Phi\right)+H_s\frac{\partial^2 Y}{\partial X^2}=\rho A\frac{\partial^2 Y}{\partial t^2},$$

where e_0a is the nonlocal constant, K is the shear factor, Φ is the rotation angle of the nanotube, $E*I*$ is the effective flexural rigidity, X is the longitudinal coordinate, and t is time. H_s is the constant parameter determined by the residual surface tension and shape of the cross section. Here, the parameters $E*I*$ and H_s are defined as

$$E^*I^*=EI+\pi E^s(R_i^3+R_o^3),$$

$$H_s=4\tau_s\,(R_i+R_o),$$

where E^*I^* is the flexural rigidity of the tube's cross-section; R_i and R_o are the internal and external radii of the tube, respectively; E^s and τ_s are the surface elasticity modulus and residual surface tension per unit length on the nanotube, respectively.

Calculation (Lee and Chang, 2010) shows that the nonlocal Timoshenko model gives slightly smaller values of vibration frequency for the comparatively long (L-length) tubes, with $L/R_o > 24$, than the Euler-Bernoulli model, and significantly smaller values for short tubes with $L/R_o < 18$, especially at second and third vibration modes. The calculated vibration frequency is even smaller, if surface effects are not considered (at $e_0a/L = 0$).

The mechanics of nonlocal effects on the vibration of single-layer graphene sheets were studied by Wang and Hu (2005), and Li et al. (2015). Resonance frequency and mass identification of zeptogram-scale nanosensor (carbon nanotubes, nanowires with fundamental frequencies up to THz) were studied based on nonlocal beam theory, and the following fundamental frequency Ω_1 equations were found:

for clamped nanocantilever with additional mass m on the free end $\Omega_1=\left(\frac{11}{140}+\frac{1}{3}\beta_\gamma-\frac{1}{10}\lambda_\gamma^2\right)^{-1/2}$;

for simply supported (1) and bridged (2-doubly clamped) nanobeams with additional mass m in the middle section $\Omega_1=\Psi_i\pi^2(\Im_i+\eta_i\beta_\gamma+\mathrm{N}_i\pi^2\lambda_\gamma^2)^{-1/2}$, where

$$\Psi_1=1,\Im_1=1,\eta_1=2,\mathrm{N}_1=1;$$

$$\Psi_2 = 4, \Im_2 = 3, \eta_2 = 8, \mathrm{N}_2 = 4.$$

Here, $\beta_\gamma = m/(\rho_\gamma A_c L)$, $\lambda_\gamma = e_0 a/L$, $\Omega = \omega^2 L^2/v^2$; ω is the angular frequency, $v = \sqrt{\omega c_\gamma r_g}$ is the phase velocity, $c_\gamma = \sqrt{E/\rho_\gamma}$ denotes the elastic wave speed, $r_g = \sqrt{J/A_c}$ is the radius of gyration of the cross section with J as the second moment of cross-sectional area A_c, e_0 is the dimensionless nonlocal parameter, a is the internal characteristic length, and L is the length of the beam.

Dimensionless fundamental frequencies on this basis can be expressed through calibration constants c_k, c_m, c_n as $\Omega^{-1} = c_k^{-1}\sqrt{1 + c_m\beta_\gamma + c_n\lambda_\gamma^2}$, and the mass m of the attached particle is expressed (Li et al., 2015) through the shift of the nanosensor's fundamental frequency Δf_c as follows:

$$\frac{m}{m_b} = \frac{\Delta f_c}{c_m f_{0c}}\left[2 + 3\frac{\Delta f_c}{f_{0c}} + 4\left(\frac{\Delta f_c}{f_{0c}}\right)^2\right] - \frac{c_n}{c_m}\lambda_\gamma^2,$$

where m_b stands for the mass of beam sensor, f_{0c} is the "classical" fundamental frequency corresponding to $\beta_\gamma = \lambda_\gamma = 0$; $c_m = 4.121$ and 2.692 for nanocantilever and nanobridge, respectively, $c_n = 12.923$ for nanobridge.

If neglecting the nonlocal effect, the mass identification formula becomes the same as it is shown in Sections 2.5.3 and 3.6 for rectangular cantilevers.

3.7.2.2 Asymmetric Elasticity

If we treat the material as a system of discrete particles (individual crystals in polycrystalline materials, elements of the molecule, etc.), the displacement of their centers of gravity can be defined by a vector field, and a small rotation about the centers of gravity by a kinematically independent moment field. Furthermore, in a theory of a continuum medium, the action of the medium on a small element isolated from it is determined by the stresses, or by the forces acting on its faces, and the moment calculated in terms of these forces. However, if we treat a medium as a system of discrete elements, the action on one element from its neighboring elements is determined by independent forces and moments. On this basis, the fundamental equations of asymmetric elasticity have been presented (Pal'mov, 1964) as follows:

$$\nabla \bullet \tau_s + K = 0;: \nabla \bullet \mu + \tau_\times + c_f = 0,$$

where $\nabla = \partial/\partial \mathbf{r}$ represents the Hamiltonian differential operator, $\boldsymbol{\tau}_s$ and $\boldsymbol{\mu}$ are dyadic of stress and couple-stress, $\boldsymbol{\tau}_\times$ is the vector of the dyadic $\boldsymbol{\tau}_s$, $\boldsymbol{K}$, and $\mathbf{c_f}$ are the intensities of forces and moments at every point in the volume of the body, respectively.

These equations are the required equations of equilibrium, and the stress dyadic $\boldsymbol{\tau}_s$ is asymmetrical, since its vector is not zero. This is origin of the term "asymmetric elasticity."

Let us recall that the dsDNA molecule is a biopolymer with two complementary sugar-phosphate chains (backbones), twisted around each other to form a right-handed helix with one turn per 10 basepairs. Each chain is a linear polynucleotide consisting of four kinds of bases: two purines and two pyramidines. The two chains are joined together by hydrogen bonds (base pairs) between pairs of nucleotides. The distance between adjacent bases is about 0.34 nm in the native B-form.

Using the relations of asymmetric elasticity, Zhou and Lai (2005) have built a model of twisted dsDNA that follows from the asymmetry of the base pair stacking interactions, and shows the effective application of the helix ribbon with a noncircular cross section to the dsDNA molecule modeling. This conclusion corresponds to our effective application of a pretwisted (helicoidal) nanostrip model for the dsDNA molecule mechanical and thermomechanical properties calculations.

3.8 INTERNAL AND EXTERNAL DAMPING

It was found in an experimental study that the Q-factor of silicon triangular cantilevers, and their arrays with lengths ranging from 500 nm to 100 μm, and thickness ranging from 30 nm to 100 nm, is as low as 5 in the air, and as high as 10^4 in vacuum. The typical mechanical characteristics of such cantilevers include a bending spring constant within 8–15 N/m and a natural frequency within 4–10 MHz. Damping effects (Chen et al., 1994) in a gas change the resonance frequency of rectangular and V-shaped cantilevers insignificantly (within 0.2–1.5%), but their quality factor changes up to 280% for rectangular cantilevers, and up to 230% for V-shaped cantilevers in argon, air, and helium media. However, the natural frequency of V-shaped cantilevers can change within 180% in different liquids such as hexane, ethanol, water, and hexadecane.

The cantilever's resonant frequency depends on the Q-factor as follows

$$f_{0,\mathrm{Q}} = f_0 \frac{\sqrt{Q_{\mathrm{fac}}^2 - 1/4}}{Q_{\mathrm{fac}}},$$

where $Q_{\mathrm{fac}} = m\omega_0/C_y$, $f_0 = \omega_0/(2\pi)$ is the cantilever's fundamental frequency without damping, m is an effective mass, and $C_y = 2C_\gamma\, m$ is the damping coefficient of the cantilever. If $Q_{\mathrm{fac}} > 1$, then $f_{0,\mathrm{Q}} \approx f_0\,[1 - 1/(8Q_{\mathrm{fac}}^2)]$. Photiadis and Judge (2004) predicted that the contribution to the loss factor for the fundamental mode of a typical cantilever beam attached to the substrate is less than $Q_{\mathrm{fac}}^{-1} \leq 10^{-7}$. Therefore, the corresponding change in the fundamental resonant frequency of the cantilever, from the attachment losses in this case, is less than $(1/8)10^{-14} f_0$. More accurate predictions may be done with the formulae for Q_{fac}^{-1}

presented in the referenced source, where its dependence on the width, thickness, and length of the oscillating cantilever is shown, and the thickness of the corresponding substrate in relation to the elastic oscillations wavenumber. In order to overcome the difficulties of resonant cantilever operation in liquids, cantilever transducers are used as a part of self-oscillating system with positive feedback. Enhancement of the quality factor, via driving force control, is useful in frequency modulation and self-excitation techniques (Albrecht et al., 1991).

Polymer microcantilevers may have larger deformations without fracture. Their quality factor is in the range from 9 to 30, at the resonant frequencies from 2 kHz to 10 kHz. However, tensile tests of polymer micro- and nanoscale fibers show that their stress–strain relationships can be nonlinear (in some cases with strain in the third power).

The low Young's modulus and high-mechanical strength of SU-8 makes it a highly suitable material for the fabrication of cantilever-based sensors, as the low Young's modulus ensures an increased sensitivity (Nordström et al., 2008).

3.8.1 Ambient Influence and Protection

Thermal and ambient-induced (vibration, relative humidity, exposure to mercury vapor, etc.) deflections of AFM cantilevers may cause change of their resonance frequency as well (Thundat et al., 1994; Jeon et al., 2004). Detailed analysis of the composite multilayer cantilever beams (silicon substrate with oxide, nitride, and polymer dielectric layers) is done by Lange et al. (2002) whose research of the temperature, pressure, and Q-factor influence on the cantilever's resonant frequency is informative. The combined inverse quality factor Q_{fac}^{-1} is, therefore, equal to the sum of inverse internal and external quality factors: $Q_{\text{fac}}^{-1} = \sum_i Q_i^{-1}$.

Temperature influence revisited. Models considered by Lange et al. (2002) are helpful and satisfactory enough, except for the model of the influence of temperature, in which Poisson's ratio is assumed to be constant with the changes of temperature. This is correct only for the isothermal conditions. The adiabatic moduli of elasticity differ from the isothermal ones (Landau and Lifshitz, 1970) and Poisson's ratio, in this case, depends on temperature. A deformation is adiabatic if there is no exchange of heat between the various parts of the body, or between the body and the surrounding medium. Therefore, the resonant frequency model in more general temperature conditions should be represented as follows:

$$f(T) = \frac{\alpha_n^2}{\sqrt{12}} \sqrt{\frac{E(T)\, h(T)^3 b(T)}{[1 - v_{\text{p}}(T)^2]\, M_g L(t)^3}},$$

where α_n is the constant of a frequency equation for the rectangular cantilever with $\alpha_{n=1} = 0.1875$, $\alpha_{n=2} = 4.694$, $\alpha_{n=3} = 7.855$, $\alpha_{n=4} = 10.996$, $\alpha_{n=k>4} = \pi(2k-1)/2$; M_g is the total mass of the cantilever; $h(T)$, $b(T)$ are the cantilever temperature; T is the dependent thickness and width; $E(T) = E(T_n)[1 + \alpha_E E(T - T_n)]$, $\nu_p(T) = \nu_p$ at isothermal conditions; $E(T) = E_{ad}(T) = E(T_n) + E^2 T \alpha_L^2 / 9C_p$, $\nu_p(T) = \nu_{pad}(T) = \nu_p + (1 + \nu_p) E T \alpha_L^2 / 9C_p$ are the temperature dependent adiabatic Young's modulus and Poisson's ratio, respectively. C_p represents the specific heat of the elastic material, α_E and α_L are the temperature coefficients of Young's modulus and linear dimensions (Table 3.7).

However, the difference between isothermal and adiabatic Young's moduli for most metals is less than two percent at normal temperature.

3.8.1.1 Minimization of the Temperature Influence. Temperature Specified Parameters

The necessary and sufficient parameters for the characterization of thermal field and temperature deformations of an elastic structure can be defined from the Fourier equation for the heat transmission:

$$\partial u_T / \partial t = a_T \frac{\partial^2 u_{Ti}}{\partial X_i^2}, \tag{3.36}$$

where $a_T = K_h / (C_p \rho_\gamma)$ is the coefficient of temperature conductivity (thermal diffusivity) that is equal to the ratio of the thermal conductivity K_h and the heat capacity $C_p \rho_\gamma$. It follows from (3.36) that the speed of the influence temperature field changes in time – $\partial u_T / \partial t$ – and the functions of its distribution in space $\partial^2 u_{Ti} / \partial X_i^2$ should be specified.

Normal temperature revision. The nominal of reference temperature (normal temperature) is commonly taken to be $t_{Tn} = 20°C$ (Comptes Rendus des Séances, 1961), that is, $T_{Tn} = 293.15\ K \approx 293\ K$. It can be shown that increasing temperature $(t_{Tn} + \Delta t_T)$ in precision elastic structures can lead to an increase in the thermodynamic fluctuations of their elements' dimensions, and an increase of their length standard uncertainty $\Delta\sigma_{\Delta T}$, such that

$$\Delta\sigma_{\Delta T} = \sqrt{\frac{\Delta t_T}{T_{Tn}}}\sigma_{xa}(T_{Tn}),$$

where $\sigma_{xa}(T_{Tn})$ is a standard uncertainty in the system element's thermodynamic dimensional fluctuations under established normal temperature T_{Tn}. For example, the expanded uncertainty of thermodynamic fluctuations increases by 40%, with an increase of temperature by 5 K only because

$$3\Delta\sigma_{\Delta T} = 3\sigma_{xa}(T_{Tn})\sqrt{5/293} \approx 0.40\,\sigma_{xa}(T_{Tn}).$$

Similarity. The analytical calculation methods for thermoelastic problems are based on the use of the so-called similarity criteria (dimensionless numbers): Fourier's thermal homochronism, Bio, Nusselt's, Prandtl's, Predvoditelev's, Reynolds', and Grashof's, etc. (Marchenko, 1965).

TABLE 3.7 Thermal Properties of Elastic Materials

Material, profile	Designation	Temperature limits, °C		Specific heat, C_p $J \cdot kg^{-1} \cdot K^{-1}$	Thermal conductivity, K_h, W/·m·K	$\frac{\alpha^{\circ}_{TE}}{\alpha^{\circ}_{TL}}$
		Melting	Application			
Microstrips	Br4Sn-3Zn**, (550)*	1020	*/3	385	390	1.0/1.0
	Br Be2*** (320)*	915	*/2	420	118	0.67/0.99
	PtAg20****	1200	<200	165	143	0.56/0.93
Steel	4340	1420	*/2	460	24–33	0.77/0.77
Stainless steel strips	420 (315)*	1370	*/2	460	25	0.77/0.7
Steel wire	–	–	–	–	32.9	–
Invar	Fe-Ni36	1427	< (315)*	515	11	1.1/0.05^^
Copper alloys	70Cu-30Zn	870	…	375	120–389	1.07
Copper alloys, hard	90Cu-10Zn					1.28/1.17
Brass-strip	12Zn-Cu	1050	…	375–393	116	…/1.21
Aluminum alloys	7075-T6 6061-T6	476	450	962 596	167 (237 for *Al*)	…/ 1.48
Titanium alloy	Ti-6Al-4V (700)*	1500	*/3	502–795	7.2–16.9	…/0.6
Nitinol	Ti-55Ni ^TEP = 9–13	1310	70–130	470–620	10–18	…/0.65
Carbon clathrates whisker, nanotubes	C ^TEP = 22	3652	<750	~600	14–22; 86–200>2000	…/0.6/0.04

(Continued)

TABLE 3.7 Thermal Properties of Elastic Materials (*cont.*)

Material, profile	Designation	Temperature limits, °C		Specific heat, C_p J·kg^{-1}· K^{-1}	Thermal conductivity, K_h, W/·m·K	$\frac{\alpha^{\circ}_{TE}}{\alpha^{\circ}_{TL}}$
		Melting	Application			
Boron carbide	B_4C	2450	<600	950	27–90	.../0.36
Silicon single crystal	Si (110)	1415	<1100	707–799	125–156	0.22/0.17
Fused quartz Silica	SiO_2	1710	<1000	703–825	1.4–1.5 4.6	–0.22/ 0.032
Zerodur	K20		<0–50 <700	800 ...	1.46 ...	/0.0013 /0.13
Silicon carbide	SiC	2700	<600	715–975	42–150	.../0.3
Silicon nitride	Si_3N_4	1800–1900	<1000	700	19–50	.../0.18to/ 0.21
Polysilicon	Si	1415	<1050	707.6	30–140	... /0.18
Tungsten Graphene	W WC Monolay C	3410 2777	... ...	132.7	108–178 84–120 >3000	... /0.29 /0.32

*Tempering temperature;
**Thermal electromotive force (TEMF) with copper TEMF = 2.0 μV/K, resistance temperature coefficient $\alpha_{T\rho} = 9.5 \times 10^{-4}$ K^{-1}, elasticity modulus temperature coefficient $\alpha_{TE} = -4.5 \times 10^{-4}$ K^{-1}; temperature coefficient of linear expansion $\alpha_L = 15.5 \times 10^{-6}$ K^{-1};
***TEMF = 1.0 μV/K, $\alpha_{T\rho}$=15.5 × 10^{-4} K^{-1};
****TEMF = 8.0 μV/K, $\alpha_{T\rho}$=10.5 × 10^{-4} K^{-1};
^TEP is the thermoelectric power in μV/K;
^^$\alpha_{TL}{}^{\circ}$= 0.003 for Super-Invar; Zerodur has $\alpha_{TL}{}^{\circ}$ = 0.008 (for others, see Tseytlin (2006)). Superscript ° indicates relative values; the ... sign shows that this parameter is not applicable, or the author does not have reliable information and recommends to verify corresponding data in other credible sources.

Distribution. The temperature distribution with the presence of internal heat sources is described by the following Fourier thermal diffusion equation,

$$\frac{\partial \theta_T}{\partial t} = a_T \nabla^2 \theta_T + \frac{W_{\text{H.S}}}{c_\text{p} \rho_\gamma}, \tag{3.37}$$

and the thermoelastic potential is described by the Laplace equation,

$$\nabla^2 \psi_T = \frac{1+\nu_\text{p}}{1-\nu_\text{p}} \alpha_{TL} \Delta \theta_T, \tag{3.37a}$$

where $W_{\text{H.S}}$ is a specific power of the internal sources of heat, and θ_T represents the temperature field.

The mean temperature of the body is calculated with an integral over internal volume of the body.

Precise solution. The precise analytical solutions of thermoelastic problems have been obtained only for semibounded bodies of so-called classical or basic forms: infinite plate, infinite cylinder, and sphere. A semibounded body is one in which a heat exchange is carried out only through one surface separating it from the medium.

Approximation. Several approximate methods have been developed for the estimation of temperature distribution in bodies with complicated forms, including: (i) a method based on the property of "heat flow stability;" (ii) a method of basic bodies; (iii) a finite element analysis. Methods (i) and (ii) have an uncertainty of, usually, more than 50%. The finite element method provides more accurate calculations, with an uncertainty of less than 35%. However, the accuracy of this method depends on the proper simulation of boundary conditions for the heat transfer and material features (isotropic, anisotropic, and orthotropic). Verification of the calculation results may be performed by experimental-calculation methods.

Area, volume, mass. The transient temperature function for elastic displacements has an exponential character,

$$\theta_{\text{Tei}} = h(t) = K_\tau [1 - e^{-(t-\tau_{\text{lag}})} / T_\tau],$$

where $T_\tau = K^* V_\Omega / (A_{\text{cs}} \times \ln 10)$ is a time constant, K_τ is a transition coefficient, K^* is a coefficient that depends on the conditions of cooling (heating) the body, V_Ω (m) is the volume (mass) of the body, A_{cs} is the area of the body's surface, and τ_{lag} is the lag time for the temperature transient function in the precision elastic structure.

The change of the transient function value during the precision system's working time t_m should be restricted by the information criterion of uncertainty negligibility η_U (Tseytlin, 2006) that corresponds to the following expression:

$$1 - \exp[-(t_\text{m} - \tau_{\text{lag}})\, A_\text{c} \times \ln 10 / K^* V_\Omega] \le \eta_U.$$

By subtracting 1 from both sides of this expression, we find that

$$\exp[-(t_{\mathrm{m}} - \tau_{\mathrm{lag}})\, A \times \ln 10 / K^{*} V_{\Omega}] \geq (1 - \eta_{U}).$$

After taking the logarithm of both sides of the latter expression, we have the relationship

$$(t_{\mathrm{m}} - \tau_{\mathrm{lag}}) \leq \frac{K^{*} V_{\Omega}[-\ln(1 - \eta_{U})]}{A_{\mathrm{cs}} \times \ln 10}. \qquad (3.38)$$

It is clear from (3.38) that the increase of the working time, within which the transient function's change is a small one, compared to the criterion $\eta_U \leq 0.35$ of negligibility, is possible by increasing the ratio between the volume V_Ω and the area A_{cs} of an elastic body. In this case, we suppose that K^* = constant. The volume of a solid body is approximately proportional to its mass, fact that corresponds to the positive effect on the thermal stabilization from the increase of the body's mass. Therefore, we should use in the structural design the optimal forms of elastic bodies with the largest possible relations V_Ω/A_{cs}. It is known that the largest V_Ω/A_{c} ratio is in the spherical body, a relative extremum is provided by a cylinder.

Adaptive measuring systems. These systems have an extended range of reference conditions for the influence quantities (*IQ*), because they are automatically self-adjustable, and can decrease the influence errors Δ_{inf} by certain automatic corrections. Let us remember that it is important to have an adaptive measurement system whose reaction is stable, and adequate to the reaction of the measurand to the same changes of *IQ*. It is useful to have standard elements with known influence functions in the dimensional chain of measuring instruments or precision machines. Differential measurements are preferred for the compensation of base displacement by gravitational field and other *IQ*.

Invariant measuring systems. These precision systems are effective if the measurand is insensitive ($\Delta_{\mathrm{inf}} \equiv 0$) to the changes of *IQ*, for example, the measurand is of a material with approximately zero temperature expansion coefficient; its quantity is invariant to the changes of temperature, as well as force, mass, etc.

3.8.1.2 Heated Tip AFM Cantilevers

The mechanical properties of polymeric nanomaterials depend strongly on temperature, and the temperature dependence of those properties can differ significantly between the macroscale and nanoscale measurand, due to localized chemical and confinement effects. Achieving the temperature-dependent measurements of mechanical properties requires means of heating the sample, or tip-sample contact. There exist two primary methods to control the temperature of the tip-sample contact in the AFM: (i) global heating of the entire measurand sample, and (ii) local heating of the cantilever tip.

The first method is useful for studying nanoscale polymer dynamics, such as observing recrystallization. Its advantage is in the ease of measuring, and in controlling the temperature. The drawbacks to global sample heating are in the significant thermal drift, and possible detrimental interaction with AFM components, as well as slow heating rates. The second approach to control temperature involves local heating of the cantilever tip (Toffoli et al., 2013; Killgore et al., 2014). In this case, the use of U-shaped cantilevers is recommended. But these cantilevers have multimode vibrations at the local heating of the cantilever tip (up to 13 modes). The first five are in-phase flexural modes, while the higher flexural modes may have anticlastic curvature (i.e., curvature perpendicular to the main flexural axis, but with opposite sign).

3.8.1.3 Microcantilever Tip Thermal Oscillation

The differential equation of the forced motion for the mechanical harmonic oscillator with viscose damping and the Langevin's stochastic thermal force $F_{\text{th}}(t) = j_k U_{\text{th}}(t)$ may be presented (Majorana and Ogawa, 1997) as

$$\ddot{x} + \frac{\omega_0}{Q_{\text{fac}}}\dot{x} + \omega_0^2 x = j_k U_{\text{th}}(t) / m,$$

where m is the mass of the tip on the cantilever–oscillator. Let us recall that, for a mechanical harmonic oscillator with viscose damping,

$$X_a^2 / 2R_{\min} = B_\omega / j_k \omega_0 Q_{\text{fac}}, \tag{3.39a}$$

where Q_{fac} is the Q-factor, j_k is the stiffness of the cantilever, B_ω is the bandwidth of the measurement (Albrecht et al., 1991), and ω_0 is the resonant frequency of the oscillator. The expression for the limit of expanded uncertainty in thermal fluctuations is as follows:

$$U_{\text{th}} = (4K_B T_T B_\omega / Q_{\text{fac}} j_k \omega_0)^{1/2}. \tag{3.39b}$$

If the thermal fluctuations correspond to the harmonic excitement with the stationary amplitude U_{th} and an angular frequency ω, then the amplitude of the oscillator fluctuations in the bandwidth $B_\omega = (\omega - \omega_0)$ is equal to:

$$U_{\text{th},\omega} = U_{\text{th}} \frac{\omega_0^2}{\sqrt{(\omega_0^2 - \omega^2)^2 + (\omega\omega_0)^2 / Q_{\text{fac}}^2}}. \tag{3.39c}$$

Substituting (3.39b) into (3.39c), we obtain the equation for the mechanical oscillator thermal fluctuations in the bandwidth B_ω as follows:

$$U_{\text{th},\omega} = (4K_B T_T B_\omega / Q_{\text{fac}} j_k \omega_0)^{1/2} \frac{\omega_0^2}{\sqrt{(\omega_0^2 - \omega^2)^2 + (\omega\omega_0)^2 / Q_{\text{fac}}^2}}, \tag{3.39d}$$

where $\frac{\omega_0^2}{\sqrt{(\omega_0^2-\omega^2)^2+(\omega\omega_0)^2/Q_{\text{fac}}^2}}=\mu_{\omega/\omega_0}$ is the transfer function (or the dynamic response factor). Expression (3.39d) is similar to that presented by Lavrik et al. (2004) for the resulting thermal noise density spectrum.

3.8.2 Methods for the Design of Vibro-Isolation Means

The quality of the transmission of a working force or motion may be defined by the following functional

$$I_s=\int_0^\infty [x(t)-u(t)]^2 dt, \tag{3.40}$$

where $x(t)$ is the actual displacement, and $u(t)$ is the specified one. Thus, the problem of synthesizing the optimal vibro-isolating system may be formulated as the reproduction of the useful signal u under the present of vibration disturbances. In this case, the frequency-response function $H\ (j\omega)$ of the dynamic system, consisting of the structure and damping, should minimize the integral estimator (3.40).

Dynamic vibro-damper. In the two-mass system with certain relations between masses m_1, m_2 (Fig. 3.8a) and stiffness j_1, j_2 of springs, one can obtain a

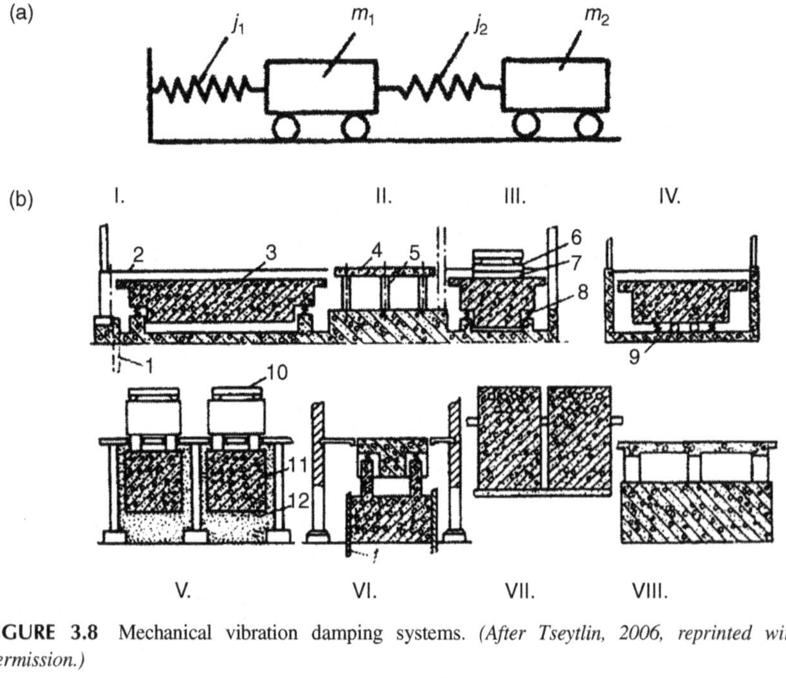

FIGURE 3.8 Mechanical vibration damping systems. *(After Tseytlin, 2006, reprinted with permission.)*

so-called antiresonance with which the point of application of the periodic force remains fixed (motionless).

The equations of the motion of mass m_1 and m_2 may be expressed in the following form:

$$-j_1 y_1 + j_2(y_2 - y_1) + F_0 \sin \omega t = m_1 \ddot{y}_1; \tag{3.40a}$$

$$-j_2(y_2 - y_1) = m_2 \ddot{y}_2, \tag{3.40b}$$

where $y_1 = y_1\ (t)$, $y_2 = y_2\ (t)$ are the displacements of corresponding masses in time. The system of simultaneous equations Eq. (3.40a) and (3.40b) is satisfied by the solutions

$$y_1 = a_1 \sin \omega t;\ y_2 = a_2 \sin \omega t, \tag{3.41}$$

which show that the oscillations have the same frequency as that of the exciting force with amplitude F_0. This partial solution describes the stationary part of the process. Substituting (3.41) into (3.40a, b) yields two equations with two unknown amplitudes, $-a_1 m_1 \omega^2 + j_1 a_1 - j_2(a_2 - a_1) = F_0$; $-a_2 m_2 \omega^2 + j_2(a_2 - a_1) = 0$.

Hence,

$$\begin{aligned} a_1 &= \frac{F_0(j_2 - m_2\omega^2)}{(j_1 + j_2 - m_1\omega^2)(j_2 - m_2\omega^2) - j_2^2}; \\ a_2 &= \frac{F_0 j_2}{(j_1 + j_2 - m_1\omega^2)(j_2 - m_2\omega^2) - j_2^2}. \end{aligned} \tag{3.42}$$

The resonance of this system occurs when the denominators in (3.42) are equal to zero, and amplitudes a_1 and a_2 tend to infinity. In this case, the resonant frequencies are equal to the natural frequencies of the system, and their number is always equal to the system's number of degrees of freedom. At antiresonance, the numerator of the first expression in (3.42) is equal to zero. It is necessary in this case to have the forcing frequency $\omega = \sqrt{j_2 / m_2}$. With this frequency value, we obtain from formulae (3.40a, b) that $a_1 = 0$ and $a_2 = -F_0 / j_2$. Hence, the first mass remains immovable, despite the application of the periodical exciting force $F = F_0 \sin \omega t$. The possibility of antiresonance is used practically in the dynamic damper of oscillations. The additional mass on the elastic link should be added to the structure for this reason. A certain adjustment of the vibro-damper provides the cancellation of vibrations in the main structure, while the additional mass m_2 vibrates intensively. This damper works properly if the forcing frequency is constant. Otherwise, the ratio of stiffness j_2 and mass m_2 should have feedback adjustment.

Passive vibro-isolators. For special precision manufacturing needs and precision measurement centers, it is possible for the whole room or the entire building to be vibro-isolated. Figure 3.8b (Tseytlin, 2006) depicts foundations with:

(i) spring damper, (ii) round columns, (iii) additional plate, (iv) rubber-spring dampers, (v) heavy block in the sand, (vi) concrete columns, (vii) with the ventilation channels for heating, and (viii) the forced concrete columns, where 1 is a metallic rabbet, 2 is the hanging floor, 3 is a concrete block, 4 is the concrete plate, 5 is a metallic or wood column, 6 is a damping insert, 7 is the plate of cast iron, 8 are the spring dampers, 9 are the rubber dampers, 10 is the precision structure, 11 is a solid concrete block, 12 is the sand foundation. The low vibration noise of the local foundation may be reduced by a pneumatic system, with damped magnetic and air springs. Such solution is used even for the improvement of stability of known Besocke-type scanners in scanning tunneling microscopes. Layered structures (gaskets) from flexible graphite, neoprene rubber, zinc-aluminum alloy (ZA-27) are also useful for vibration damping. The details of these solutions are beyond the scope of this text.

Active vibro-protective systems with automatic control and general problems of vibration in machines with elastic transmission mechanisms are discussed in Kolovskii (1976), Kolovsky et al. (2000).

Impact. The action of impact on the precision elastic structure is a complicated one. This action that does not cause damage to a structure may be estimated by equalizing the potential energy of the impact, and the energy of the caused elastic deflection. For example, one can estimate the largest deflection s_δ of an elastic element with flexural rigidity j_s at the impact by some free falling mass m_δ from height h_δ above this element, on the basis of the following equation:

$$m_\delta g(h_\delta + s_\delta) = \tfrac{1}{2} j_s s_\delta^2.$$

Therefore, $s_\delta = \dfrac{m_\delta g}{j_s} + \sqrt{\left(\dfrac{m_\delta g}{j_s}\right)^2 - \dfrac{2m_\delta g h_\delta}{j_s}}$.

3.8.2.1 Acoustic Noise

Noise influence. It was shown that the uncertainty level of the metrological large range scanning probe microscope is approximately 3 nm, and is principally due to the vibrational noise originating from the building, and other sources of acoustic noise. A noise reduction requires the use of improved acoustic isolation. To reduce the influence of the ground vibration, the inherent frequency ω_0 should be less than $1/\sqrt{2}$ of ω_0. Zheng et al. (2014) developed and studied a vibro-isolation system for measuring small forces in a range less than 10^{-7} N, based on an electrostatic force generated, by injecting air damping. A pair of coaxial cylindrical capacitors generates the electrostatic force when a voltage is applied across the inner and outer electrodes. The outer electrode is clamped to the frame and the inner electrode that is clamped at the end of an elastic parallelogram mechanism, with four cylindrical notch hinges (Section 2.6), and moves freely in the vertical direction. The air damper operates between the outer and inner cylinders.

A displacement of the inner cylinder is measured by the laser interferometer. However, we should note that the utilized elastic parallelogram cannot have a rectilinear motion contrary to the assertion of Zheng et al. (2014). A double parallelogram (Tseytlin, 2006) or "orthoplanar" (Howell, 2001) should be used, instead, to obtain the quasi-rectilinear motion. See also Qin et al. (2013), on the elimination of parasitic motions in symmetric flexure hinges.

3.8.2.2 Dust Content and Settling

Dust can influence the behavior of elastic systems by adding some mass to a thin elastic element, and increasing damping in small functional gaps. Soil erosion, wear of asphalt pavement, aero plankton, microorganisms, and all production processes are sources of dust in the air.

Particles and aerosols. The air contains particles and the vapors of petroleum oil and other organic liquids that may be referred to as aerosols. Dust particles may arbitrarily vary in shape and chemical content. Air within production areas contains numerous fine fibers, including metallic fibers. In modern industrial cities, dust is comprised of mineral (70%) and organic (30%) substances. The fractional composition of dust in the air is shown in Figure 3.9a. A steep descent of the curve with an increase in the particle size – a_{ch} – is due to their rapid settling. Calculations based on Stock's formula for the free fall of a body in perfectly still air yield the dust with dimensions $a_{ch} \leq 1$ μm settling velocity is equal to $v_{ach} \leq 3.5 \times 10^{-5}$ m/s. The particles of at least 10 μm in diameter settle down rapidly enough, with the velocity $v_{ach} > 0.003$ m/s. The concentration of the dust particles in air increases as their size decreases.

With an increase of dust particle concentration N_{ach}, however, the probability of their mutual collision and coagulation under the action of attractive forces (Van der Waals and electric type) also increases. These forces are dependent on the size of particles. The binding energy of two ball-shaped particles is estimated as 10^{-17} J. Such binding energy substantially exceeds that of atoms in chemical compounds. Consequently, when in contact, the particles tenaciously

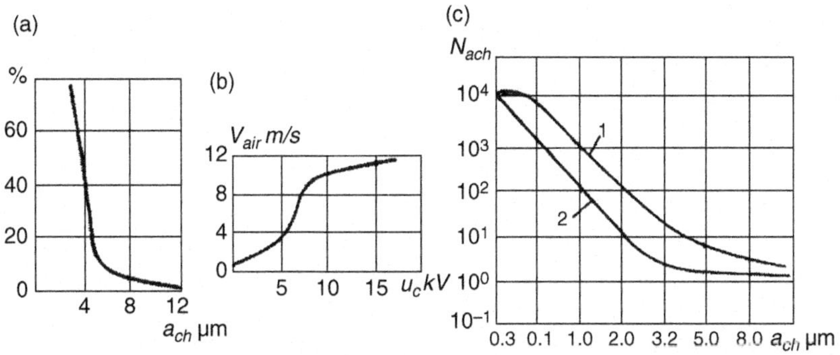

FIGURE 3.9 Features of dust particles in air. *(After Tseytlin, 2006, reprinted with permission.)*

adhere to one another. According to aerosol mechanics, the coagulation velocity of dust particles is proportional to their concentration. Therefore, with an arbitrarily large concentration of dust at the instant of its formation, even in a short length of time, the concentration may not exceed 10^6 particles in 1 liter of air. The coagulation of dust particles is speeded up in moist air, since the absorbed film of moisture increases the binding forces of particles in collision. Therefore, the concentration of particles in rain (curve 2 in Fig. 3.9c) is less than in dry air (curve 1).

The dissimilar charges of particles also encourage coagulation. The limits for the splitting of dust particle cover the range down to molecular size. Particles of 0.5–1.0 nm in size were detected with the help of an electronic microscope.

Cyclone, moisture, adhesion. Another mechanism of airborne dust settling is due to the inertia of dust particles (cyclone effect), when eddy (cyclonic) currents of air pass in the neighborhood of objects. In this case, the dust particles are thrown onto the objects and adhere to them. Under substantial humidity, the particle-to-surface binding forces abruptly increase. The surface adhesion of the particles of 1–2 μm in diameter that have settled on the object features such tenacity that the particles cannot be blown off by a compressed air jet at a velocity of 200 m/s.

Charge. Charge accumulation by particles is due to a variety of causes: the absorption of ions from air, friction of particles against objects, photoelectric effect, thermal electron emission, etc. The charge value is a function of the airflow velocity V_{air} (Fig. 3.9b) and gas composition. As can be seen from Figure 3.9b, the charge of dust particles $u_{c,}$ kV, can reach several hundred volts, even when the velocities are comparatively low. The polarity of dust particles can be both positive and negative. The attraction of the charged particles to objects is caused by either the induced forces of "mirror reflection," or the charges on the object's surface. The term "dust-repelling material" that may be illustrated by referring to films, paints, and coatings is only taken to signify minimum dust-retaining properties, that is, the materials absorbing water vapor poorly, and discharging dust particles in collision.

Density of air. The following approximate formula for air density $\rho_{\gamma a}$ is derived from the equation for the state of a real gaseous mixture:

$$\rho_{\gamma a} = p_{at.a}\, M_a R_{g.c}^{-1} T_T^{-1} k_z^{-1} [1 + (\varepsilon_{v.v} - 1)\varphi_e / p_{at.a}],$$

where $p_{at.a}$ is the summary air pressure; $\varphi_e = \varphi_{Ha}$ is the partial pressure of the water vapor; M_a is the molar mass of the dry air that is equal to: $M_a = M_{a0.4} + 12.011\,[(V_{CO_2} / V) - 0.0004] \times 10^{-3}$, and for the molar content of $CO_2 = V_{CO_2} / V$, which is equal to 0.04% in the normal reference air, $M_{a0.4} = 28.9635 \times 10^{-3}$ kg/mol; $R_{g.c} = 8.31451(1 \pm 8.4 \times 10^{-6})$ J/molK is the gas constant; T_T is the temperature (in K) of air and water vapor; $\varepsilon_{v.v}$ is the ratio of the molecular mass of water vapor and dry air equal to 0.6220, if the content of CO_2 equals 0.04%, and $k_z \approx 0.99960$ is an air compression coefficient at temperature T_T = 293.15 K, $p_{at.a}$ = 101325 Pa (~760 mm of the mercury column), and $\varphi_H = 50\%$. Let us recall that the value of k_z is less dependable on humidity at lesser temperatures. For calculation purposes in practice, the following expression is employed:

$$\rho_{\gamma a} = 0.12027\, p_{at.a} M_a T_T^{-1} k_z^{-1}(1 - 0.3780\, c_{pt}\varphi_e / p_{at.a}),$$

where $c_{pt}\varphi_e/p_{at.a} = d_H$ is the mole fraction of water vapor, and $c_{pt} \approx 1.004$ is a correction factor. The density of the normal air is therefore equal to $\rho_{\gamma a} = 1.203\ kg/m^3$ with air temperature T_T = 293.15 K (20°C), pressure $p_{at.a}$ = 101325 Pa, $\varphi_H = 58\%$, corresponding to the partial pressure of water vapor φ_e = 1333.22 Pa, and the content of carbon dioxide (CO_2) of 0.04%. The random component of the relative uncertainty in the estimation-experimental appraisal of air density equals $\pm 3 \times 10^{-5}$ with the bias being within $\pm 4 \times 10^{-4}$. The CO_2 content in the laboratories and shops is variable, and a change of 0.01% in the CO_2 content produces a relative change of 4×10^{-5} in the air density that calls for a check on this quantity, when high-precision measurements of mass are involved. At the same time, the sum of the CO_2 and oxygen concentrations is assumed to be sufficiently constant, and equal to 0.20979. For practical applications in mass measurement and air damping estimation in precision mechanical oscillators, the conditional density of normal air $\rho_{\gamma a}$ = 1.2 kg/m^3 can be used. For more details on the determination of the air density with International Committee equation (1981/1991) (see Davis, 1992).

Refractive index. Air viscosity, its density, and Reynolds number are important for the calibration of AFM microcantilevers' flexural and torsional spring constants.

The refractive index n_λ is one of the most important parameters employed in optical interference measurement that enables a transition from the wavelengths in the vacuum λ_{vac} to their values in the ambient air:

$$\lambda_{air} = \lambda_{vac} / n_\lambda.$$

To determine refractive index $n_{\lambda,p,t,\varphi}$ in air, under any values of pressure p, temperature t_T, and humidity φ_H, we employ the corrected Edlén (1953) formula, such that

$$(n_{\lambda,p,t,\varphi} - 1) = (n_{\lambda s1} - 1) A_\lambda C_\lambda^{-1} (1 + t_T \alpha_T)^{-1} - \varphi_{Ha} B_\lambda,$$

where $A_\lambda = 1 + 10^{-8}(0.601 - 0.00972 t_T)p$; $B_\lambda = [3.7345 - 0.0401\, \tilde{v}^2\,] \times 10^{-10}$, $C_\lambda = 96095.43$, $\alpha_T = 0.003661$.

The pressure p is here in Pa, temperature in °C, $\tilde{v} = 1/\lambda_{vac}$. The index $n_{\lambda s1}$ of standard air refraction, as corrected for the content of CO_2, is equal to:

$$(n_{\lambda s1} - 1) = \left(1 + 0.0004 \frac{n_{CO_2} - n_{\lambda s}}{n_{\lambda s} - 1}\right)(n_{\lambda s} - 1).$$

Experimental data show that the behavior of quantities ($n_{\lambda,p,t,\varphi}$–1) is the same as that for air densities, that is, such quantities follow Boyle-Marriott's and Gay-Lusaak's laws, and the index of air refraction minus one is obtained by adding together quantities (n_λ–1) for the individual components. As defined by standards, the reference air under normal atmospheric conditions is comprised of 20.9476% oxygen, 78.087% nitrogen, 0.917–0.934% argon, and 0.04% carbon dioxide. The standard air contains no humidity data. The effect of CO_2 content on the change Δn_{CO_2} of the refractive index can be determined from the following relationship:

$$\Delta n_{CO_2} \times 10^6 = (0.0145 + 0.00011\tilde{\nu}^2)(k_{CO_2} - 3),$$

where $\Delta n_{CO_2} = n_{CO_2} - n_{\lambda s}$; $k_{CO_2} \times 10^{-4}$ is the content of CO_2 in the air.

Refractometer. The most accurate values for the index refraction in air are provided by direct measurements taken with the help of a refractometer.

3.8.2.3 Reference Conditions for Length's Measurement

Noncontact interference measurements of length are the most accurate in this field, but they are very sensitive to the conditions of their execution, if the difference between the interference rays' paths is more than 1 mm (Tseytlin, 1981). The wavelength of a primary reference radiation of krypton 86 in normal air is $\lambda_1 = 605.61574$ nm, with the refractive index $n_{\lambda 1} = 1.00027159$; the wavelength of the secondary reference radiation is $\lambda_2 = 546.07819$ nm, with the refractive index $n_{\lambda 2} = 1.00027259$, and the wavelength of the laser stabilized with the iodine cell is $\lambda_{las} = 0.632819505$ μm, with the refractive index $n_\lambda = 1.00027128$.

Also significant for interferometric instruments and other elastic precision structures are the limitation of the speed of air motion (can cause refraction, deflection, vibration), its dust content, the strength of electromagnetic fields, spatial orientation, etc.

3.8.2.4 Electrostatic Field. Source and Action

An electrostatic field is created by the friction of nonconducting elements, sources of electrical discharge in nonconducting media, ambient air ionization, etc. The action of this field on the precision elastic system is produced by force that occurs between the electrostatic field and electric charges accumulated on the dielectric elements. The influence force of an electric field with a strength of 1 V/m per the 1 C charge equals 1 N. This force is used, for example, in electrostatic chucks, for clamping the semiconductor wafers in microfabrication.

3.8.2.5 Electromagnetic Field. Protection Against LF and HF

The relative permeability μ_{mag} for the tape material from 0.02 mm to 2.5 mm thick, of iron-nickel alloys, ranges from 10,000 to 35,000. In practice, complete magnetic saturation for such alloys occurs with the field strength of 600 A/m to 800 A/m. Ferromagnetic screens are only used for protection against low-frequency (LF) interference carrying fields, since at high frequencies (HF) the actual cross section of ferromagnetic walls becomes negligible, as a result of surface skin effect. For the purpose of protection against HF interference carrying fields with at least 1 kHz frequency, the conducting shields of copper, aluminum, and other nonmagnetic materials are used.

The strength of electric field $\boldsymbol{H}_E$, the magnetic component $\boldsymbol{H}_{mag}$, the magnetic inductance $\boldsymbol{B}_{mag}$, the electric bias $\boldsymbol{D}_E$, the density of electric current j_I, and the density of electric charges σ_{pI} are related by Maxwell equations for the electromagnetic field. This field is a vertical one.

3.9 CALIBRATION

A variety of methods have been proposed to determine the cantilever spring constant. The most commonly employed methods are the thermal noise, the added mass methods, and electromagnetic balance (Song et al., 2015) that determine the spring constant for a point load at the end of the cantilever. These methods are applicable for the calibration of bending and torsional spring constants. Some authors show that both constants can be calibrated simultaneously. It may be true, from the practical standpoint, but there are some theoretical concerns because the twisting of the cantilever changes its bending spring constant, as in the pretwisted strips.

3.9.1 Quantum Effects

A mechanical resonator may reveal either quantum or classical behavior, depending on the decoherence-inducing environment (Wei et al., 2006). If the energy $2\pi\hbar\nu$ of the vibration with frequency ν quanta is larger than the thermal energy k_BT, then the mechanical oscillation could be regarded as a quantum mechanical. Let us recall Heisenberg's uncertainty principle in accordance with which it is impossible to specify or determine simultaneously both the position Δx and velocity ΔV of a particle with "full accuracy," because $\Delta x \Delta V \geq \hbar/2m$ and the standard quantum limit $\Delta x_{SQL} = \sqrt{\dfrac{\hbar}{2m\omega_0}}$, where ω_0 is angular resonant frequency, and m is the mass of a simple harmonic oscillator. Meanwhile, the recent research on the basis of Casimir force measurements proves some change in the Newtonian constant on the distances that less than 300 nm (Decca et al., 2003).

Quantum fluctuations create intermolecular forces that go through macroscopic bodies. At molecular separations of a few nanometers or less, these interactions are part of the van der Waals forces. However, at larger distances, and between macroscopic condensed media, the sign of the force can be changed from attractive to repulsive by suitable choice of interacting materials immersed in a liquid. Repulsive Casimir-Lifshitz forces could allow quantum levitation of objects in a fluid, and lead to ultra-low static friction (Munday et al., 2009; Derjaguin et al., 1956).

A number of authors have proposed methods of calibration of the AFM cantilever spring constant involving the hydrodynamic drag of the cantilever. The force acting on the cantilever is equal to $F_{\text{dr}} = \frac{1}{2} c_{\text{d}} \rho_{\gamma} v^2 L w$, where c_{d} is the drag coefficient, ρ_{γ} is the density of fluid, v is its velocity, L and w are the length and width of the cantilever.

Li et al. (2006) developed a new diamagnetic levitation spring system with magnetically suspended body for the lateral force calibration of an atomic friction force microscope (FFM). A wide recognition in FFM calibration is given to the so-called "wedge" method (Ogletree et al., 1996; Varenberg et al., 2003), using two facet slopes of grating sample with a known angle position difference.

A microfabricated array (MARS) of reference spiral-springs (Cumpson et al., 2003) has limited options for the calibration of AFM cantilevers' spring constant that is more than 0.05 N/m. The spiral spring-spoke (see Tseytlin, 2006, Fig. 8.9f) elastic behavior is a complicated one, and similar to that in the curved beam that is loaded perpendicular to its plane. The precision limitation of this calibration device for application in the measurement of nanonewton and piconewton forces may be imposed by vacuum fluctuations (Casimir force), and van der Waals molecular forces (attraction change in inverse ratio to the seventh power of distance between molecules) acting on the spiral springs, as well.

The calibration of the cantilever spring constant is a very important problem for the measurement and imaging by atomic force microscopy. There are many effective methods developed for this purpose. However, most of these methods are dedicated to the calibration of the spring constant for the cantilever in a quasistatic state with one free end, or the regime of its vibration with the natural (the first) eigenfrequency. In the meantime, atomic force microscopes with the cantilevers vibrating at the higher eigenmodes are more sensitive to the position of the probe, an additional mass, and application of force to be measured, because their virtual spring constants are significantly larger. The proper calibration in these cases can be done using thermal noise (Butt and Jaschke, 1995; Hutter and Bechhoefer, 1993; Lübbe et al., 2013; Frentrup and Allen, 2011), the reference cantilever (array of cantilevers) methods, and using laser Doppler vibrometry (Gates et al., 2013, 2015). However, it is useful for atomic force microscopy application and calibration, especially in the case with reference cantilevers, to have a preliminary evaluation of an expected transformation in the spring constant value at higher eigenmodes for the micro- and nanocantilevers with free end (tapping mode), and the end in a contact with the soft elastic or hard object. These evaluations for a rectangular cantilever with the free end were made by different methods: the equivalent point mass model, the solution of the differential equation of motion, the thermal noise estimation, and the kinetostatic presentation of nodal points (Tseytlin, 2005, 2006). Only the latter method showed analytically the influence of an effective section position (the probe or additional mass) on the cantilever spring constant transformation that is important for the higher eigenmodes, and for the natural mode of vibration, as well. An accuracy of this calibration is within 15%. The method of nondestructive experimental determination of a biomaterial rectangular cantilever spring constant in water was developed by Snow et al. (2008), with the heating of water. This method has uncertainty within 15–20% that is still under requirements of the information criterion, η_U, the measurement uncertainty negligibility.

Calibration of the torsional and lateral constants of cantilever sensors by interaction with a flow of fluid was studied by Parkin and Hähner (2014). Sahin (2007) studied the calibration of time-varying tip-sample force measurement with harnessing bifurcations in tapping-mode AFM. Borysov et al. (2014) presented a theoretical framework for the dynamic calibration of the higher eigenmode parameters (stiffness and optical lever invers responsivity) of an AFM cantilever, by using tip-surface interaction method. Forchheimer et al. (2014) used bimodal and multimodal AFM for determining surface properties. Eslami et al. (2015) showed that contrast optimization in tapping mode is most effective in bimodal AFM in which the higher eigenmode is not a subject to the topographical acquisition control loops, but it can be useful for single-mode imaging. The measurements of surface features were provided on the samples of soft Nafion polymer thin films.

3.9.2 Information Criterion of Expended Uncertainty Negligibility

Let us recall that this criterion η_U (Tseytlin, 2006; Section 6-1) is equal to:

$$\eta_U \le \frac{A_э}{\delta_p} e^{H_э(x)},$$

where $A_э = 0.5$ is a normalizing factor for symmetrical distributions of the measuring quantities and $A_э = 1$ for asymmetrical ones; $H_э(x) = -\int_{-\infty}^{+\infty} \varphi(x) \ln \varphi(x)\, dx$ is the entropy or measure of variable's x uncertainty; $\varphi(x)$ is the probability density function of the random variables distribution; δ_p is the tolerance for the object-value x of the measurand. If the statistical distribution of measurand x and measuring uncertainties are normal, then $\eta_U = 35\%$ at $\delta_p/\sigma = 6$ (corresponds to 99.7% of distribution), and the negligibility of component uncertainty is 12%, or may be estimated as $\eta_v = \eta_U / \sqrt{\aleph}$, where $\aleph$ is the number of measurement uncertainty quasi-equal components (calibration, instrument system, methodical, and measurand object).

REFERENCES

Albrecht, T.R., Gütter, P., Horne, D., Rugar, D., 1991. Frequency modulation detection using high-Q cantilevers for enhanced force microscopy sensitivity. J. Appl. Phys. 69, 668–673.

Anczykowski, B., Krüger, D., Babcock, K.L., Fuchs, H., 1996. Basic properties of dynamic force spectroscopy with the scanning force microscope in experiment and simulation. Ultramicroscopy 66, 251–259.

Basak, S., Raman, A., 2007. Dynamics of tapping mode atomic force microscopy in liquids: theory and experiments. Appl. Phys. Lett. 91, 064107.

Bauza, M.B., Hocken, R.J., Smith, S.T., Woody, S.C., 2005. Development of a virtual probe tip with an application to high aspect ratio microscale features. Rev. Sci. Instrum. 76, 095112.

Bhat, B.R., Wagner, H., 1976. Natural frequencies of a uniform cantilever with a tip mass slender in the axial direction. J. Sound. Vib. 45, 304.

Borysov, S.S., Forchheimer, D., Haviland, D.B., 2014. Dynamic calibration of higher eigenmode parameters of a cantilever in atomic force microscopy by using tip-surface interactions. Beilstein J. Nanotech. 5, 1899–1904.

Braun, D., 2011. Ultimate quantum bounds on mass measurements with nano-mechanical resonator. EPL 94, 68007.

Burra, R.K., Vankara, J., Reddy, D.V., 2012. Effect of added mass using peak shifting technique. J. Micro/Nanolith. MEMS. MOEMS. 11 (2).

Butt, H.J., Jaschke, M., 1995. Calculation of thermal noise in atomic force microscopy. Nanotechnology 6, 1.

Chang, W.-J., Lee, H.-L., Chen, T.Y.-F., 2008. Study of the sensitivity of the first four flexural modes of an AFM cantilever with a sidewall probe. Ultramicroscopy 108, 619.

Chen, G.Y., Warmack, R.J., Thundat, T., Allison, D.P., Huang, A., 1994. Resonance response of scanning force microscopy cantilever. Rev. Sci. Instrum. 65, 2532–2537.

Comptes Rendus des Séances, 1961. XI Conference Generale des Poids et Measures.

Cumpson, P.J., Hedley, J., Zhdan, P., 2003. Accurate force measurement in the atomic force microscope: a microfabricated array of reference springs for easy cantilever calibration. Nanotechnology 14, 918–924.

Davis, R.S., 1992. Equation for the determination of the density of moist air (1981/91). Metrologia 29, 67–70.

Decca, R.S., Fishbach, E., Klimchitskaya, G.L., Krause, D.E., Lopez, D., Mostepanenko, V.M., 2003. Improved tests of extra-dimensional physics and thermal quantum field theory from new Casimir force measurements. Phys. Rev. D 68 (15), 116003.

Derjaguin, B.V., Abrikosova, L.I., Lifshitz, E.M., 1956. Direct measurement of molecular attraction between solids separated by a narrow gap. Q. Rev. Chem. Soc. 10, 295–329.

Edlén, B., 1953. The dispersion of standard air. JOSA 43 (5), 339–344, idem. ibid., 1966. The refractive index of air. Metrologia 2, 71–80.

Eringen, A.C., 1983. On differential equations of nonlocal elasticity and solutions of screw dislocations and surface waves. J. Appl. Phys. 54 (9), 4703.

Eslami, B., López-Guerra, E.A., Diaz, A.J., Solares, S.D., 2015. Optimization of the excitation frequency for high probe sensitivity in single-eigenmode and bimodal tapping-mode AFM. Nanotechnology 26, 165703, (12p.).

Forchheimer, D., Borysov, S.S., Platz, D., Haviland, D.B., 2014. Determining surface properties with bimodal and multimodal AFM. Nanotechnology 25, 485708, (8p.).

Frentrup, H., Allen, M.S., 2011. Error in dynamic spring constant calibration of atomic force microscope probes due to nonuniform cantilevers. Nanotechnology 22 (29), 295703.

Gates, R.S., Osborn, W.A., Pratt, J.R., 2013. Experimental determination of mode correction factors for thermal method spring constant calibration of AFM cantilevers using laser Doppler vibrometry. Nanotechnology 24 (25), 255706.

Gates, R.S., Osborn, W.A., Shaw, G.A., 2015. Accurate flexural spring constant calibration of colloid probe cantilevers using scanning laser Doppler vibrometry. Nanotechnology 26 (23), 235704.

Hähner, G., Parkin, J.D., 2011. Mass determination and sensitivity based on resonance frequency changes of the higher flexural modes of cantilever sensors. Rev. Sci. Instrum. 82, 035108.

Howell, L.L., 2001. Compliant Mechanisms. Wiley, New York, NY.

Huang, Q.X., Hou, Q., 2011. Higher eigenmodes vibration cantilever's resonance frequency and cantilever spring constant. Appl. Mech. Mat. 44–47, 489–493.

Hutter, J.L., Bechhoefer, J., 1993. Calibration of atomic-force microscope tips. Rev. Sci. Instrum. 64, 1868.

Ilic, B., Krylov, S., Graighead, H.G., 2010. Young's modulus and density measurements of thin atomic layer deposited films using resonant nanomechanics. J. Appl. Phys. 108, 044317.

Jeon, S., Braiman, Y., Thundat, T., 2004. Cross talk between bending, twisting, and buckling modes of three types of microcantilever sensors. Rev. Sci. Instrum. 75, 4841–4844.

Killgore, J.P., Tung, R.C., Hurley, D.C., 2014. Characterizing the free and surface-coupled vibrations of heated-tip atomic force microscope cantilevers. Nanotechnology 25, 345701.

Kolovskii, M.Z., 1976. Automatic Control of Vibro-Protective Systems. Nauka, Moscow, (in Russian).

Kolovsky, M.Z., Evgrafov, A.N., Semenov, Yu.A., Slousch, A.V., 2000. Advanced Theory of Mechanisms and Machines. Springer, Berlin, Transl. by L. Lilov.

Krylov, A.N., 1936. Vibrations of Ships. Gostekh, Leningrad, (in Russian).

Landau, L.D., Lifshitz, E.M., 1970. Theory of Elasticity, second English ed. Pergamon Press, Oxford, Transl. from Russian by J.B. Sykes and W.H. Reid.

Lange, D., Brand, O., Batles, H., 2002. CMOS Cantilever Sensor Systems: Atomic Force Microscopy and Gas Sensing Applications. Springer, Berlin.

Lavrik, N.V., Sepaniak, M.J., Datskos, P.G., 2004. Cantilever transducers as a platform for chemical and biological sensors. Rev. Sci. Instrum. 75, 2229–2253.

Lee, H.-L., Chang, W.-J., 2010. Surface effects on frequency analysis of nanotubes using Timoshenko nonlocal beam theory. J. Appl. Phys. 108, 093503.

Li, Q., Kim, K.-S., Rydberg, A., 2006. Lateral force calibration of an atomic force microscope with diamagnetic levitation spring system. Rev. Sci. Instrum. 77, 065105.

Li, X.-F., Tang, G.-J., Shen, Z.-B., Lee, K.Y., 2015. Resonance frequency and mass identification of zeptogram-scale nanosensor based on the nonlocal beam theory. Ultrasonics 55, 75–84.

Lübbe, J., Temmen, M., Rahe, P., Kühnle, A., Reichling, M., 2013. Determining cantilever stiffness from thermal noise. Beilstein J. Nanotechnol. 4, 227–233.

Majorana, E., Ogawa, Y., 1997. Mechanical thermal noise in coupled oscillators. Phys. Lett. A. 233, 162–168.

Maranganti, R., Sharma, P., 2007. Length scales at which classical elasticity breaks down for various materials. Phys. Rev. Lett. 98, 195504.

Marchenko, V.M., 1965. The Temperature Fields and Stresses in the Design of Flying Apparatus. Mashinostroenie, Moscow, (in Russian).

Melcher, J., Hu, S., Raman, A., 2007. Equivalent point-mass models of continuous force microscope probes. Appl. Phys. Lett. 91, 053101.

Morse, P.M., 1948. Vibration and Sound. McGraw-Hill, New York, NY.

Munday, J.N., Capasso, F., Parsegian, V.A., 2009. Measured long-range repulsive Casimir-Lifshitz forces. Nature 457 (8), 170–173.

Neumeister, J.M., Ducker, W.A., 1994. Lateral, normal, and longitudinal spring constants of atomic force microscopy cantilevers. Rev. Sci. Instrum. 65, 2527–2531.

Nordström, M., Keller, S., Lillemose, M., Boisen, A., et al., 2008. SU-8 cantilevers for Bio/chemical sensing; fabrication, characterization and development of novel read-out methods. Sensors 8 (3), 1595–1612.

Ogletree, D.F., Carpick, R.W., Salmeron, M., 1996. Calibration of frictional forces in atomic force microscopy. Rev. Sci. Instrum. 67, 3298.

Pal'mov, V.A., 1964. Fundamental equations of the theory of asymmetric elasticity. PMM 28 (3), 401 (in Russian).

Panovko, Y.G., 1968. Natural and forced vibrations of beams and beam systems. In: Birger, I.A., Panovko, Y.G. (Eds.), Strength, Stability, Vibrations. Handbook in Three Volumes, 3, Mashinostroenie, Moscow, pp. 285–346 (in Russian).

Parkin, J.D., Hähner, G., 2014. Calibration of the torsional and lateral spring constants of cantilever sensors. Nanotechnology 25, 225701.

Parlak, Z., Tu, Q., Zausher, S., 2014. Liquid contact resonance AFM: analytical models, experiments, and limitations. Nanotechnology 25, 445703, 11p..

Photiadis, D.M., Judge, J.A., 2004. Attachment losses of high *Q* oscillators. Appl. Phys. Lett. 85, 482–484.

Poncharal, P., Wang, Z.L., Ugarte, D., de Heer, W.A., 1999. Electrostatic deflection and electromechanical resonance of carbon nanotubes. Science 283, 1513–1516.

Pradhan, S.C., Murmu, T., 2010. Application of nonlocal elasticity and DQM in the flapwise bending vibration of a rotating nanocantilever. Physica E 42, 1944.

Qin, Y., Shrinzadeh, B., Zhang, D., Tian, Y., 2013. Compliance modeling and analysis of statically indeterminate symmetric flexure structures. Precise Eng. 37 (2), 415–424.

Rast, S., Wattinger, G., Gysin, U., Meyer, E., 2000. Dynamics of damped cantilevers. Rev. Sci. Instrum. 71, 2772.

Sader, J.E., 2003. Susceptibility of atomic force microscope cantilever to lateral forces. Rev. Sci. Instrum. 74, 2438–2443.

Sahin, O., 2007. Harnessing bifurcations in tapping-mode atomic force microscopy to calibrate time-varying tip-sample force measurements. Rev. Sci. Instrum. 78, 103707.

Sahin, O., Quate, C.F., Solgaard, O., Atalar, A., 2004. Resonant harmonic response in tapping-mode atomic force microscopy. Phys. Rev. B 69, 165416.

Snow, D.E., Weeks, B.L., Kim, D.J., Pitchimani, R., Hope-Weeks, L.J., 2008. Nondestructive experimental determination of biomaterial rectangular cantilever spring constant in water. Rev. Sci. Instrum. 79 (8), 083706.

Song, Y., Wu, S., Xu, L., Fu, X., 2015. Accurate calibration and uncertainty estimation of the normal spring constant of various AFM cantilevers. Sensors 15, 5865–5883.

Souayeh, S., Kacem, 2014. Computational models for large amplitude nonlinear vibrations of electrostatically actuated carbon nanotube-based mass sensors. Sens. Actuat. A 208, 10–20.

Stark, R.W., Drobek, T., Heckl, W.M., 2001. Thermomechanical noise of a free v-shaped cantilever for atomic-force microscopy. Ultramicroscopy 86, 207.

Thundat, T., Warmack, R.J., Chen, G.Y., Allison, D.P., 1994. Thermal and ambient-induced deflections of scanning force microscope cantilevers. Appl. Phys. Lett. 64, 2894–2896.

Toffoli, V., Carrato, S., Lee, D., Jeon, S., Lazzarino, M., 2013. Heater-integrated cantilevers for nano-samples thermogravimetric analysis. Sensors 13, 16657–16671.

Tseytlin, Y.M., 1981. Normal Conditions of Measurement in the Mechanical Engineering. Mashinostroenie, Leningrad (in Russian).

Tseytlin, Y.M., 2005. High resonant mass sensor evaluation: an effective method. Rev. Sci. Instrum. 76, 115101.

Tseytlin, Y.M., 2006. Structural Synthesis in Precision Elasticity. Springer, New York, NY.

Tseytlin, Y.M., 2007. Nanostep and film-coating thickness traceability, Proceedings of 53 IIS, ISA, Tulsa, OK, USA, 470 Papertp007iis006.

Tseytlin, Y.M., 2008. Atomic force microscope cantilever spring constant evaluation: a kinetostatic method. Rev. Sci. Instrum. 79, 025102.

Tseytlin, Y.M., 2010. Kinetostatic model of spring constant ratio for an AFM cantilever with end extended mass. Ultramicroscopy 110, 126.

Turner, J.A., Wiehn, J.S., 2001. Sensitivity of flexural and torsional vibration modes of atomic force microscope cantilevers to surface stiffness variations. Nanotechnology 12, 322.

Varenberg, M., Etsion, I., Halperin, G., 2003. An improved wedge calibration method for lateral force in atomic force microscopy. Rev. Sci. Instrum. 74, 3362.

Vogel, M., Mooser, C., Karrai, K., 2003. Optically tunable mechanics of microlevers. Appl. Phys. Lett. 83, 1337–1339.

Wang, L., Hu, H., 2005. Flexural wave propagation in single-walled carbon nanotubes. Phys. Rev. B 71, 195412.

Weaver, W., Timoshenko, S.P., Young, D.H., 1990. Vibration Problems in Engineering, fifth ed. Wiley, New York, NY.

Wei, L.F., Liu, Yu-xi, Sun, C.P., Nori, F., 2006. Probing tiny motion of nanomechanical resonators. Phys. Rev. Lett. 97, 237201.

Zang, J., Liu, F., 2008. Modified Timoshenko formula for bending of ultrathin strained bilayer films. Appl. Phys. Lett. 92, 021905.

Zheng, Y., Song, L., Hu, G., Zhao, M., Tian, Y., Zhang, Z., Fang, F., 2014. Improving environmental noise suppression for micronewton force sensing based on electrostatic by injecting air damping. Rev. Sci. Instrum. 85, 055002.

Zhou, Z., Lai, P.-Y., 2005. On the asymmetric elasticity of twisted dsDNA. Phys. A 350, 70.

Chapter 4

Kinematics and Nonlinear Motion Transformation in Elastic Helicoids

4.1 HELIX PARAMETERS

The strain-stress state of the helicoidal (pretwisted) beam differs in several ways from the corresponding state of the prismatic beams. First, the pretwisted beams have torsion deformation (twist–untwist) under the stretching load applied along their axis. Second, their length changes under the applied torque; normal stresses occur with twisting, and tangent stresses with stretching. These special features are significant at larger ratios $\Lambda_\lambda = (b/h)^2 = \lambda^{-1}$ between the width $b = 2b_1$ and the thickness $h = 2h_1$ of the beam's cross-section.

The translational motion in the pretwisted helicoids has a functional relation to the rotation through the helix parameter κ_p. Let the cross-section profile corresponds to the following equation

$$r_a = [r(\vartheta)\cos\vartheta]\,\mathbf{i_a} + [r(\vartheta)\sin\vartheta]\,\mathbf{j_a}, \tag{4.1}$$

where $r(\vartheta)$ is the equation for the polar curve of the cross-section profile's contour. Let this profile conduct the helical motion with parameter κ_p that corresponds to the certain translational motion along the axis, at the profile rotation for one radian. Parameter κ_p has a positive sign if the helical motion is right-handed, and a negative sign if the helical motion is a left-handed one. The equation of the helicoid side surface in matrix form is as follows:

$$r_1 = M_{1a}\, r_a, \tag{4.2}$$

where

$$\mathrm{M}_{1a} = \begin{bmatrix} \cos\vartheta & \sin\vartheta & 0 & 0 \\ -\sin\vartheta & \cos\vartheta & 0 & 0 \\ 0 & 0 & 1 & \kappa_p^{-1}\vartheta \\ 0 & 0 & 0 & 1 \end{bmatrix} \tag{4.3}$$

Advanced Mechanical Models of DNA Elasticity

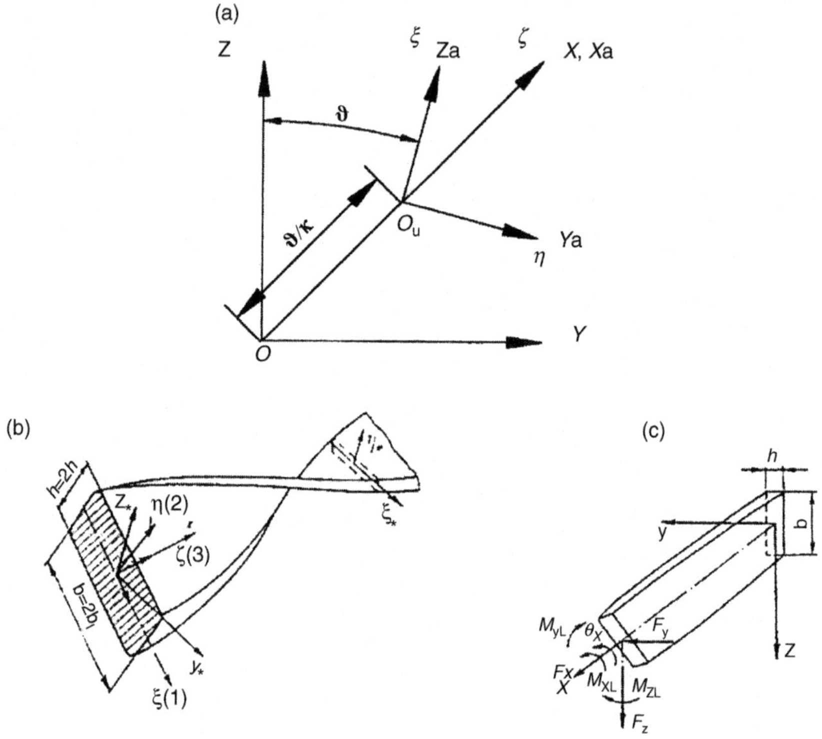

FIGURE 4.1 Fixed and following helix coordinate systems (a), for the helix motion (b), and helicoid loading (c). *(After Tseytlin, 2006, reprinted with permission.)*

and ϑ is the angle of the cross-section's rotation (Fig. 4.1a) around the helicoids longitudinal axis ζ.

On the basis of Eqs (4.1–4.3), we finally obtain the equation of the helicoids' side surface

$$r_1 = [r\cos(\nu+\vartheta)]\mathbf{i} + [r\sin(\nu+\vartheta)]\mathbf{j} + \kappa_p^{-1}\vartheta\mathbf{k},$$

where i_a, j_a, i, j, k are the unit-vectors of the corresponding coordinate axes.

The important feature of the helicoidal surface may be presented by the equation

$$Y n_z - Z n_y = \kappa_p n_x,$$

where n_x, n_y, n_z are the normal vector's projections on the Cartesian coordinate axes X, Y, Z. The initial helix parameter is equal to the relative twist of the unloaded naturally pretwisted (helicoidal) beam:

$$\kappa_{p0} = d\vartheta_0/d\varsigma. \tag{4.4}$$

The beam is pretwisted uniformly if κ_{p0} = constant. Then, $\kappa_{p0} = \vartheta_L/L$, where ϑ_L is the change in the angle position of the cross-section in the range of the beam's length (L). The corresponding points of different cross-sections are positioned on the helix lines, with helix angles that are equal to:

$$\beta_0 = \tan^{-1}(\kappa_{p0}R), \tag{4.5}$$

where $R = [(\xi - \xi_j)^2 + (\eta - \eta_j)^2]^{0.5}$ is the distance of a helical surface point to axis ς_j, and ξ_j and η_j are the coordinate axes moving with the cross-section. If $\beta_0 \ll 1$, we can assume with necessary accuracy that

$$\beta_0 = \kappa_{p0}R. \tag{4.5a}$$

The internal force F_ζ and moment M_ξ, M_η, M_ζ factors related to the axes shown in the subscripts for the helicoidal beam are expressed (Tseytlin, 1965; Shorr, 1968; Tseytlin, 2006) through the coefficients of rigidity in the following matrix form:

$$\begin{bmatrix} F_\zeta \\ M_\xi \\ M_\eta \\ M_\zeta \end{bmatrix} = \begin{bmatrix} j_\zeta & j_{\varepsilon\xi} & j_{\varepsilon\eta} & j_{\varepsilon\theta} \\ j_{\xi\varepsilon} & j_\xi & j_{\xi\eta} & j_{\xi\theta} \\ j_{\eta\varepsilon} & j_{\eta\xi} & j_\eta & j_{\eta\theta} \\ j_{\theta\varepsilon} & j_{\theta\xi} & j_{\theta\eta} & j_\theta \end{bmatrix} \begin{bmatrix} \zeta' \\ \aleph_\xi \\ \aleph_\eta \\ \theta' \end{bmatrix}, \tag{4.6}$$

where $j_{kl} = j_{lk}$ ($k, l = \varepsilon, \xi, \eta, \theta$); $\aleph_\xi$, $\aleph_\eta$ are the curvatures; $\zeta' = \varepsilon$ is the elongation strain, and θ' is the torsion per beam's unit length. The coefficients in the matrix of rigidity for moderately pretwisted beams are defined through formulae, shown in Table 4.1 version A.

The notations in Table 4.1 are as follows: $T = T_o + 2(1 + \nu_p)\,\kappa_{p0}^2(J_r^0 - T_r^0)$, $J_\xi = \int_{Ac} \eta^2\, dA_c$; $J_\eta = \int_{Ac} \xi^2\, dA_c$ are the axial moments of the cross-section, A_c is the area of the cross-section, $J_p = \int_{Ac} R^2\, dA_c$ is the polar moment, $J_{p\xi} = \int_{Ac} R^2\eta\, dA_c$, $J_{p\eta} = \int_{Ac} R^2\xi\, dA_c$ are the polar-axial moments of inertia; $J_r^0 = \int_{Ac} R^4\, dA_c$, $T_o = \int_{Ac} (R^2 + \varphi'_\psi)\, dA_c$; $T_\xi = \int_{Ac} (R^2 + \varphi'_\psi)\eta dA_c$, $T_\eta = \int_{Ac} (R^2 + \varphi'_\psi)\xi dA_c$, $T_r^0 = \int_{Ac} [R^4 + (\varphi'_\psi)^2] dA_c$; $\varphi'_\psi = \varphi'_\eta(\xi - \xi_j) - \varphi'_\xi(\eta - \eta_j)$, and $\varphi'_\eta = \partial\psi^o/\partial\eta$; $\varphi'_\xi = \partial\psi^o/\partial\xi$, where ψ^o and T_o are the function of rotation and torsional constant for non-twisted beam with the same cross-section, respectively.

Let us recall that the torsional rigidity for the solid beam with the complicated contour of solid cross-section may approximately be estimated by Saint-Venant's formula, $T_o \approx A_c^4/(40\, J_p)$, where A_c is the area, and J_p is the polar moment of inertia for the corresponding cross-section. However, this formula may be not accurate for cross-sections with the concave and very stretched contours.

The coefficients in the reciprocal to (4.6) matrix of compliance for helicoidal beams are defined through the formulae shown in Table 4.1 version B, where $\nu_\theta = [GT_0(1 + \beta_p^2)]^{-1}$; $\beta_p^2 = 2(1+\nu_p)k_{p0}^2(J_r - T_r)/T_o$; $J_r - T_r = J_r^0 - T_r^0 - [(J_p - T_o)^2/A_c] - [(J_{p\xi} - T_\xi)^2/J_\xi] - [(J_{p\eta} - T_\eta)^2/J_\eta]$.

TABLE 4.1 The Coefficients in Matrix of Rigidity (A) and Matrix of Compliance (B) for Moderate Helicoidal Beams

Coordinate	Version	$\zeta'=\varepsilon$	ξ	η	θ
$\zeta'=\varepsilon$	A	$j_\zeta = j_\varepsilon = EA$	$j_{\varepsilon\xi} = 0$	$j_{\varepsilon\eta} = 0$	$j_{\varepsilon\theta} = \kappa_{p0}E(J_p - T_o)$
	B	$\nu_{\varepsilon} = \frac{1+\nu_\theta\, j_{\varepsilon\theta}^2/j_\varepsilon}{j_\varepsilon}$	$\nu_{\varepsilon\xi} = \frac{\nu_\theta\, j_{\varepsilon\theta}\, j_{\xi\theta}}{j_\varepsilon j_\xi}$	$\nu_{\varepsilon\eta} = \frac{\nu_\theta\, j_{\varepsilon\theta}\, j_{\eta\theta}}{j_\varepsilon j_\eta}$	$\nu_{\varepsilon\theta} = -\frac{\nu_\theta\, j_{\varepsilon\theta}}{j_\varepsilon}$
ξ	A	$j_{\xi\varepsilon} = 0$	$j_\xi = EJ_\xi$	$j_{\xi\eta} = 0$	$j_{\xi\theta} = \kappa_{p0}\, E(J_{p\xi} - T_\xi)$
	B	$\nu_{\xi\varepsilon} = \frac{\nu_\theta\, j_{\varepsilon\theta}\, j_{\xi\theta}}{j_\varepsilon j_\xi}$	$\nu_{\xi} = \frac{1+\nu_\theta\, j_{\xi\theta}^2/j_\xi}{j_\xi}$	$\nu_{\xi\eta} = \frac{\nu_\theta\, j_{\eta\theta}\, j_{\xi\theta}}{j_\eta j_\xi}$	$\nu_{\xi\theta} = -\frac{\nu_\theta\, j_{\xi\theta}}{j_\xi}$
η	A	$j_{\eta\varepsilon} = 0$	$j_{\eta\xi} = 0$	$j_\eta = EJ_\eta$	$j_{\eta\theta} = -k_{p0}E(J_{p\eta} - T_\eta)$
	B	$\nu_{\eta\varepsilon} = \frac{\nu_\theta\, j_{\varepsilon\theta}\, j_{\eta\theta}}{j_\varepsilon j_\eta}$	$\nu_{\eta\xi} = \frac{\nu_\theta\, j_{\eta\theta}\, j_{\xi\theta}}{j_\eta j_\xi}$	$\nu_{\eta} = \frac{1+\nu_\theta\, j_{\eta\theta}^2/j_\eta}{j_\eta}$	$\nu_{\eta\theta} = -\frac{\nu_\theta\, j_{\eta\theta}}{j_\eta}$
θ	A	$j_{\theta\varepsilon} = \kappa_{p0}E(J_p - T_o)$	$j_{\theta\xi} = \kappa_{p0}\, E(J_{p\xi} - T_\xi)$	$j_{\theta\eta} = -k_{p0}E(J_{p\eta} - T_\eta)$	$j_\theta = GT$
	B	$\nu_{\theta\varepsilon} = -\frac{\nu_\theta\, j_{\varepsilon\theta}}{j_\varepsilon}$	$\nu_{\theta\xi} = -\frac{\nu_\theta\, j_{\xi\theta}}{j_\xi}$	$\nu_{\theta\eta} = -\frac{\nu_\theta\, j_{\eta\theta}}{j_\eta}$	*

*$\nu_\theta = [j_\theta - (j_{\varepsilon\theta}^2/j_\varepsilon) - (j_{\xi\theta}^2/j_\xi) - j_{\eta\theta}^2/j_\eta]^{-1}$. *Some indices may be interchangeable in different publications, for example, $\varepsilon \leftrightarrow \zeta$ because $\varepsilon = \zeta'$ and different orientation of coordinate axes.*

TABLE 4.2 The Compliance Coefficients in the Cartesian Coordinates

Coordinate	x	z	y	θ
x	$v_x = v_\varepsilon$	$v_{xz} = v_{\xi\varepsilon} \cos \vartheta_0 x - v_{\eta\varepsilon} \sin\vartheta_0 x$	$v_{xy} = v_{\xi\varepsilon} \sin \vartheta_0 x + v_{\eta\varepsilon} \cos \vartheta_0 x$	$v_{x\theta} = v_{\varepsilon\theta}$
z	$v_{zx} = v_{\xi\varepsilon} \cos \vartheta_0 x - v_{\eta\varepsilon} \sin\vartheta_0 x$	$v_z = v_\xi \cos^2 \vartheta_0 x + v_\eta \sin^2 \vartheta_0 x - v_{\xi\eta} \sin 2\vartheta_0 x$	*	$v_{z\theta} = v_{\xi\theta} \cos\vartheta_0 x - v_{\eta\theta} \sin\vartheta_0 x$
y	$v_{yx} = v_{\xi\varepsilon} \sin \vartheta_0 x + v_{\eta\varepsilon} \cos \vartheta_0 x$	*	$v_y = v_\xi \sin^2 \vartheta_0 x + v_\eta \cos^2 \vartheta_0 x + v_{\xi\eta} \sin 2\vartheta_0 x$	$v_{y\theta} = v_{\xi\theta} \sin\vartheta_0 x + v_{\eta\theta} \cos\vartheta_0 x$
θ	$v_{\theta x} = v_{\varepsilon\theta}$	$v_{\theta z} = v_{\xi\theta} \cos \vartheta_0 x - v_{\eta\theta} \sin \vartheta_0 x$	$v_{\theta y} = v_{\xi\theta} \sin\vartheta_0 x + v_{\eta\theta} \cos \vartheta_0 x$	v_θ

*$v_{zy} = v_{yz} = (v_\xi - v_\eta) \sin \vartheta_0 x \cos \vartheta_0 x + v_{\xi\eta} \cos 2\vartheta_0 x$; $\vartheta_0 = k_{p0}$. *Some indices may be interchangeable in different publications, for example,* $\varepsilon \leftrightarrow \zeta$ *because* $\varepsilon = \zeta'$ *and different coordinate axes.*
Source: After Tseytlin (2006), reprinted with permission.

Matrix of strain. The components of the beam deformation in the X, Y, Z system may be expressed in matrix form as the product of the matrix of compliance v_{ij} by the column matrix of loads $F_i(F_x, M_x, M_y, M_z)$ in the fixed Cartesian coordinates:

$$\begin{bmatrix} S'_x \\ -S''_y \\ S''_z \\ \theta'_x \end{bmatrix} = \begin{bmatrix} v_x & v_{xz} & v_{xy} & v_{x\theta} \\ v_{zx} & v_z & v_{zy} & v_{z\theta} \\ v_{yx} & v_{yz} & v_y & v_{y\theta} \\ v_{\theta x} & v_{\theta z} & v_{\theta y} & v_\theta \end{bmatrix} \begin{bmatrix} F_x \\ M_z \\ M_y \\ M_x \end{bmatrix} = [v_{ij}][F_i], \tag{4.7}$$

where v_{ij} are the coefficients of compliance in the corresponding directions X, Y, Z, θ_x (Fig. 4.1a,c); S_x, S_y, S_z, θ_x are the elastic displacements in the directions of the coordinate axes X, Y, Z and the angle of rotation around axis X; S'_x, S''_y, S''_z, θ'_x are the derivatives of the displacements with respect to the coordinate x. The values of compliance coefficients v_{kl} in the "following" coordinates ξ_*, η_*, ζ_* with directions ξ, η, ζ, $\vartheta = \theta$ are related to coefficients v_{ij} in the Cartesian coordinates X ,Y, Z, and θ by the equations presented in Table 4.2.

The state of symmetry corresponds to the equality of the certain compliance's components: $v_{kl} = v_{lk}$; $v_{ij} = v_{ji}$.

There is a simplification for the thin beams with a profile that is longer in the ξ direction, because the shift constituent is a very small one in the η direction. As a result, the derivative of the function of rotation for such beams is equal to $\varphi'_\psi = R^2$, where $R^2 = \xi^2 + \eta^2$. Let us open matrix (4.7) and conduct the necessary integration to find the elastic displacement components at the bending of a cantilever helicoidal beam:

$$S_x = \int_0^L (v_x F_x + v_{xz} M_z + v_{xy} M_y + v_{x\theta} M_x)\,dx, \tag{4.8}$$

$$S_y = -\int_0^L \int_0^x (v_{zx} F_x + v_z M_z + v_{zy} M_y + v_{z\theta} M_x)\,dx^2; \tag{4.9}$$

$$S_z = \int_0^L \int_0^x (v_{yx} F_x + v_{yz} M_z + v_y M_y + v_{y\theta} M)\,dx^2; \tag{4.10}$$

$$\theta_x = \int_0^L (v_{\theta x} F_x + v_{\theta z} M_z + v_{\theta y} M_y + v_\theta M_x)\,dx. \tag{4.11}$$

There should be preliminary substitution into (4.8) through (4.11) of the expressions for the coefficients of compliance v_{ij} in the fixed system of coordinates, through the corresponding coefficients v_{kl} in the "following" system of coordinates. The bending moments in the beam section (Fig. 4.1c) are equal to:

$$M_z = M_{zL} - F_y(L - X) - F_x S_y, \tag{4.12}$$

$$M_y = M_{yL} - F_z(L - X), \tag{4.13}$$

$$M_x = M_{xL} + F_z S_y, \tag{4.13a}$$

where moments M_{xL}, M_{yL}, M_{zL}, and forces F_x, F_y, and F_z are applied to the end of the beam. The moment $F_x S_y$ may be neglected, if the displacement S_y and the longitudinal force F_x are small. Then, the solution of the integrals in (4.8–4.11) is simplified.

4.2 MOTION STABILITY

If the displacement S_y is significant, and the moment $F_x S_y$ cannot be neglected, a differential equation of the second order is obtained after the opening of matrix (4.7):

$$\begin{aligned} -S_y'' &= [M_{zL} - F_y(L - x) - F_x S_y](v_\xi \cos^2 \vartheta_0 x + v_\eta \sin^2 \vartheta_0 x) \\ &\quad + (v_\xi - v_\eta)\sin \vartheta_0 x \cos \vartheta_0 x [M_{yL} + F_z(L - x)] + v_{\xi\theta} \cos \vartheta_0 x [M_{xL} + F_z S_y], \end{aligned}$$

TABLE 4.3 Loss of Beam's Stability Under Compression and Torsion

Type of load	Schematic of loading	The critical value of load
Compression and torsion of flat beam	 $Q = F_x$	$\theta_{tr} = \frac{4}{3}\frac{n_2 - n_1}{(1-n_1)(1-n_2)}$ Elliptical cross-section: $n_1 = 1 - \frac{2}{(1+\nu_p)(1+h_1^2 b_1^{-2})}$; $n_2 = 1 - \frac{2}{(1+\nu_p)(1+h_1^{-2} b_1^2)}$ Rectangular cross-section: $n_1 = 1 - \frac{6\nu_{kch}}{1+\nu_p}$; $n_2 = 1 - \frac{6\nu_{kch}}{1+\nu_p} b^2 h^{-2}$; $^*\nu_{kch} = \frac{H_k}{Gh^3 b}$.
Compression of elastic helicoid cantilever	See Fig. 4.1	$-F_x = -\nu_{tr}\frac{\pi^2}{4L^2\nu_\eta}$; $\nu_\xi < \nu_\eta; \vartheta_0 L > 2\pi; \nu_{tr} \approx \frac{2\nu_\eta}{\nu_\eta + \nu_\xi}$

*H_k is the section torsional rigidity with load F_x.
Source: After Tseytlin (2006), reprinted with permission.

which transforms in the corresponding nonhomogeneous differential equation

$$\begin{aligned}
&S_y'' - F_x S_y(\nu_\xi \cos^2\vartheta_0 x + \nu_\eta \sin^2\vartheta_0 x) + F_z S_y \nu_{\xi\theta}\cos\vartheta_0 x = \\
&-(M_{zL} - F_y L)\nu_\xi - [F_y \nu_\xi + (M_{yL} + F_z L)(\nu_\xi - \nu_\eta)\vartheta_0]x \\
&+[(M_{zL} - F_y L)(\nu_\xi - \nu_\eta)\vartheta_0^2 + F_z(\nu_\xi - \nu_\eta)\vartheta_0]x^2 + [F_y(\nu_\xi - \nu_\eta)\vartheta_0^2 \\
&+(2/3)(M_{yL} + F_z L)(\nu_\xi - \nu_\eta)\vartheta_0^3]x^3 - (2/3)F_z(\nu_\xi - \nu_\eta)\vartheta_0^3 x^4 - M_{xL}\nu_{\xi\theta}\cos\vartheta_0 x.
\end{aligned} \tag{4.14}$$

The right-hand side of Eq. (4.14) can be approximated by a polynomial of the fourth degree of x. The left-hand side of this equation, after corresponding transformations, can be presented in the form of Mathieu's equation (McLachlan, 1964). The final equation at negative force F_x and $F_z = 0$ may be presented as follows:

$$S_y'' + S_y F_x \frac{\nu_\xi + \nu_\eta}{2}\left[1 - \frac{\nu_\eta - \nu_\xi}{\nu_\xi + \nu_\eta}\cos\vartheta_0 x\right] = 0, \tag{4.15}$$

where F_x, ν_ξ, ν_η are the parameters; cos $\vartheta_0 x$ is the periodic function. It is known, however, that Mathieu's equations have fields with both stable and unstable solutions. The physical essence of these specifics is important for the beam shape stability at bending. It means that the pretwisted (helicoidal) thin beam can lose stability at bending, if the beam is loaded by a longitudinal compressing load (Table 4.3).

This loss of stability is followed by large changes in the deformation at a small change in the load on the boundary of the critical field. A similar effect is also possible for initially plane straight beams that are loaded with a compressing force and twisting moment. The critical compressing force in these cases is much smaller than the one for compressed straight beams, in accordance with the known Euler's equations.

Specific rigidities. The relations between displacement $s_\gamma = s/L$, rotation $\theta_\gamma = \theta/L$, the stretching longitudinal force F_Q, and torque M at the helicoidal beam, with length L, extension and twisting, can be expressed through the three rigidity features:

$$F_Q = j_{kk}\, s_\gamma - j_{\tau k}\theta_\gamma; \tag{4.16}$$

$$M = j_{k\tau}\, s_\gamma - j_{\tau\tau}\, \theta_\gamma, \tag{4.17}$$

where j_{kk} is the extensional rigidity at extension deformation; $j_{\tau k} = j_{k\tau}$ is the torsional rigidity at the extension deformation, or vice versa; $j_{\tau\tau}$ is the torsional rigidity at the twist deformation.

The stress components in the helicoidal beam correspond to the stress tensor

$$\sigma^{\xi\eta} = \sigma_F^{\xi\eta} S_\gamma + \sigma_M^{\xi\eta}\theta_\gamma, \tag{4.18}$$

where $\sigma_F^{\xi\eta}$ are the stress tensor contravariant components at the longitudinal deformation, and the absence of the external twisting moment ($M = 0$); $\sigma_M^{\xi\eta}$ are the stress tensor components at the torsion deformation and the absence of the external longitudinal stretching force ($F_Q = 0$). The stress tensor in matrix form for coordinates 1, 2, 3 (Fig. 4.1b) is equal to:

$$\hat{\sigma}_{1T} = \begin{bmatrix} \sigma^{11} & \sigma^{12} & \sigma^{13} \\ \sigma^{21} & \sigma^{22} & \sigma^{23} \\ \sigma^{31} & \sigma^{32} & \sigma^{33} \end{bmatrix}. \tag{4.19}$$

The indices show the direction of the corresponding component; the first number denotes the normal to the plane, where the stress component has action, the second one denotes the direction to which the corresponding stress component is parallel.

4.3 TRANSMISSION NONLINEARITY

All of the concepts discussed earlier belong to the linear helicoidal beam behavior. But, for elastic helicoids, it is very important to study the possible nonlinear relations between beam longitudinal stretch and its angle of rotation. This will allow us to estimate the possible quasi-linear working range of the helicoid, and the systematic deflection in that function from nonlinearity. We study these problems with the variation method, on the basis of the potential

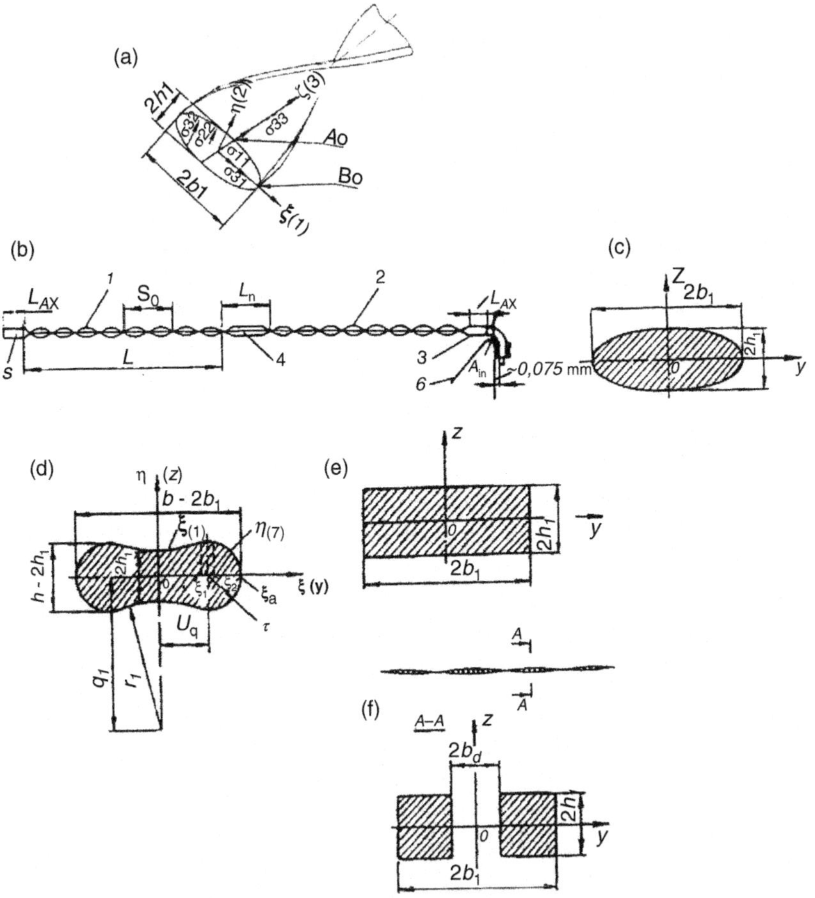

FIGURE 4.2 Pretwisted thin strip (a,b), its typical stress components (b), and cross-sections: (c) ellipse, (d) oval with waist, (e) rectangle, (f) punched. *(After Tseytlin, 2006, reprinted with permission.)*

energy equation by comparing its linear and nonlinear tensor components. The first step is to get a system of equations in order to find a linear kinematics relation between the longitudinal strain $s_\gamma = s/L$ and the untwist rotation $\theta_\gamma = \theta/L$ of the elementary helicoid with unit length $L = 1$. Let us recall that several systems of coordinates (Figs 4.1a,b; and 4.2a) are used to simplify the solution: the Cartesian coordinates X, Y, Z, the non-Cartesian helical coordinates $\xi(1)$, $\eta(2)$, $\zeta(3)$, and ξ_* and η_* as "following" coordinates that are rigidly connected to the initial contour of the beam's cross-section. Figure 4.1b represents a pretwisted (helicoidal) beam as an envelope at the cross-section contour's uniform rotation and uniform sliding, with respect to the straight line $\zeta(3)$. The "following" Cartesian coordinates have a fixed axis X (Fig. 4.1a,b) that coincides with the straight line ζ and two axes $Y_a = \xi_*$ and $Z_a = \eta_*$ rotating around this line.

The total potential energy with respect to the length unit for a pretwisted (helicoidal) beam with linear elastic material following Hooke's law may be expressed through the tensor of deformation (strain) components

$$U_v = G\iint_{A_c}[\varepsilon_{11}^2+\varepsilon_{22}^2+\varepsilon_{33}^2+2(\varepsilon_{12}^2+\varepsilon_{13}^2+\varepsilon_{23}^2)+v_p\Theta^2/(1-2v_p)]dA_c - F_Q s_\gamma - M\theta_\gamma, \tag{4.20}$$

where G is the shear modulus, Θ is an invariant of the tensor of deformation, A_c is the area of the beam's cross-section, F_Q and M are a stretching force and a twisting moment (torque), respectively; s_γ and θ_γ are the stretch strain and the angle of twist per unit of the beam's length, respectively. The constituents of deformation in this case correspond to the differential equation

$$\frac{\partial s_u^{(2)}}{\partial y} - \frac{\partial s_u^{(1)}}{\partial z} = 2\theta_\gamma x + F_\vartheta(z,y),$$

where $s_u^{(1)}, s_u^{(2)}$ are the displacements in the directions of coordinates 1 and 2 in cross-sectional plane, θ_γ is an angle of rotation per unit length of the twisted body, $F_\vartheta(z,y)$ is a function that reflects local rotations of the elements of the pretwisted (initially, "naturally" twisted, helicoidal) beam's cross-section.

The system of five simultaneous equations is obtained with the differentiation of potential energy U_v. Their structure is a complicated one, and can be shown by blocks, as follows:

$$\begin{aligned}&\partial U_v/\partial\Gamma_a = f_1(S_\gamma;\theta_\gamma;\kappa_{p0};J_p;T_0);\ \iint_{A_c}(\partial\psi^0/\partial\vartheta)^2\,dA_c;\\ &\iint_{A_c}[(\partial\psi^0/\partial\xi)^2+(\partial\psi^0/\partial\eta)^2]dA_c;\\ &\iint_{A_c}(\partial\psi^0/\partial\vartheta)(\xi^2-\eta^2)dA_c;\ \iint_{A_c}\xi\rho\eta(\partial\psi^0/\partial\rho)dA_c;(\Gamma_a;\Gamma_b;\Gamma_c)=0;\end{aligned} \tag{4.21}$$

$$\partial U_v/\partial\Gamma_b = f_2(S_\gamma;A_c;\kappa_{p0};J_p;T_0;I_{20};I_{02};\Gamma_a;\Gamma_b;\Gamma_c)=0; \tag{4.22}$$

$$\begin{aligned}&\partial U_v/\partial\Gamma_c = f_3(S_\gamma;\kappa_{p0};I_{20};I_{02};I_{22};I_{40};I_{04});\\ &\iint_{A_c}(\partial\psi^0/\partial\vartheta)(\xi^2-\eta^2)dA_c;\iint_{A_c}\xi\rho\eta(\partial\psi^0/\partial\rho)dA_c;(\Gamma_a;\Gamma_b;\Gamma_c)=0;\end{aligned} \tag{4.23}$$

$$\partial U_v/\partial\theta_\gamma = f_4(\theta_\gamma;J_p;T_o;\Gamma_a) = -M/f_7[G(J_p-T_0)]; \tag{4.23a}$$

$$\partial U_v/\partial s_\gamma = f_5(S_\gamma;\kappa_{p0};A_c;J_p;T_o;\Gamma_a;\Gamma_b) = F_Q/f_6(4GA_c). \tag{4.24}$$

Here, ψ^0 is the warping function in torsion of a plane straight beam with the same cross-section as for the pretwisted one (initial contour); ϑ and ρ are the contour polar coordinates (ρ is a radius-vector); $I_{qm} = \int_{A_c}\xi^q\eta^m\,dA_c$ are the two-multiple (double) integrals; $J_p = I_{20} + I_{02}$ is the polar moment of the cross-section inertia;

$T_0 = \int_{A_c} (\partial\psi^\circ/\partial\vartheta)dA_c + J_p$ is the geometric rigidity at the twist of the plane straight beam with the same cross-section; Γ_a, Γ_b, and Γ_c are the constants that should be defined from the solution of the system of simultaneous equations (4.21–4.24); M is the twisting moment (torque); F_Q is the longitudinal extension acting force. S_γ and θ_γ values for the quasi-linear system are determined from the solution of (4.21–4.24) by substituting the warping function ψ^0 and expressions of the double integrals $I_{qm} = \int_{A_c} \xi^q \eta^m \, d\xi \, d\eta$ for the corresponding cross-sections. The indices in the integrals are respectively equal to $q = 0, 1, 2, 4, 6$; $m = 0, 1, 2, 4, 6$.

The nonlinear system should have the "following" Cartesian coordinates Z_a, Y_a rotated with variable speed, because its helix parameter $\kappa_p = \kappa_{p0} + \theta_\gamma$ will change at the beam section rotation, unlike that of the constant helix parameter κ_{p0} in the linear system. It should also be remembered that, at the stretching of the pretwisted beam, the angle of untwist rotation is negative, $(\theta/L) = \theta_\gamma < 0$. As a result, we will find the relationship between components ε_{ik} of strain tensor for a nonlinear body in the "following" system of coordinates, and the known strain tensor components $\varepsilon^0{}_{ik}$ for a linear body (noted by zero superscript indices) of the same pretwisted beam:

$$\varepsilon_{11} = \varepsilon^0_{11}; \varepsilon_{22} = \varepsilon^0_{22}; \varepsilon_{12} = \varepsilon^0_{12} = 0; \varepsilon_{33} = \varepsilon^0_{33} - \theta_\gamma \Gamma_a \partial\psi^0/\partial\vartheta; \tag{4.24a}$$

$$\varepsilon_{13} = \varepsilon^0_{13} - 2\theta_\gamma \Gamma_b \xi^2 \eta; \varepsilon_{23} = \varepsilon^0_{23} - 2\theta_\gamma \Gamma_c \xi \eta^2, \Theta = \Theta^0 - \theta_\gamma \Gamma_a \partial\psi^0/\partial\vartheta, \tag{4.24b}$$

where 1, 2, 3 are the coordinate axes numbers in Figures 4.1b and 4.2a.

The potential energy at the stretching of the elementary (with length $L = 1$) nonlinear pretwisted beam without an external twist moment ($M_{xL} = 0$) can be expressed as

$$\begin{aligned} R_v = G \iint_{A_c} \{ & [\varepsilon^{0\,2}_{11} + \varepsilon^{0\,2}_{22} + (\varepsilon^0_{33} - \theta_\gamma \Gamma_a \partial\psi^0/\partial\vartheta)^2] + 2[(\varepsilon^0_{13} - 2\theta_\gamma \Gamma_c \xi^2 \eta)^2 \\ & + (\varepsilon^0_{23} - 2\theta_\gamma \Gamma_c \xi^2 \eta)^2] + (\Theta^0 - \theta_\gamma \Gamma_a \partial\psi^0/\partial\vartheta)^2 \} \, dA_c - F_Q S_\gamma = R^0_v + R'_v, \end{aligned} \tag{4.25}$$

where $R_v{}^0 = U_v$ is the known expression (4.20) of the potential energy at a small deformation of the pretwisted beam, and $M_{xL} = 0$, R'_v is an additional member that shows the influence of the large degree of beam untwist. By subtracting (4.20) from (4.25) with the substitution, $M = M_{xL} = 0$, we finally obtain

$$\begin{aligned} R'_v = G \iint_{A_c} 2\{ & (-\theta_\gamma \Gamma_a \partial\psi^0/\partial\vartheta)[2S_\gamma - 2\kappa_{p0} \Gamma_a (\partial\psi^0/\partial\vartheta) + 4\Gamma_b \\ & - 4\Gamma_c (\xi^2 - \eta^2)] + (\theta_\gamma \Gamma_a \partial\psi^0/\partial\vartheta)^2 \\ & - 4\Gamma_c \theta_\gamma \xi \eta \left[(\Gamma_a/2) \left(\xi \frac{\partial\psi^0}{\partial\xi} + \eta \frac{\partial\psi^0}{\partial\eta} \right) - 2\Gamma_c \kappa_{p0} \xi \eta (\xi^2 + \eta^2) \right] \\ & + 4\Gamma_c^2 \theta_\gamma^2 \xi^2 \eta^2 (\xi^2 + \eta^2) \} \, dA_c. \end{aligned} \tag{4.26}$$

The constants Γ_a, Γ_b, Γ_c can be defined with the principle of stationary potential energy, when deriving the solution by the method of variations at $\kappa_p = \kappa_{p0} + \theta_\gamma$:

$$\partial R_v/\partial \Gamma_b = (\partial R_v^0/\partial \Gamma_b); \tag{4.27}$$

$$\partial R_v/\partial \Gamma_c = (\partial R_v^0/\partial \Gamma_c); \tag{4.28}$$

$$\partial R_v/\partial \theta_\gamma = (\partial R_v^0/\partial \theta_\gamma) + (\partial R'_v/\partial \theta_\gamma); \tag{4.29}$$

$$\partial R_v/\partial \Gamma_a = (\partial R_v^0/\partial \Gamma_a); \tag{4.30}$$

$$\partial R_v/\partial S_\gamma = (\partial R_v^0/\partial S_\gamma). \tag{4.31}$$

Then, the following five equations at $M = 0$ are found:

$$S_\gamma + 6\Gamma_b + 2b_1^2\Gamma_c + (1/3)(\kappa_{p0} + \theta_\gamma)\Gamma_a b_1^2(1-3\lambda) = 0; \tag{4.32}$$

$$-S_\gamma - 6\Gamma_b + 3.9b_1^2\Gamma_c - (3/5)b_1^2(1-2\lambda)(\kappa_{p0} + \theta_\gamma)\Gamma_a = 0; \tag{4.33}$$

$$\begin{aligned} &2(\kappa_{p0} + \theta_\gamma)(S_\gamma + 2\Gamma_{b_b}) - (\theta_\gamma/2) + \Gamma_a[½ + (6/5)b_1^2(1-\lambda)(\kappa_{p0} + \theta_\gamma)^2] \\ &-(12/5)\Gamma_c(1+\lambda)(\kappa_{p0_{p0}} + \theta_\gamma)b_1^2 = 0; \end{aligned} \tag{4.34}$$

$$\begin{aligned} &\Gamma_a - \theta_\gamma(1+4\lambda) - 45S_\gamma\Gamma_a - (12/5)b_1^2\kappa_{p0}\Gamma_a^2 - 8\Gamma_a\Gamma_b \\ &-(24/5)b_1^2\Gamma_a\Gamma_c - (12/5)\theta_\gamma b_1^2\Gamma_a^2 = 0; \end{aligned} \tag{4.35}$$

$$S_\gamma + 2\Gamma_b + (1/3)(\kappa_{p0} + \theta_\gamma)\Gamma_a b_1^2(1-3\lambda) - (2/3)b_1^2\Gamma_c = (2/3)k_Q^0. \tag{4.36}$$

The solution of these simultaneous equations relative to k_Q^0, Γ_c, Γ_a, Γ_b, S_γ yields the nonlinear equation for the average transformation ratio $i_{\theta\text{av}}$, in °/μm, of the flexible helicoids as follows (Tseytlin, 2006)

$$i\theta\,\text{av} = |\theta_\gamma/2S_\gamma| = 0.0286(\kappa_{p0} + \theta_\gamma) \frac{E[1 + \theta_\gamma(\kappa_{p0} + \theta_\gamma)^{-1} - 3(\kappa_{p0} + \theta_\gamma)b_1^2\theta_\gamma]}{4G\lambda[1 + \nu_{\text{if}}(\kappa_{p0} + \theta_\gamma)^2 b_1^2\lambda^{-1} + 3\theta_\gamma b_1^2\lambda^{-1}(\kappa_{p0} + \theta_\gamma)/8]}, \tag{4.37}$$

where ν_{if} is the transformation coefficient of cross-section form that depends on the I_{qm} integrals over the cross-section area. Note that this transformation ratio is effective for the typical double elastic helicoids with length $2L$ that consists of two pretwisted beams 1 and 2, each of length L with the left and right helix, respectively, positioned on one axis and connected through the rigid rotatable body 4 (Fig. 4.2b).

On the basis of (4.37), we can find that the relationship between the transformation ratios for the different forms of pretwisted strip's cross-section with different transformation coefficients ν_{if1} and ν_{if2} may be expressed as follows:

$$\frac{i_{\theta\text{av}}(\nu_{\text{if1}})}{i_{\theta\text{av}}(\nu_{\text{if2}})} = \frac{\lambda b_1^{-2} + \nu_{\text{if2}}(\kappa_{p0} + \theta_\gamma)^2 + 3\theta_\gamma(\kappa_{p0} + \theta_\gamma)/8}{\lambda b_1^{-2} + \nu_{\text{if1}}(\kappa_{p0} + \theta_\gamma)^2 + 3\theta_\gamma(\kappa_{p0} + \theta_\gamma)/8}. \tag{4.37a}$$

The substitution of $\theta_\gamma = 0$ and $\chi_{kp} = \kappa_{p0}^2 b_1^2 \lambda^{-1}$ into (4.37) yields the formula of the transformation ratio for the linear pretwisted strip system, with the different cross-sections

$$i_\theta = 0.0286\kappa_{p0} E/[4G\lambda(1+\nu_{if}\chi_{kp})]. \quad (4.37b)$$

We can also show that (4.37) can be expressed as a cubic equation relative to the angle θ_γ (in rad/mm) with the subsequent notations:

$$A_3\theta_\gamma^3 + A_2\theta_\gamma^2 + A_1\theta_\gamma + A_0 = 0, \quad (4.38)$$

where $A_0 = -0.00715(ES_\gamma/G\lambda)\kappa_{p0}$;
$A_1 = 2[14.325\nu_{if}\chi_{kp} - 0.00715(ES_\gamma/G\lambda) + 14.325] + 0.02145\chi_{kp} ES_\gamma/G$;
$A_2 = 28.65\chi_{kp}[(2\nu_{if} + 0.375) + 0.0015 ES_\gamma/G]\kappa_{p0}^{-1}$;
$A_3 = ½ b_1^2[57.3(\nu_{if} + 0.375) + 0.0429 ES_\gamma/G]\lambda^{-1}$.

The instantaneous transformation ratio can be expressed as a partial derivative of untwist function θ_γ with respect to stretch of the strip S_γ per its unit length:

$$\partial\theta_\gamma^\circ/2\partial S_\gamma = i_\theta = -1.229(E/G)(b_1^2/4\lambda)\times \frac{2\theta_\gamma^3 + 4\kappa_{p0}\theta_\gamma^2 + [2\kappa_{p0}^2 - (4/3)b_1^{-2}]\theta_\gamma - (2/3)b_1^{-2}\kappa_{p0}}{3\theta_\gamma^2 A_3 + 2\theta_\gamma A_2 + A_1}, \quad (4.39)$$

where S_γ is in μm/mm

It is also possible to obtain a nonlinear equation for the untwist angle of the pretwisted beam, if in (4.14) one substitutes the coefficients of compliance with $\vartheta_0 = \kappa_{p0}$ by $\vartheta_0 = (\kappa_{p0} + \tau)$. In this case, the corresponding cubic equation will be (Shorr, 1968) as follows:

$$\vartheta^3 + 3\vartheta^2\beta_\rho + 2(1 + \mu_Q + \beta_\rho^2)\vartheta = 2(\vartheta_m - \mu_Q\beta_\rho), \quad (4.40)$$

where $\mu_Q = (1/GT_o)[(J_p/A_c)F_\zeta - (J_{p\eta}/J_\eta)M_\eta + (J_{p\xi}/J_\xi)M_\xi]$;
$\beta_\rho^2 = 2(1 + \nu_p)\kappa_{p0}^2 J_r/T_0$; $\vartheta_m = (\beta_\rho/\kappa_{p0}) M_\zeta/G T_0$; $\vartheta = (\beta_\rho/\kappa_{p0})\tau$; and $\tau = \theta'$.

Internal moments M_η, M_ξ, M_ζ, and force F_ζ correspond to the matrix (4.6).

Eq. (4.40) is a cubic one, as is (4.38). However, (4.38) corresponds directly to the cubical kinematical relationship between the angle of rotation and the elongation of the stretching beam, unlike (4.40), where the coefficients contain the variable force expressions. Analysis shows that (4.40) can have one root at $\mu_Q \ll 1$, when its discriminant $\Delta = \beta_\rho^2 + (2-\beta_\rho^2)^3/27 > 0$; it can also have three roots at $\beta_\rho^2 > 7.2$ that leads to $\Delta < 0$. The latter condition still corresponds to the moderate pretwisted beam field, because $\beta_o^2 \ll 1$. Note that (4.38) has only one value for a certain angle of rotation, and correspondingly yields an unambiguous kinematic solution. The members with ϑ^3 and ϑ^2 in (4.40) can be omitted if β_o, β_ρ, and ϑ are small. In this case, $\vartheta = (\vartheta_m - \mu_Q\beta_\rho)/(1 + \mu_Q + \beta_\rho^2)$, and the angle of rotation with the average stress $\sigma_{av} = F_\zeta/A_c$ at a pure extension is equal to:

$$\theta = G^{-1}T_0^{-1}[1+(F_\zeta A_c^{-1}J_p/GT_0)+\beta_\rho^2]^{-1}LF_\zeta J_p\kappa_{p0}/A_c, \tag{4.41}$$

where L is the length of the helicoidal beam, and A_c is the area of its cross-section. This equation still shows the nonlinear relationship between the angle of rotation and the stretching force. However, this nonlinearity is small, at a small initial extension.

The torsional rigidity at the pretwisted beam extension $j_\theta = F_\zeta/\theta^o$ in cN/(angular degree) for a linear system can also be expressed (Tseytlin, 1965) through the coefficient of form rigidity ν_{jf}, as follows

$$j_\theta = 69.6GA\lambda(1+\nu_{jf}\chi_{kp})/L\kappa_{p0}. \tag{4.42}$$

We estimate the values of form coefficients ν_{if} and ν_{jf} for the cross-sections of a helicoidal (pretwisted) beam with different shapes in the following section.

4.4 MOTION TRANSFORMATION IN A THIN HELICOIDAL STRIP

Structure and shape. The greatest motion transformation in the double helicoidal strip sensor occurs between the rotation of its "output" (middle) section and the strip's longitudinal extension, or at the change of distance between the strip's ends. The thin helicoidal strip practically does not have transverse rigidity. It usually consists of two sequentially connected pretwisted strip segments 1, 2 (Fig. 4.2b), with opposite direction of helix lines, and the length L of each, flat crosspiece 4 between them, and flat segments 3, 5 on both ends. End 3 is connected to the input 6 in point A. One of the pretwisted segments has the left-handed helix lines, and another one has right-handed helix lines. The crosspiece element 4 is the rotational output of the strip's sensor to which a pointer, mirror, or the secondary transducer can be connected. The cross-section form of such strips depends on the method of their manufacturing.

These forms include: ellipse (Fig. 4.2c) at the rolling of round wires by the cylindrical rolls, oval with a waist (Fig. 4.2d) at the rolling of a thin wire by the balls, and a rectangular one (Fig. 4.2e) at the slitting of strips from rolled thin, wide tapes. The punched cross-section (punched, perforated strip) is also available (Fig. 4.2f). The small accurate holes in the narrow and thin strip can be provided with the laser ray or EDM. Other methods of such helicoids manufacturing are also possible. The geometric features of elliptical and rectangular contour are widely known. The oval with waist and punched strip geometric features are more complicated, and may be expressed with the quartic equation of the fourth degree similar to oval of Cassini:

$$(y^2/b_1^2)+(z^2/h_c^2)+\varsigma(h_c^{-2}-b_1^{-2})(y^2-z^2)-\varsigma(y^4-6y^2z^2+z^4)/(b_1^2h_c^2)=1-\varsigma.$$

Here, $Ç$ is a constant defined under the condition $h_1^2/(b_1^2+h_c^2) > Ç > -½(\sqrt{2}-1)$. For a punched section (punched or perforated strip), there should be an additional condition, $h_c^2 < 0$. Let us recall that the Cassini oval is an anallagmatic curve that is invariant under inversion. Such oval contour, for the simplification of calculations, may also be approximated with the arcs of a circle and segments of straight lines. The equation of the first arc with radius r_1 (Fig. 4.2d) can be expressed as: $\xi_{(1)} = [-\eta^2 + 2\eta q_1 + (r_1^2 - q_1^2)]^{0.5}$. The equation of the second arc r_2 is as follows:

$\eta_{(2)} = [\xi^2 + 2\xi U_q + (r_2^2 - U_q^2)]^{0.5}$. The I_{qm} integrals over oval with waist cross-sections approximated by arcs and segments of straight lines converge with Abel's integrals. Vesicles with the shape similar to oval Cassini (dumbbell, pancake) are found in the mechanics of the cell, as well (see Boal, 2002; Figs 7.6–7.10).

Materials. The materials used for the microstrips under consideration are shown in Table 4.4. Other materials are also applicable (Chapter 1, Section 1.7).

Form coefficients.

Let us evaluate the coefficients of form for rigidity ν_{jf} and for the transformation ν_{if} of different cross-section profiles. The rigidity of the pretwisted beam under extension related to the untwist angle, on the basis of (4.41), is equal to:

$$j_\theta = F_ç/\theta = GT_0 A_c [1 + (\sigma_{av} J_p/GT_0) + \beta_\sigma^2]/(L J_p \kappa_{p0}). \tag{4.43}$$

Rigidity. If the strip's cross-section is an ellipse with a ratio of thickness $h = 2h_1$ to width $b = 2b_1$ that equals

$\sqrt{\lambda} = h/b \ll 1$, then $J_p/T_0 = (1+\lambda)^2/4\lambda$; $J_r/T_0 = (b_1^2/16)(1 - 2/3\lambda + \lambda^2) \times (1+\lambda)/\lambda \approx b_1^2/16\lambda$; and $\beta_p^2 = 2(1+\nu_p)\, \kappa_{p0}^2\, J_r/T_0 \approx (1+\nu_p)\, \kappa_{p0}^2\, b_1^2/8\lambda$.

The function J_r for the elliptical cross-section is equal to:

$J_r = \pi b^6(\sqrt{\lambda})(1 - 2/3\lambda + \lambda^2)/1024 \approx \pi b^6(\sqrt{\lambda})/1024$. Substitution of these expressions into (4.43) yields

$$j_\theta \approx 4G\lambda A_c \{[1 + \sigma_{av}(1+\lambda)^2/4G\lambda] + [(1+\nu_p)\kappa_{p0}^2 b_1^2/8\lambda]\}/[L\kappa_{p0}(1+\lambda)^2].$$

Hence, if $\sigma_{av} \to 0$ at the beginning of the stretch of the pretwisted strip with an elliptical cross-section and the expression of j_θ in cN/o, we finally have an equation similar to Eq. (4.42):

$$j_{\theta o} \approx 69.6 G A_c \lambda (1 + \nu_{jf} \chi_{kp})/L\kappa_{p0}, \tag{4.44}$$

where $\nu_{jf} = \nu_{jfel} = (1+\nu_p)/8$; area $A_c = \pi b_1 h_1$; the λ ratio in the sensor usually corresponds to $\lambda^{0.5} \le 1/15\ldots1/25$, and if $\nu_p = 0.333$, as for many materials, $\nu_{jfel} \approx 0.17$.

If the strip's cross-section is a rectangular one with similar geometric characteristics, then

$T_0/J_p \approx 4\lambda/(1+\lambda)$; $J_r = 0.00555 b^6(\sqrt{\lambda})/[1 + 1.067(0.5)^2]$; $J_r/T_0 = 0.0525 b_1^2/\lambda$; $\beta_p^2 = 0.1052(1+\nu_p)\kappa_{p0}^2\, b_1^2/\lambda$; and area $A_c = 4b_1 h_1$.

TABLE 4.4 Mechanical Properties of Typical Elastic Materials (for more detail see Tseytlin (2006))

Material					Modulus of elasticity				
Profile	Designation Dim., t_T, load	$\sigma_t°$ ($\sigma_f°$)	$\sigma_e°$	$\sigma_{en}°$ max	$E°$ $(E/\rho_\gamma)°$	$G°$ $(\nu_p°)$	*F.T.*	$\rho_\gamma°$	σ_r^0, MYS
Micro-strips	Br4Sn-3Zn *	1.0	1.0	1.0	1.0 (1.0)	1.0 (1.00)	...	1	1
	Br Be 2(320)**	1.39	1.25	2.0	1.2(1.2)	1.1 (1.26)	...	0.96	1
	PtAg20	1.74	2.00	0.5	1.5(0.7)	1.5 (0.9)	...	2.17	...
Fused silica quartz oriented	SiO_2, including microfibers	0.05 – 5 0.97	0.07 11.4	0.02	0.62 – 0.95	0.70 (0.50 – 0.64)	0.9	0.26 – 0.30	...
Polysilicon	Si, tension bending	1.38 – 2.6 (2.5)	2.29 – 4.28	2.88 – 3.27	1.42	(0.79)	0.85 – 1.9	0.26	...
Graphene	Monolayer C			115	9.7	10 (0.68 – 0.97)		0.15	

*Young's modulus $E = 112.8$ GPa; shear modulus $G = 44.1$ GPa; Poisson's ratio $\nu_p = 0.279$; tensile (UTS) strength $\sigma_t = 1.13$ GPa; yield stress $\sigma_e = 0.785$ GPa; endurance limit $\sigma_{en} = 0.260$ GPa (approximately, on the basis of 10 million cycles); elasticity temperature coefficient $\alpha_E = -4.5 \times 10^{-4}\ K^{-1}$ (approximately); temperature coefficient of linear expansion $\alpha_L = 15.5 \times 10^{-6}\ K^{-1}$, $E/\rho_\gamma = 12.84$ MN·m/kg $= 12.84 \times 10^6\ m^2/s^2$; specific gravity $\rho_\gamma = 8.8\ g/cm^3$ (8800 kg/m^3), and MYS (material yield system) $\sigma_r = 300$ MPa. *F.T.* represents fracture toughness in MPa $m^{1/2}$, and $\sigma_f°$ is the relative fracture strength σ_f/σ_t for brittle material. Superscript ° indicates relative values with the reference to accepted data for the microstrips from bronze Br4Sn-3Zn; recommended factor of safety is not less than 1.3.

**Quenching temperature t_T is shown in parenthesis with two asterisks in °C; all properties at 20°C.

In this case, at $\sigma_{av} \to 0$, the rigidity form coefficient for the rectangular cross-section of the helicoid strip extension is $\nu_{jf} = \nu_{jfrc} = 0.1052(1 + \nu_p)$, and if $\nu_p = 0.333$, then $\nu_{jfrc} = 0.14$; at $\nu_p = 0.279$ for the zinc bronze (Table 4.4) $\nu_{jfrc} \approx 0.135$.

Transformation. The pretwisted (helicoidal) strip with a thin rectangular cross-section and constant thickness h has compliance that for the "following" coordinates with directions ξ, η, ζ, and ϑ are shown (Tseytlin, 2006; Shorr, 1968) in Table 4.5.

The corresponding compliances ν_x, $\nu_{x\theta}$, and ν_θ in Cartesian coordinates have direct correlation with compliances in the "following" coordinates:

$\nu_x = \nu_\varsigma = (EA)^{-1}\,\{1 + [(0.0417(1+\nu_p)b^4\vartheta_o^2)/(h^2 + 0.0263\,(1+\nu_p)\,b^4\vartheta_o^2)]\}$;

$\nu_{\theta x} = \nu_{\varsigma\vartheta} = -\,(GT_o)^{-1}\,[0.0833\,b^2\,\vartheta_o\,h^2/(h^2 + 0.0263\,(1+\nu_p)\,b^4\vartheta_o^2)]$;

$\nu_\theta = \nu_\vartheta = (GT_o)^{-1}\,\{h^2/[h^2 + 0.0263\,(1+\nu_p)\,b^4\vartheta_o^2]\}$.

The matrix Eq. (4.7) for the pretwisted beam under the pure extension yields, in this case, expressions for the derivatives of displacements $S'_x = S_\gamma$ for extension and $\theta'_x = \theta_\gamma$ for untwist, as follows

$$S_\gamma = \nu_x F_Q = E^{-1}A^{-1}\left(1+\frac{0.0417(1+\nu_p)b^4\kappa_{po}^2}{h^2+0.0263(1+\nu_p)b^4\kappa_{po}^2}\right)F_Q; \tag{4.45}$$

$$\theta_\gamma = \nu_{\theta x}F_Q = -\frac{0.08333\,b^2h^2\,\kappa_{po}}{GT_0[h^2+0.0263(1+\nu_p)b^4\,\kappa_{po}^2]}F_Q, \tag{4.46}$$

where $T_0 = bh^3/3$ for very thin rectangle ($b/h > 10$).

Hence, the transformation ratio in (angular degree)/μm for the pretwisted strip in linear deformation state (remember that the thin strip is a double one, with left-hand and right-hand helix segments) with a rectangular cross-section in accordance with (4.45) and (4.46), after elementary cancellations, is equal to:

$$\begin{aligned} i_\theta = 57.295\cdot 10^{-3}\theta_\gamma/2S_\gamma &= -\frac{57.295\cdot 3(0.0833)E\kappa_{po}}{2G\lambda[1+0.272(1+\nu_p)b_1^2\kappa_{po}^2/\lambda]1000} \\ &= -\frac{0.0286E\kappa_{po}}{4G\lambda(1+\nu_{\text{ifrc}}\chi_{kp})}, \end{aligned} \tag{4.47}$$

where the kinematic transmission form coefficient $\nu_{ifrc} = 0.272\,(1 + \nu_p)$; and, for example, at $\nu_p = 0.279$ for the zinc bronze (Table 4.4), $\nu_{ifrc} \approx 0.345 \approx 29/84$. Eq. (4.47) is the same expression that was presented in Tseytlin (2006). We can also find the transformation form coefficient ν_{ifel} for an elliptical cross-section, where

$$\nu_\varepsilon = \frac{4}{E\pi b^2\sqrt{\lambda}}\left\{1+\frac{(1+\nu_p)\beta_m^2(1-\lambda)^4}{8\lambda(1+\lambda)}\left[1+\frac{(1+\nu_p)\beta_m^2\,(1-\lambda)^2}{8\lambda\,(1+\lambda)}(1+2/3\lambda+\lambda^2)\right]^{-1}\right\}$$

TABLE 4.5 The Compliance v_{ij} of the Helicoidal Strips with Rectangular Cross-Section

	ζ	ξ	η	ϑ
ζ	$\frac{1}{EA}\left[1+\frac{0.0417\,(1+v_p)\,b^4\vartheta_0^2}{h^2+0.0263\,(1+v_p)\,b^4\vartheta_0^2}\right]$	$\frac{1}{GT_0}\frac{0.0111\,b^4\theta_0^2 h}{1.267\,h^2+0.0333\,(1+v_p)\,b^4\vartheta_0^2}$	0	$-\frac{0.0833\,b^2\vartheta_0\,h^2}{GT_0[h^2+0.0263\,(1+v_p)\,b^4\vartheta_0^2]}$
ξ		$\frac{1}{EJ_\xi}\left[1+\frac{0.0089\,(1+v_p)\,b^4\vartheta_0^2}{1.267\,h^2+0.0333\,b^4\vartheta_0^2}\right]$	0	$\frac{1}{GT_0}\frac{0.133\,b^2\vartheta_0 h}{1.267\,h^2+0.0333\,(1+v_p)\,b^4\vartheta_0^2}$
η	0	0	$\frac{1}{EJ_\eta}$	0
ϑ		$\frac{1}{EJ_\xi}\left[1+\frac{0.0089\,(1+v_p)\,b^4\vartheta_0^2}{1.267\,h^2+0.0333\,b^4\vartheta_0^2}\right]$	0	$\frac{1}{GT_0}\frac{h^2}{[h^2+0.0263\,(1+v_p)\,b^4\vartheta_0^2]}$

Note: $i = \zeta, \eta, \xi, \vartheta$; $j = \zeta, \eta, \xi, \vartheta$; Cell $\zeta\zeta$ contains expression for v_ζ; cell $\xi\zeta$ contains $v_{\xi\zeta}$ etc. $v_{ij} = v_{ji}$.
Source: After Tseytlin (2006), reprinted with permission.

$$v_{\varepsilon\theta} = -\frac{2\beta_m(1-\lambda)^2}{G\pi\, b^3\lambda^{3/2}}\left[1+\frac{(1+v_p)\beta_m^2(1-\lambda)^2}{8\lambda(1+\lambda)}(1+2/3\lambda+\lambda^2)\right]^{-1};\, i_\theta = v_{\varepsilon\theta}/2v_\varepsilon,$$

and i_θ, in angular degree/μm, similar to (4.47) after certain substitutions and cancellations is equal to:

$$i_\theta = -\frac{0.0286\,E\kappa_{po}}{4G\lambda(1+v_{\text{ifel}}\,\chi_{kp})},$$

where the form transformation coefficient for the elliptical cross-section is equal to $v_{\text{ifel}} \approx (1 + v_{\text{p}})/4$; and at $v_p = 0.250\ldots0.4$, as for many materials, with $v_{\text{ifel}} = 0.32 - 0.35$.

More on rigidity. The expression for the extension rigidity of a pretwisted strip with a rectangular cross-section related to the untwist angle (in cN/angular degree) can be found by using (4.46) as well. In this case,

$$j_\theta = (1000/57.295)P_Q/(\theta_\gamma L) = \frac{17.453GT_o[h^2+0.0263(1+v_p)b^4\kappa_{po}^2]}{0.0833b^2h^2\kappa_{po}},\quad \text{and}$$

after elementary cancellations and substitutions, we finally have the same expression as in (4.44):

$$j_\theta = 69.6GA_c\lambda(1+v_{\text{jfrc}}\,\chi_{kp})/L\kappa_{p0}. \tag{4.48}$$

For the cross-section of the oval with waist, the rigidity form coefficient v_{jfoc} is theoretically and practically close to one for the elliptical cross-section, but its area is close to the rectangular cross-section. This is because the oval with a waist cross-section of thin strips practically has a very shallow slope (less than 0.0025) on the long side, and rounding with the radius equal to half of the thickness on the short side. The depth of a "crater" on the long side of such a cross-section is not more than 3–5 interference fringes, or less than 1.5 μm (60 μin.). We recall that the rigidity coefficient for the pretwisted (helicoidal) strip with elliptical cross-section is $v_{\text{jfel}} \approx 0.17$. Therefore, we accept that $v_{\text{jfoc}} = 5/28 \approx 0.178$, a value that corresponds to the experimental data of the book's author.

The longitudinal rigidity of the pretwisted strip increases slightly after its comparatively large untwist $\theta°$ when the actual helix pitch

$S_{\text{a}} = 360L/[360(L/S_0) - \theta°]$ increases, and the pretwist helix parameter value $k_p = 2\pi/S_{\text{a}}$ correspondingly decreases, in comparison with its initial value k_{p0}.

4.4.1 Stress Parameters, Dangerous Points

The stress state in the most critical of the elliptical cross-section points A_o and B_o (Fig. 4.2a) may be approximately presented by the following expressions (see Tseytlin, 2006) at the absence of an external twisting moment:

$$\sigma_{QAo}^{11} \approx \sigma_{QAo}^{22} \approx -(9/112)\chi_{kp} C\sigma^{-1}\sigma_0^{33}; \sigma_{QAo}^{12} = 0; \tag{4.49}$$

$$\sigma_{QAo}^{13} = (h_1\kappa_{p0}/2\lambda)C_\sigma^{-1}\sigma_0^{33}; \sigma_{QAo}^{23} = 0; \tag{4.50}$$

$$\sigma_{QAo}^{33} = [1+(69/224)\chi_{kp}]C_\sigma^{-1}\sigma_0^{33}, \tag{4.51}$$

where $C_\sigma = 1 + 5/28\ \chi_{kp}$; $\sigma^{33}{}_0 = F_Q/A_c$, and $\chi_{kp} = b_1^2\ \kappa_{p0}^2/\lambda$.

This system may be represented by the tensor

$$\hat{\sigma}_{2T} = \begin{bmatrix} \sigma_{QA_0}^{11} & 0 & \sigma_{QA_0}^{13} \\ 0 & \sigma_{QA_0}^{22} & 0 \\ \sigma_{QA_0}^{31} & 0 & \sigma_{QA_0}^{33} \end{bmatrix} \tag{4.52}$$

The component $\sigma_{QA_0}^{22}$ of tensor $\hat{\sigma}_{2T}$ (as follows from the tensor's view) is one of the three principal components of the stress tensor: σ_{qAo}, σ_{iAo}, and σ_{jAo}. If $\sigma_{QA_0}^{22} = \sigma_{qAo}$, the two other principal components are defined with the known formula

$$\sigma_{i,j} = ½(\sigma^{11}+\sigma^{33}) \pm [(1/4)(\sigma^{11}-\sigma^{33})^2 + (\sigma^{13})^2]^{0.5}. \tag{4.53}$$

Hence, after substituting into (4.53) corresponding components, and eliminating small members, in comparison with one, we will have

$$\sigma_{iAo} = -(16/224)\chi_{kp} C_\sigma^{-1}\sigma_0^{33}; \tag{4.54}$$

$$\sigma_{jAo} = [1+(67/224)\chi_{kp}]C_\sigma^{-1}\sigma_0^{33}. \tag{4.55}$$

We can conclude that, on the basis of the comparison values σ_i, σ_j, and σ_q, the following sequence $\sigma_j = \sigma_1 > \sigma_i = \sigma_2 > \sigma_q = \sigma_3$ appears, and therefore the stress status in point A_o is characterized by the expression

$$\sigma_{1Ao} - \sigma_{3Ao} = [1+(85/224)\chi_{kp}]C_\sigma^{-1}\sigma_0^{33}. \tag{4.56}$$

In point B_0, $\eta = 0$; $\xi = b_1$. Then, the stress tensor in this point will have the following components:

$$\sigma_{qBo} = \sigma_{QBo}^{11} = (19/112)\chi_{kp} C_\sigma^{-1}\sigma_0^{33}; \tag{4.57}$$

$$\sigma_{iBo} = (6/56)\chi_{kp} C_\sigma^{-1}\sigma_0^{33}; \tag{4.58}$$

$$\sigma_{jBo} = [1+(69/224)\chi_{kp}]C_\sigma^{-1}\sigma_0^{33}. \tag{4.59}$$

We can conclude that on the basis of comparison of values σ_i; σ_j, and σ_q, the following sequence $\sigma_j = \sigma_1 > \sigma_i = \sigma_2 > \sigma_q = \sigma_3$ appears, and the stress status in point B_o is characterized by expression

$$\sigma_{1Bo} - \sigma_{3Bo} = [1 + (63/224)\chi_{kp}]C_\sigma^{-1}\sigma_o^{33}. \tag{4.60}$$

Hence, the stress state in point A_o is more dangerous because

$$(\sigma_{1Ao} - \sigma_{3Ao}) > (\sigma_{1Bo} - \sigma_{3Bo}).$$

Effective cross-section. Derived for the elliptical cross-section, the conclusion is effective for a rectangular cross-section, as well. However, the most critical point (with the largest tangential stress) in the cross-section of the oval with waist type (Fig. 4.2d) under torsion is not in the middle of a long side. This point, in accordance with the Saint-Venant theory, may be split in two and positioned at the distance of $\sim½\, b_1$ from the center of cross-section profile that distances it from the center fiber with the largest normal stress. The reader can find a more detailed study of this problem in Filon (1900).

4.4.2 Strip Parameters Deflection

If the nonuniformity of the helicoid (pretwisted strip) parameters is a small and random one (with normal distribution as the result of many factors' influence), then the motion transformation in the group of pretwisted strips with the same nominal parameters may be presented in the following form:

$$\theta = \theta_\gamma\, L \pm 1/2\sqrt{\sum_k (\Delta\theta_k L_k)^2}, \tag{4.61}$$

where $\Delta\theta_k$ is the range of possible deflection of the untwist angle on each kth segment of the pretwisted strip; L_k is the length of the kth strip's segment; $\Delta\kappa_p$, ΔE_k, ΔG_k, Δb_{1k}, Δh_{1k} are the corresponding deflections of the parameters on each segment L_k of the strip from the nominal value. The calculation and the manufacturing practice (Table 4.6) show that the random deflections of all parameters, in the range of 5–15% from nominal, change the transformation ratio by up to 5–10%. The strongest influence on the transformation ratio comes from the deflection of thickness, because thickness has the smallest absolute value.

Pendulum application. There were attempts to use the torsional pendulum method for the identification of the flat blank strips, similar to that used successfully for the torsion suspensions. However, this simple method is not representative for the pretwisted strips sensors.

Table 4.7 data show that pretwisted strips made from flat strips (blank No 2 and 4) with very close periods in the torsion pendulum have different transformation ratios at the pretwist parameter $k_p = \pi$ mm^{-1}. At the same time, the pretwisted strips with close transformation ratios were made from flat blanks No. 3 and 4, with the different period values. Blank No. 5 and No. 4 with the

TABLE 4.6 The Deflections of Helicoidal Strip Parameters (Example)

Nominal value of parameter	Manufacturing deflection of parameters	Deflection of transmission ratio* Δi_θ, o/μm, from nominal value	
		Experimental	Calculated
$2h_1 = 0.008$ mm	$\Delta h = \pm\ 0.0009$ mm	+ 0.2	+ 0.28
$2b_1 = 0.12$ mm	$\Delta b = \pm\ 0.01$ mm		
$\kappa_p = \pi$ rad mm^{-1}	$\Delta\kappa_p = \pm\ 0.5$ rad mm^{-1}	– 0.3	– 0.28
$G = 44$ GPa	$\Delta G = \pm\ 3$ MPa		

**The nominal transformation ratio $i_\theta = 3.4$ o/μm.*
Source: After Tseytlin (2006), reprinted with permission.

TABLE 4.7 Period T_τ of the Torsion Pendulum Oscillations with the Strip Suspension

Blank No.	Dimensions of microstrip in mm					
	$2b_1$	$2h_1$	m_w, g	R_{cyl}, mm	T_τ, s	i_θ, °/μm
1	0.08	0.004	14	11	140	7.25
2	0.1	0.006	18	11	105	4.76
3	0.12	0.008	54	15	130	3.45
4	0.122	0.006	18	11	90	3.82
5	0.122	0.006	54	15	169	3.82

Source: After Tseytlin (2006), reprinted with permission.

same cross-section dimensions have different periods in the pendulum, with different mass m_w. Let us look at this problem in more detail. The oscillation period T_τ of the flat straight strip depends on its torsion rigidity

$$T_\tau = 2\pi\sqrt{I_w L_{pm}/(GJ_{tr})}. \tag{4.62}$$

The torsion rigidity J_{tr} of the strip with a rectangular cross-section is equal to:

$$J_{tr} = [(16/3)\,b_1 h_1^3 + (1/3G)F_a b_1^2]G = (16/3)\,b_1 h_1^3 G + (1/3)\,F_a b_1^2. \tag{4.62a}$$

$$I_w = (1/2)\,m_{wg}\,R_{cyl}^2, \tag{4.63}$$

where $F_a = m_{wg}\,g$ is the force of longitudinal tension that is produced by mass m_w of a cylindrical torsion pendulum with radius R_{cyl}, and L_{pm} is the length of the pendulum suspension. The comparison of (4.62) with any expression for the transmission ratio or the longitudinal rigidity of the pretwisted strip shows that the condition of T_τ = constant does not necessarily mean that i_θ = constant or j_θ = constant; and conversely, the condition of the constant working features does not require one certain value of torsion rigidity. This is evident because the pretwisted strip is characterized by the three different rigidity features (j_{kk}, $j_{\tau k}$, and $j_{\tau\tau}$). Moreover, there is no representation of the pretwist factor in (4.62). Hence, the identification of the blank strips to obtain the specified transformation ratio and longitudinal rigidity should be based on the measurement of the strip's dimensions, by objective methods. Part of these parameters indeed depends on the specifics of the manufacturing procedure: constant thickness and variable width at the cutting of strips from the sheet with a certain thickness h and constant squared relation λ at the rolling of the wire by cylindrical or spherical rolls. It is impossible to draw conclusions about the quality of the blanks, on the basis of the torsion pendulum described above. However, the torsion pendulum method is helpful in some cases. The following equation reflects the elliptical cross-section parameters relations at $\lambda^{-1} \gg 1$:

$$(4\lambda)^{-1} = (J_p/T_0) - 1/4, \tag{4.64}$$

where J_p is the polar moment of cross-section inertia; T_0 is the geometric torsion rigidity of a flat strip with the same cross-section as for the pretwisted one. The transformation ratio, in this case, may be expressed as follows:

$$|i_\theta| = \frac{\kappa_p E}{G[(J_p/T_0) - 1] + (29/21)(b_1^2 \kappa_p^2 G)}. \tag{4.65}$$

There are effective relationships for the relatively thick strips and beams ($\chi_{kp} < 1$) with any shape of cross-section whose parameters may be defined with the torsional pendulums

$$|i_\theta| = \kappa_p (E/G)[1 - (J_p/T_0)]; \tag{4.66}$$

$$F_Q = EAS_\gamma + E(J_p - T_0)\kappa_p \theta_\gamma. \tag{4.67}$$

It is known, from the theory of torsion pendulum oscillations, that $M/\theta = I_w (2\pi/T_{\tau 1})^2$, where M is the torque, θ is the amplitude angle of rotation at oscillations, $I_w = ½\, m_{wg} R_{cyl}^2$ is the dynamic moment of inertia of a cylindrical pendulum with radius R_{cyl}, mass m_{wg}, and $T_{\tau 1}$ is the period of pendulum oscillations. As a result, $J_{tr} = ML_{pm}/(\theta G) = I_w (2\pi/T_{\tau 1})^2 L_{pm}/G$. The rigidity at pure torsion GT_0 of the strip can be defined from the known J_{tr} in accordance with Eq. (4.62a), after subtraction of a part, $(1/3)F_a b_1^2$, caused by the pendulum mass action (longitudinal tension $F_a = gm_{wg}$). The best way to subtract is to use a pendulum with a small mass, whose influence may be neglected. The value of G in suspension wire material is determined from the system's period $T_{\tau 2}$ of torsional oscillation as

$$G = 16(2\pi/T_{\tau 2})^2 L_f m_w R_{cyl}^2/(\pi d^4). \tag{4.68}$$

The estimation of polar moment J_p for any symmetric section can also be done with the torsion pendulum if necessary (Tseytlin, 1965).

Important second member. Eq. (4.62a) has two members, one of which is dependent on the strip shear modulus, and another one is independent of strip material elastic parameters, but depends on the longitudinal force applied to the pendulum (Tseytlin, 2006). This problem may even have a negative influence in the micromechanical system device with hanged mass (Cumpson et al., 2003) for the calibration of AFM cantilever spring constants, because the elastic element of the beam rotation is not round, unlike the carbon fiber with diameter 7 μm in their preceding model. However, the second member of an equation similar to Eq. (4.62a) was successfully used in novel torsion balance for the measurement of the Newtonian gravitational constant (Quinn et al., 1992, 1997). This example shows how many useful sides the nature of precision elasticity has when the used model is correct.

4.5 COILED RIBBON HELICOIDS

History. The device was first proposed by Ayrton and Perry (1884), and may be called after their names. This spring is usually considered as a helical bar. But this model for double-helicoid wide coiled ribbon strips produces incorrect relationships between extension $s = \Delta$ and the angular untwist θ of the crosspiece.

Parameters and features. As shown in Fig. 4.3a, the double-coiled helicoid strip has two parts (1 and 2) coiled in opposite directions, straight ends (3 and 5) and the crosspiece (4) that connects the coiled parts and serves as the

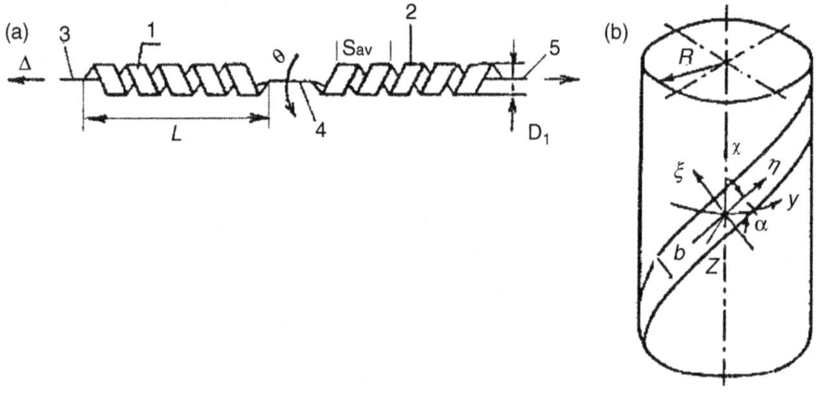

FIGURE 4.3 Coiled strip schematic (a), and its middle surface (b). *(After Tseytlin, 2006, reprinted with permission.)*

output element. These helicoids have a large helix angle α ranging from 40° to 70°, and a large ratio between width b and height h of the strip's cross-section, such that b/h = 30–70. From a theoretical standpoint, these coiled strips are studied as cylindrical shells (Biderman, 1963; Tseytlin, 2006) that provide more correct formulae for the transmission function in two stages of deformation.

Let us first look at the initial stage of deformation in one part of the coiled strip. Prior to deformation, the middle surface of the strip coincides with the surface of a cylinder of radius R_0, and the edges of the ribbon form the angle α_o with respect to the plane of the cylinder's parallel circle. Figure 4.3 shows a coiled strip schematic (a), and the middle surface presentation (b). Following deformation, the radius of the cylinder and the angle of the helical line take the values R and α. The coordinate system x, y, z (Fig. 4.3b) is connected to a point of the shell's middle surface, and $y = R\varphi$, where φ is a central angle. Along with these coordinate axes, orthogonal axes η and ξ are used, and they are tangent to the helix line, and directed as a binormal to it, respectively.

The components of displacement for the points on the middle surface in directions x, y, z are denoted as u, v, w. The whole deformation of the spring-shell, in accordance with the proposal made by Biderman (1963), has two constituents: (i) deformation without change to the cylindrical shell and without extension of the middle surface; and (ii) deformation without change to the pitch and helix angle, but with extension of the middle surface.

Thus, $u = u_1 + u_2$; $v = v_1 + v_2$; and $w = w_1 + w_2$. The values of R and α (in accordance with Biderman et al., 1970) are associated with the spring longitudinal extension s and the mutual rotation angle θ of the ends, through relationships that follow from the condition that the developed length L_r of the strip remains constant:

$$s = L_r(\sin\alpha - \sin\alpha_0), \tag{4.69}$$

$$\theta = -L_r[(\cos\alpha/R) - (\cos\alpha_0/R_0)], \tag{4.70}$$

where

$$L_r = L_{r0} = 2\pi R_0\, n_{r0}/\cos\alpha_0, \tag{4.71}$$

$n_{r0} = L_r/S_{av}$ is the number of coils in one part of the coiled strip, with certain direction of helix lines, and average pitch S_{av}. In the first type of deformation, the initial curvatures of the lines x and y are

$$\kappa_1 = \sin^2(\alpha - \alpha_0)/R_0, \quad \kappa_2 = \cos^2(\alpha - \alpha_0)/R_0. \tag{4.72}$$

The expressions for stresses in the first constituent of deformation are as follows (Biderman, 1963):

$$\sigma_{x1} = \frac{E}{1-\mu^2}(\varepsilon_{x1} + \mu\varepsilon_{y1}) = -\frac{E}{1-\mu^2}\mu\frac{\Delta}{R^2}z, \tag{4.73}$$

$$\sigma_{y1} = \frac{E}{1-\mu^2}(\varepsilon_{y1} + \mu\varepsilon_{x1}) = -\frac{E}{1-\mu^2}\frac{\Delta}{R^2}z, \tag{4.74}$$

$$\tau_{xy1} = \frac{E}{2(1+\mu)}\gamma_{xy1} = -\frac{E}{1+\mu}\frac{\omega}{R}z, \tag{4.75}$$

where $\mu = \nu_p$.

And the displacements are equal to

$$u_1 = \omega y, v_1 = -\frac{\Delta}{R}y - \omega x, w_1 = \Delta, \quad s = 2\pi Ri\omega, \quad \theta = 2\pi i\left(\frac{\Delta}{R} + \omega\tan\alpha\right). \tag{4.76}$$

Displacement w_2 on the middle surface of the coiled spring (second constituent) satisfies the following differential relationship

$$w_2^I + 4m_w^4 w_2 = 0, \tag{4.77}$$

where $m_w = [3(1-\nu_p^2)/R^2h^2]^{1/4}\cos\alpha$.

Eq. (4.77) has a similar form with the equation of the axis-symmetrical deformation of the thin cylindrical shells whose solution may be expressed through Krylov's (Table 3.1) beam functions $Y_1(m_w\xi)$, $Y_2(m_w\xi)$, $Y_3(m_w\xi)$, and $Y_4(m_w\xi)$. These functions are connected through the following differential relations $Y_1' = -4\,m_w Y_4$, $Y_2' = m_w Y_1$, $Y_3' = m_w Y_2$, and $Y_4' = m_w Y_3$.

The displacement w_2 (ξ) is symmetrical with relation to the middle of the coil's turn $\xi = 0$, therefore it was represented by the formula

$$w_2 = A_w Y_1(m_w\xi) + B_w Y_3(m_w\xi). \tag{4.78}$$

The coefficients A_w and B_w were defined in order to satisfy the boundary conditions.

Finally (Tseytlin, 2006, p. 205–206), the "twisting" M^t and "flexural" M^f moments are equal to:

$$\begin{aligned} M^f = {} & (Db/\cos\alpha)\{(1/R_0 - 1/R)[1 - K_\xi(\sin^2\alpha + \nu_p\cos^2\alpha)^2] \\ & -(1-\nu_p)R_0^{-1}\sin(\alpha-\alpha_0)[\sin(\alpha-\alpha_0) \\ & -K_\xi(\sin^2\alpha + \nu_p\cos^2\alpha)\sin(\alpha+\alpha_0)]\}, \end{aligned} \tag{4.79a}$$

$$\begin{aligned} M^t = {} & (Db/\cos^2\alpha)\{(1/R - 1/R_0)\{1 - K_\xi[1-(1-\nu_p)^2\cos^4\alpha]\}\sin \\ & \alpha + (1-\nu_p)R_0^{-1}\sin(\alpha-\alpha_0)\{2\cos\alpha\cos(\alpha-\alpha_0) + \sin\alpha\sin(\alpha-\alpha_0) \\ & -K_\xi\sin\alpha\sin(\alpha+\alpha_0)[1+(1-\nu_p)\cos^2\alpha)]\}. \end{aligned} \tag{4.79b}$$

Besides, $M^t = F_pR$, where F_p is the longitudinal stretching force. If flexural moment $M^f = 0$ under spring tension with free rotation of ends (or, equivalently, the middle section), then one can find that

$$\frac{1}{R}=\frac{1}{R_0}+\frac{1-\nu_p}{R_0}\sin(\alpha-\alpha_0)\frac{K_\xi(\sin^2\alpha+\nu_p\cos^2\alpha)\sin(\alpha+\alpha_0)-\sin(\alpha-\alpha_0)}{1-K_\xi(\sin^2\alpha+\nu_p\cos^2\alpha)^2}. \quad (4.79c)$$

If quantity λ_{co} is large enough, that is, the corresponding expression $b_{co}\alpha/\sqrt{Rh}$ is also large, then $K_\xi \sim 1/\lambda_{co} \to 0$ and

$$\theta=-L_r\left[\frac{\cos\alpha-\cos\alpha_0}{R_0}-\frac{1-\nu_p}{R_0}\cos\alpha\sin^2(\alpha-\alpha_0)\right]. \quad (4.79d)$$

Now, we assume that with a small longitudinal displacement s, the change $\Delta\alpha$ in helix angle α is also small. Therefore, $\alpha=\alpha_0+\Delta\alpha$ and $s=L_r[\sin(\alpha_0+\Delta\alpha)-\sin\alpha_0]$, where $\Delta\alpha \ll 1$.

Hence,

$$s=L_r\Delta\alpha\cos\alpha_0 \quad \text{and} \quad \Delta\alpha=\frac{s}{L_r\cos\alpha_0}. \quad (4.79e)$$

In addition, $(\cos\alpha-\cos\alpha_0)=-\Delta\alpha\sin\alpha_0$ and $\sin^2(\alpha-\alpha_0)=(\Delta\alpha)^2$. Substituting these expressions into (4.79d), we obtain the following formula for the untwist angle θ at a small longitudinal stretch of the coiled ribbon strip: $\theta=\frac{s\sin\alpha_0}{R_0\cos\alpha_0}=\frac{s}{R_0}\tan\alpha_0$, because the limit of the second member in brackets of (4.79d) has a very small value, and $\lim_{\Delta\alpha\to 0}\frac{1-\nu_p}{R_0}(\Delta\alpha)^3\sin\alpha_0^2=0$. Let us also recall that $\tan\alpha_0=\frac{S_{av}}{\pi D_{r0}}$, where S_{av} is the average helix pitch and $D_{r0}=2R_0$ is the initial diameter of the helix middle surface.

Hence, we derive a very simple formula for the transmission ratio under a small extension s of the coiled ribbon $i_\theta=\frac{\theta}{s}=-\frac{2S_{av}}{\pi D_{r0}^2}$ that corresponds entirely to the similar formula derived by Biderman (1963) with the stipulation that the radius of the helix surface does not change under small extension, that is, $R-R_0=0$ (the first stage of deformation). For double-coiled ribbon strip, this formula transforms to:

$$i_\theta=-\frac{S_{av}}{\pi D_{r0}^2}. \quad (4.80)$$

This expression corresponds well to experimental data at the beginning of the extension of the spring. However, the transformation function at a larger extension becomes a nonlinear one, corresponding to the quasi-elliptical expression (Tseytlin, 2006, p. 207):

$$[\theta^2/(2L_r/D_r)^2]+[(L+s)^2/2L_r^2]=1, \quad (4.81a)$$

where D_r is the helicoid's outside diameter; and L is the length of each coiled part.

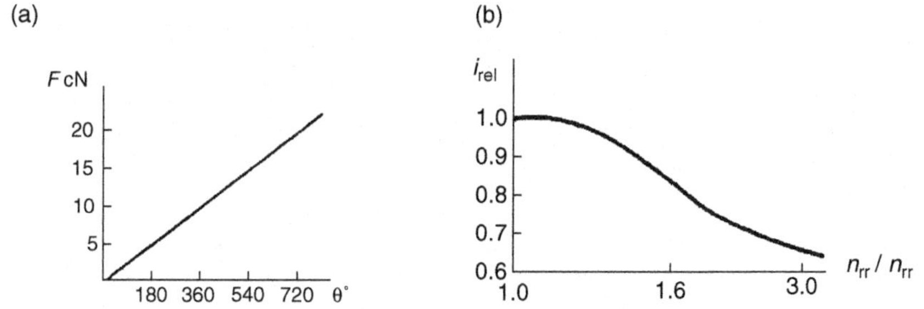

FIGURE 4.4 Coiled strip force function (a), and kinematic transmission at different ratio of coils numbers (b).

Expression (4.81a) is the canonical equation of an ellipse with semi-axes $\sqrt{2}L_r$ and $2L_r/D_r$ in the Cartesian coordinates. At the beginning of the crosspiece rotation, the mean transmission ratio (in agular degrees per micrometer) is equal to $i_{\theta m} \approx -0.0573 S_{av}/\pi D_r^2$, corresponding well to the theoretical formula (4.80) shown earlier. Figure 4.4a shows the type of relationship between the tension force F and the angle of rotation $\theta°$ for the double-coiled helicoid that is close to linear with stiffness $j_{\theta'} = 6.3\dfrac{bh^3E}{D_rL}$. The limit of the crosspiece's elastic untwist for the strip of iron-nickel-chrome-titanium (Ni41ChrTi) corresponds to the empirical expression $\theta_{\lim} = n_r\ (0.15–0.33)$.

The transmission ratio of the double-coiled helicoid becomes more linear, with an increase in the number n_r of coils. If the number of coils in the part with the right-hand helix n_{rr} and in the part with the left-hand helix n_{rl} is different, then the mean transmission ratio decreases as

$$i_{\theta m} = -4\sqrt{n_{rr} n_{rl}}\ S_0/[(n_{rr} + n_{rl})\pi\ D_r^2]. \tag{4.81b}$$

Figure 4.4b shows the relative transmission ratio of double-coiled strip with the different ratio of the coils on the left- and right-hand sides, n_{rr}/n_{rl}. The largest transmission ratio has a double-coiled strip with $n_{rr} = n_{rl}$. Small deflections from 1 in the ratio n_{rr}/n_{rl} changes the coiled strip's transmission ratio insignificantly. For example, if $n_{rr}/n_{rl} = 0.8$, the transmission ratio decreases only by 0.8%. This frees us from the necessity to keep the exact equality between the number of coils with the left-handed and right-handed helices. The combined coiled helicoid consists of a strip coiled in one direction, and a flat strip attached to its end. The transmission ratio for this system can be expressed as

$$i_{\theta m} \approx -(2S_{om}/\pi D_r^2)/(1 + j_{k2}/j_{k1}), \tag{4.82}$$

where $j_{k1} = \dfrac{Ebh^3 \cos\alpha}{12\pi(1+v_p)Rn_r} \dfrac{1}{1+(1-2v_p)\cos^2\alpha}$ at $K_\xi = 0$ and $j_{k2} = Gbh^3/3L$ are the torsion stiffness coefficients for the coiled and flat strips, respectively. If the ratio j_{k2} / j_{k1} decreases, the absolute value of the transmission ratio increases. Its upper limit value, in this case, is equal to the transmission ratio of the double-coiled strip.

4.6 TRANSMISSION RELATIONS

Kinematics and stiffness. Functional formula.

The elastic kinematic system (ELKS) functional formula at the input displacement s may be presented in the following form:

$$F(s) \supset F_k(s) + F_j(Q) + \delta F_j(Q) + \delta F_{kc},$$

where $F(s)$ is the exit displacement of the ELKS, $F_k(s)$ is the exit displacement of the equivalent kinematic system with rigid links or, in the ELKS straight chain without resistance to motion from sequential elastic elements, $F_j(Q)$ and $\delta F_j(Q)$ are the linear and nonlinear feedback of ELKS compliances, and δF_{kc} represents the environmental conditions that influence the ELKS kinematics. The feedback compliance shows the influence of the load opposite to motion direction, applied to the driven link of the ELKS on its kinematics. The necessary analytical solutions for the elastic double-helicoid microstrips (Table 4.8) were derived only in the 1960s (Tseytlin, 1965) on the basis of the developed methods of the theory of elasticity.

One can see the three typical transmission segments known from experimental data of the pretwisted (helicoidal) strip under extension s_1 (Fig. 4.5a) and rotation $\theta°$ (Fig. 4.5a,b): I is the initial nonlinear segment with a variable average $i_{\theta av}$ transmission ratio; II is the linear one, with the constant (at deflection within 1–2%) transmission ratio; and III is the nonlinear segment with a variable transmission ratio where the strip's material can be in the elastic-plastic state (dashed line). It is natural that calculation relationships for linear segment II are based on the linear theory of elastic helicoid deformation. The transmission ratio $i_{\theta II}$ in °/μm and driving stiffness $j_{\theta II}$ in cN/o of the sensor are equal to: $i_{\theta II} = 57.3\ \theta_\gamma/s_\gamma$ and $j_{\theta II} = F/[57.3\ (\theta_\gamma L)]$.

The values of θ_γ and s_γ correspond to the untwist angle and extension of the strip per unit length, respectively. Their functional magnitudes are found from the solution of the simultaneous equations and presented in variant 1 (Table 4.8). However, the expressions of variant 1 in Table 4.8 do not show the influence of the flat crosspiece 4 (Fig. 4.2b) and straight ends 3 and 5, which the double-helicoid strip sensor usually has.

TABLE 4.8 Transmission Relationships in Studied Helicoids

Name and type of transmission	Schematic	Calculation relationships	Notes
Helicoidal strip	Fig. 4.1	1. $\mathbf{F_k}$**(s)+** $\mathbf{F_j}$ **(Q)** (Tseytlin, 1965, 2006) $i_\theta = \frac{Ek_p}{4G\lambda}\frac{0.0286}{1+\nu_{if}\chi_{kp}}$; $\chi_{kp} = \frac{b_1^2 k_{p0}^2}{\lambda}$, (a) $\nu_{iel} = 0.345$; (b) $\nu_{irc} = 0.32–0.35$, (c) $\nu_{ipf} \approx 0.39$ $j_\theta = 69.6\frac{GA\lambda}{Lk_{p0}}\left(1+\nu_{jf}\chi_{kp}\right)$; $\nu_{jel,ow} = 0.179$; $\nu_{jrc} = 0.14$; $j_{\theta pf}^{-1} = 2^{-1}\frac{j_{\theta c}+j_{\theta rc}}{j_{\theta c}\, j_{\theta rc}}$; $\nu_{if} = 0.41$ $j_{\pi\pi} = GT_0\left(1+\nu_{if}\chi_{kp}\right)$, $j_{\theta c} = \frac{69.6G\lambda A_c}{Lk_p}\{1+\chi_{kp}[0.5-0.3(1+\mu_b)]\}$, $A_c = (b-b_d)h$, $\mu_b = b_d/b_1 < 1$; $L_c/L = 0.5$ See also (4.83), (4.84).	$\frac{L_{str}}{2L} \approx 0$ (a) $\lambda \geq 0.4\times10^{-2}$ (b) $\lambda < 0.4\times10^{-2}$ $j_{\tau k} = \frac{EAb_1^2 k_p}{4}$; $j_{kk} = EA$
		2. $\mathbf{F_k}$ **(s) +** $\mathbf{F_j}$ **(Q) +** $\boldsymbol{\delta}$**F (Q)** see ((4.37),(4.37b),(4.39))	$\nu_{if} = 0.32 - 0.39$
Coiled ribbon helicoid	Fig. 4.3a	1. $\mathbf{F_k}$ **(s)** $i_{\theta av} \approx 0.0573\frac{S_{av}}{\pi D_r^2}$, $j_\theta = 6.3\frac{bh^3E}{D_r L}$	(see Tseytlin, 2006, p. 207)
		1. $\mathbf{F_k}$**(s) +** $\mathbf{F_j}$ **(Q)** After Biderman (1963) See (4.79e)	

Index c in the punched strip parameters corresponds to the segments with cut, including their length L_c;
L_{str} is the length of straight segments.
Source: After Tseytlin (2006), reprinted with permission.

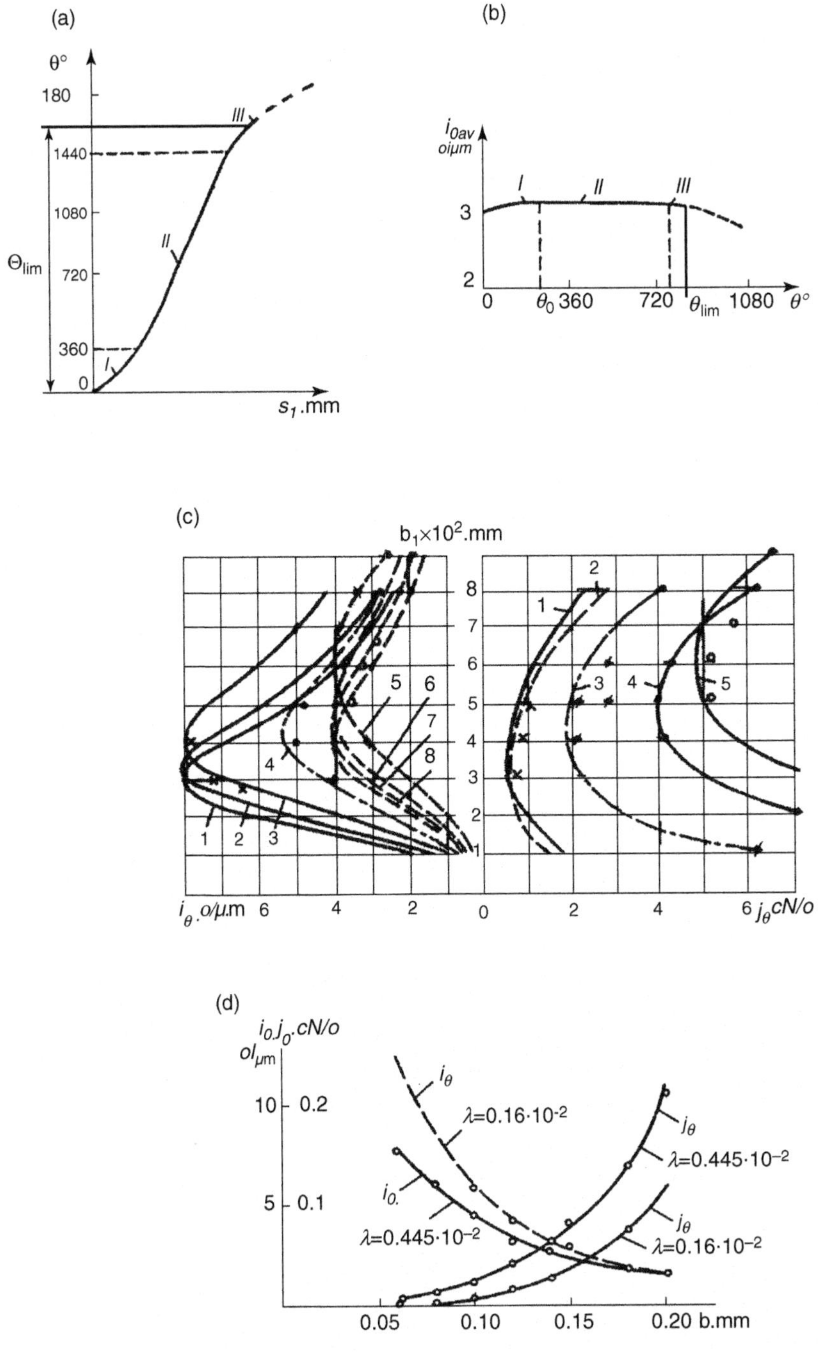

FIGURE 4.5 Double pretwisted strip's typical transmission and stiffness features. *(After Tseytlin, 2006, reprinted with permission.)*

Straight segments. With the accounting of this straight segment extension, the expressions for $i_{\theta\text{II}}$, in °/μm, and j_{II} are as follows (Tseytlin, 1965):

$$i_{\theta\text{II}} = 0.0572\left(E\kappa_p L/4G\lambda\right)\left[L_{\text{str}}(1+\nu_{\text{jf}}\chi_{\text{kp}})+2(1+\nu_{\text{if}}\chi_{\text{kp}})L\right]^{-1}, \quad (4.83)$$

$$j_{\text{II}} = EA(1+\nu_{\text{jf}}\chi_{\text{kp}})\left[(1+\nu_{\text{jf}}\chi_{\text{kp}})L_{\text{str}}+2(1+\nu_{\text{if}}\chi_{\text{kp}})L\right]^{-1}, \quad (4.84)$$

where the longitudinal stiffness $j_{\text{II}} = j_{\theta\text{II}}\, i_{\theta\text{II}}$ in cN/μm, $L_{\text{str}} = L_{\text{crp}} + 2L_{\text{send}}$, and $\chi_{\text{kp}} = b_1^2\kappa_p^2/\lambda$.

The influence of the length of crosspiece L_{crp} and straight ends L_{send} in the real device's transmission and stiffness may be up to 10%. The substitution of $L_{\text{str}} = 0$ into (4.83) converts it into the corresponding expression for i_θ shown in Table 4.8.

Comparison. We can show that for the helicoidal strips with an elliptical cross-section, where $\nu_{\text{if}} = 0.345$, the expression for the transmission ratio in quasi-linear transformation part may have another more digestible form, because

$$i_\theta = \frac{Ek_p}{4G\lambda}\frac{0.0286}{1+\nu_{\text{if}}\chi_{kp}} \approx 0.045\frac{E}{GS_0}\frac{1}{(h/b)^2+3.4(b/S_0)^2}, \quad (4.84a)$$

where S_0 is the pitch of the helix.

The charts in Figures 4.5c and 4.6, built on the basis of the relationships presented in Table 4.8, help to explain the contradictions of the conclusions obtained by Perry (1888) and other researchers with the limited experimental data (see Tseytlin, 2012). Figure 4.5c presents the calculated and experimental transmission ratio $i_\theta(b_1)$ and stiffness $j_\theta(b_1)$ in linear segment of transmission for pretwisted strips with parameters 1 – $h_1 = 0.002$ mm, $k_p = 2.1$ mm^{-1}; 2 – $h_1 = 0.002$ mm, $k_p = \pi$ mm^{-1}, × – experimental data; 3 – $h_1 = 0.002$, $k_p = 2.44$, ◆ – experimental data for j_θ; 4 – $h_1 = 0.004$, $k_p = \pi$, • – experimental data for i_θ and j_θ; 5 – $h_1 = 0.004$, $k_p = 2.1$, o – experimental data for j_θ and ◆ for i_θ; 6 – $h_1 = 0.004$, $k_p = \pi$, o – experimental data for i_θ; 7 – $h_1 = 0.004$, $k_p = 3.83$; 8 – $h_1 = 0.004$, $k_p = 4.94$. Figure 4.5d shows calculated transmission ratio $i_\theta(b)$ and stiffness $j_\theta(b)$ functions with specified parameter λ.

It is clear that most of the relationships in Figure 4.6 have a relative maximum or minimum. This prevented an unambiguous conclusion about the influence of certain parameters (except the thickness) upon the transmission ratio, or the longitudinal stiffness at the limited experimental data.

The form of the microstrip's cross-section has an influence on the transmission ratio, and especially on the longitudinal stiffness of the helicoidal microstrip. This influence on i_θ and j_θ for the solid microstrips with identical dimensions and material is not more than 10%, but the longitudinal stiffness of the perforated strips differs significantly.

The calculating relationships shown in Table 4.8 correspond well to the experimental data. The difference between calculated and experimental data obtained by many researchers and the author of this book on a wide variety of

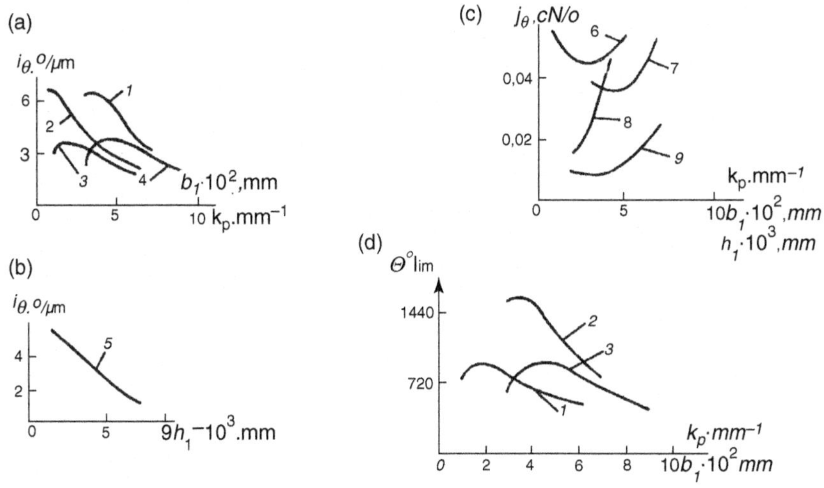

FIGURE 4.6 Helicoidal strip parameters optimization options. *(After Tseytlin, 2006, reprinted with permission.)*

helicoidal strips is not more than 10% for transmission ratio i_θ and is less than 15% for longitudinal stiffness j_θ.

Ratajczyk (1968) performed special research on the accuracy of the relationships proposed by the author of this book for the pretwisted double-helicoidal strips. The conclusions in all cases were positive (e.g., Tseytlin (2006), Table 4.9).

A large role in the original mass production of the measurement sensors with pretwisted strips – "mikrokators," in the twentieth century was played by the Swedish company Johansson Gage Co. Photoelectric transducers with pretwisted strips for sorting automata were produced (in the 1970s) at the Leningrad Instrumental Plant, in Russia. Spring measuring heads with pretwisted strips and increased measuring range by the elastic compensator at decreased amplification rate, and with multirevolution pretwisted strip structure at the double end pointer (Patent US 3,337,963, 1967; GB 1,104,460, 1968) were also developed in that Plant. Despite the decline in the production of such systems in recent years, due to the development of electronic systems with wide measurement range (inductive, capacitive), and AFM with micronanocantilevers in quasi-static deformation, natural vibration and vibration at higher modes, flexible helicoidal strips continue to be of interest and use. For example, the present author recently designed a model of DNA molecules using the same helicoidal structure (see Chapters 1, 6, 7).

Untwist limit. The limits of the elastic untwist for the helicoidal strip's crosspiece can be estimated on the basis of the ultimate stress state theory with the maximum shear stress, as follows:

$$\theta_{\lim} = \tau_e/[\pi(\delta\sigma_1 - \delta\sigma_3)], \tag{4.85}$$

TABLE 4.9 The Limiting and Equivalent Ultimate Crosspiece Untwist

				Working regime				
Parameters of the strip				Quasi-static		Cyclic, $N_c = 0.5 \times 10^7$		
$2b_1$	$2h_1$	L		θ°_{lim}		θ°_0	θ°_d	θ°_{eq}
mm			n_v	Calculated with (4.87)	Experimental			
0.1	0.006	18	8	1343	1440	270	1080	1205
0.12	0.008	18	9	971	1080	270	830	880
0.20	0.016	18	8	479	486	180	360	405

Source: After Tseytlin (2006), reprinted with permission.

where $\tau_e = \sigma_e/(1.73 - 2)$ is the shear stress that corresponds to the yield stress of material; $\delta\sigma_1$ and $\delta\sigma_3$ are the corresponding additions to the largest and the smallest principal stresses in the stress tensor, with the extension of the helicoidal strip and rotation of its crosspiece to the unit angle of 1 rad. In accordance with (4.56) for the critical point of the strip's cross-section, we have

$$(\delta\sigma_1 - \delta\sigma_3) = (4G\lambda/\kappa_p L)(1 + 0.379\,\chi_{kp}) \tag{4.86}$$

because $\sigma_0^{33} = F_Q/A_c$ and force $F_Q = 4G\lambda A_c\, C_\sigma\theta/(k_p L)$.

By substituting (4.86) into (4.85), we obtain the relationship for the limit angle of the strip's untwist at $\theta_{\lim} < Lk_{p0}/4\pi$ as

$$\theta_{\lim} = \frac{\sigma_e \kappa_p L}{8\pi G\lambda(1 + 0.379\chi_{kp})}. \tag{4.87}$$

The coefficient 0.379 insignificantly differs from the corresponding coefficient in the expression for the transmission ratio $\nu_{if} = 0.345$. Hence, we can approximately express the limit of the untwist angle through the strip's transmission ratio:

$$\theta_{\lim} \approx \frac{\sigma_e L i_\theta}{2 \times 0.0286\pi E}. \tag{4.88}$$

This expression is effective for the helicoidal strips with elliptic and rectangular cross-sections.

Optimization. The optimal parameters of the helicoidal strips can be identified from the relationships shown in Table 4.8. From this, we can find the partial effective relationships of the transmission ratio, and the longitudinal force with thickness $2h_1$, width $2b_1$, helix twist parameter κ_p of the strip, and elastic E and shear G moduli of material. Functions $i_\theta = f_1(E, G)$ and $j_\theta = f_2(E, G)$ are assumed to be linear.

We defined the position of the extreme points (Fig. 4.6a,c) on curves $i_\theta\ (b_1)$, $i_\theta(\kappa_p)$, $j_\theta\ (b_1)$, and $j_\theta(\kappa_p)$ through the partial differentiation of (4.83) and (4.84) at $L_{str} = 0$.

Figure 4.6 shows calculated relationships of (a) $i_\theta\ (b_1,\ k_p)$; (b) $i_\theta\ (h_1)$; (c) $j_\theta\ (b_1,\ k_p)$ for double helicoidal strips with parameters $1 - h_1 = 0.002$ mm, $k_p = \pi\ \text{mm}^{-1}$; $2 - h_1 = 0.002$, $b_1 = 0.06$ mm; $3 - h_1 = 0.004$, $b_1 = 0.06$; $4 - h_1 = 0.004$, $k_p = \pi$; $5 - b_1 = 0.05$, $k_p = \pi$; $6 - h_1 = 0.004$, $b_1 = 0.06$; $7 - h_1 = 0.004$, $k_p = \pi$; $8 - b_1 = 0.06$, $k_p = \pi$; $9 - h_1 = 0.0022$, $k_p = \pi$; (d) $\theta^{\circ}_{\lim}$ as a function of pretwist parameter k_p for strip 1 with $b = 0.12$ mm, $h = 0.008$ mm; as a function of the width $b_1 = 1/2\ b$ of the strip's cross-section for strip 2 with $k_p = \pi$, $h = 0.004$; and for strip 3 with $k_p = \pi$, $h = 0.008$.

For the strips with elliptical cross-sections, the relationship for the transmission ratio of width $2b_1$ has its maximum at $b_1 = 1.31\sqrt{h_1/\kappa_p}$, the same relationship of the helix twist parameter κ_p has its maximum at $\kappa_p = 1.70h_1/b_1^2$. The relationship for the longitudinal stiffness (Fig. 4.6c) of the strip's width has a minimum at $b_1 = 1.17\sqrt{h_1/\kappa_p}$ and a minimum at the helix parameter $= \kappa_p = 2.36h_1/b_1^2$. The relationships for the transmission ratio and the longitudinal stiffness of the microstrip's thickness do not have extreme points that follow from the partial derivatives of these relationships (Fig. 4.6b). Hence, with the change in the microstrip's geometrical and elastic parameters, one can choose, during the structural design sensor or model, the microstrip that effectively provides the specified transmission ratio and/or longitudinal stiffness. Moreover, the same transmission ratio or stiffness can be obtained with different combinations of parameters b_1, h_1, and k_p. If the strip's cross-section thickness ultimately diminishes to zero, the transmission ratio theoretically approaches infinity, but the longitudinal stiffness approaches zero, that excludes transformation options.

As it follows from (4.88) for $\theta_{\lim}$, the number of the crosspiece's revolutions in the elastic range of the strip's deformation is proportional to the strip's length L, its transmission ratio i_θ, and yield stress σ_e of the material, but inversely proportional to Young's modulus of elasticity, E. Graphs of $\theta^{\circ}_{\lim}$ in angular degrees that depend on the width, thickness, and parameter of the strip's pretwist κ_p are shown in Figure 4.6d, for microstrips of 4-Tin 3-Zinc bronze with $L = 18$ mm. In this case, curve 3 corresponds to the strips with $\kappa_p = \pi\ \text{mm}^{-1}$; $h = 0.008$ mm; curve 2 corresponds to the strips with $\kappa_p = \pi$ mm^{-1}; $h = 0.004$ mm, and curve 1 corresponds to the strips with $b = 0.12$ mm;

h = 0.008 mm. The position of the maximum-value points on these curves is found through the partial derivatives of (4.87):

$$\partial\theta_{\lim}/\partial h_1 = -\frac{\sigma_e h_1 k_p L/b_1^2}{4\pi G[(h_1^2/b_1^2)+0.379b_1^2k_p^2]^2};$$
$$\partial\theta_{\lim}/\partial b_1 = -\frac{\sigma_e k_p L\left(-\dfrac{2h_1^2}{b_1^3}+170b_1k_p^2\right)}{8\pi G(h_1^2/b_1^2+0.379b_1^2k_p^2)^2};\qquad(4.89)$$
$$\partial\theta_{\lim}/\partial\kappa_p = \frac{\sigma_e L(h_1^2/b_1^2-0.379b_1^2\kappa_p^2)}{8\pi\, G(h_1^2/b_1^2+0.379b_1^2\kappa_p^2)^2}.$$

By equalizing these expressions to zero, we can find that the relationship between $\theta_{\lim}$ and the cross-section width that has a maximum at $b_1 = 1.28\sqrt{h_1/\kappa_p}$, and the relationship between $\theta_{\lim}$ and pretwist parameter κ_p has a maximum at $\kappa_p = 1.623h_1/b_1^2$. The relationship between $\theta_{\lim}$ and the cross-section thickness $2h_1$ does not have extreme points. The limiting number of the working elastic untwist turns $\theta_{\lim}$ of the microstrip's crosspiece does not continuously increase with the increase of the number of initial turns on the strip's twisted segment, as one would suppose. Conversely, the limiting number of elastic turns $\theta_{\lim}$ and the transmission ratio start to decrease after the number of initial turns exceeds a value corresponding to the extreme point. The optimal value of the pretwisted turns on the strip's segment is equal to:

$$n_{\text{v.opt}} = \frac{1.8}{2\pi}\frac{h_1}{b_1^2}L+3. \qquad (4.90)$$

In this case, of course, the calculated optimal number $n_{\text{v.opt}}$ should be larger than $\theta_{\lim}/(2\pi)$, where $\theta_{\lim}$ is expressed in rad. The increase of the pretwisted turns on the strip's segments can increase the linear segment II (see Fig. 4.5a,b) of the transmission ratio's function, but this can slightly simultaneously increase the initial nonlinear segment I. As a result, part of segment II can fall in the elastic-plastic zone of the strip's deformation, because the large increase of pretwisted turns n_v will decrease $\theta_{\lim.}$

To diminish the driving force rate at the specified transmission ratio, one should use relatively thinner strips, that is, strips with smaller λ at smaller absolute thickness h. These strips have a smaller longitudinal stiffness. For example, the transmission ratio is approximately equal to the strips with cross-sections 0.006 × 0.122 mm and 0.008 × 0.12 mm. The longitudinal stiffness for the latter is 1.6 times larger.

Nonlinear transmission. Charts with initial I and quasi-linear II regions for calculated average $i_{\theta\text{av}}$ transmission ratios, in accordance with nonlinear functions (4.37) and (4.38), for a set of the helicoidal strips with dimensions of thickness × width × helix parameter (k_p) and length L = 18 mm of each pretwisted segment, are shown in Figure 4.7a. Dash-dot lines correspond to function (4.40), and dashed lines correspond to the strip's elastic-plastic stress state in the region III of transformation function.

Let us discuss the influence of the driven element's mass clamped to the helicoidal strip's crosspiece on its transmission function. The load F_N (Fig. 4.7b) from the mass clamped to the crosspiece at the horizontal strip's position is perpendicular to the strip's axis, and causes sag s_{sa}, as in any thin string. This sag can be decreased by the longitudinal preliminary stretch of the strip. The pretwisted strip will not have a constant transmission ratio with linear characteristics if the sag s_{sa} is large. This can be explained with the relationship that follows from the schematic in Figure 4.7b:

$$F_N = \frac{2Fs_{sa}}{(L+s/2)} + j(L+s/2)\frac{s_{sa}^3}{(L+s/2)^3}; \quad j = F/s_f = j_\theta i_\theta, \qquad (4.91)$$

where $s = s_0 + s_p$ is the change of the distance between the strip's ends on the straight line connecting them; s_f is the actual elongation of the strip. The relative additional strip's deformation in accordance with the triangle OBC is equal to:

$$k_\partial = \frac{s_{sa}^2}{2(L+s/2)^2}, \quad \text{but} \quad k_\partial = \frac{s_f - s}{2(L+s/2)}.$$

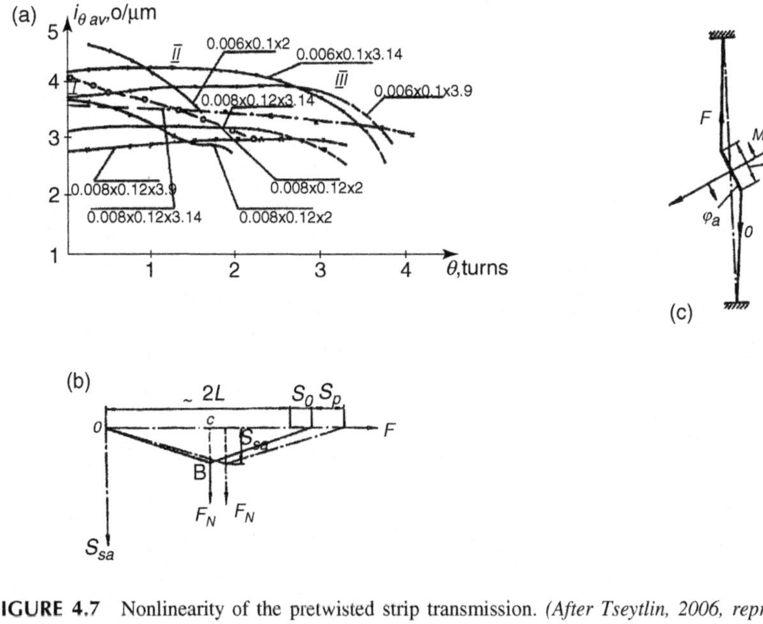

FIGURE 4.7 Nonlinearity of the pretwisted strip transmission. *(After Tseytlin, 2006, reprinted with permission.)*

Thus, we have

$$F_N = 2 j_\theta i_\theta s\sqrt{2k_\partial} + j_\theta i_\theta (L + s/2)(2k_\partial)^{3/2}. \tag{4.92}$$

The preliminary longitudinal extension (s_0) of the distance between the strip's ends should satisfy the condition

$$2L(k_{\partial 0} - k_\partial) = (s_f - s_p) \le \{\delta s_n\}, \tag{4.93}$$

where $\{\delta s_n\}$ is the permissible additional nonlinearity of the transmission caused by the strip's sag; $k_{\partial 0} = \dfrac{s_f - s_0}{2(L + s_0/2)}$; and s_p is the working relative displacement of the strip's ends. The limit of the allowable transverse load on the strip's crosspiece can be found from (4.92), with condition

$$2Lk_\partial + s_0 + s_p < \theta_{\lim}/i_\theta. \tag{4.94}$$

The torsional unbalance ΔM_∂ caused by the unbalanced load (ΔF_g) clamped to the crosspiece at the distance r_{un} yields nonlinear deflection of the transmission function in the horizontal strip, as a result of the acting additional torque proportional to the sine of the crosspiece's rotation angle θ. This leads to the additional rotation angle $\delta\theta$ of the crosspiece that equals

$$\delta\theta = \pm \Delta M_\partial L / 2 j_{\tau\tau}, \tag{4.95a}$$

where

$$\Delta M_\partial = \Delta F_g \, r_{\text{un}} \sin\theta. \tag{4.95b}$$

This component of the transmission nonlinearity is practically independent from the preliminary extension of the strip. The torque generated on the crosspiece of the microstrip is equal to: $M = j_{\tau k}\, s_\gamma - i_{\tau\tau}\theta_\gamma$.

The well-balanced mechanism has the torque on a crosspiece that is close to zero. An additional error in the transmission can be caused by the longitudinal unbalanced load M_∂ (Fig. 4.7c, shown as *Ma* and φa) that is clamped to the crosspiece. We can show this by the following approximation. The strip's equilibrium yields the equality of the moment from the pointer's longitudinal unbalance, and the moment from the strip's tension that are acting on the crosspiece with length L_{n}. This segment of the strip can be considered rigid, as a result of clamping a rigid element to it, or at a length less than $6b_1$. The equilibrium equation is $M_\partial = j_\theta \theta L_{\text{n}} \sin\varphi_\partial$, from which $\sin\varphi_\partial \approx \varphi_\partial = M_\partial/(j_\theta \theta L_{\text{n}})$. This rotation of the crosspiece causes additional deformation of the strip's pretwisted segments, whose value is equal to $\Delta_\partial = L_{\text{n}}(1 - \cos\varphi_\partial) \approx 0.5 L_{\text{n}} \phi_\partial^2$, or after substituting the expression for φ_∂ through the moment of unbalance, we get

$$\Delta_\partial = 0.5 L_n^{-1} M_\partial^2 / (j_\theta^2 \theta^2). \tag{4.95c}$$

4.7 FATIGUE OPTIONS AND STATISTICAL FLUCTUATIONS

In real conditions, the angle of the crosspiece's rotation consists of the constant component θ_0 that corresponds to the preliminary extension of the strip, and the variable component θ_d that corresponds to the working angle of the crosspiece rotation. In accordance with the theory of an asymmetric cycle loading, the influence of the constant and variable components is not the same. As shown above, the pretwisted strip has a complex stress state with dominating extension ($\sigma_Q = \sigma_Q^{33}$) and torsion ($\tau_Q = \sigma_Q^{13}$) stresses. For the fatigue limit calculation, we will use the following equation:

$$\{\tau_{en}\}^2 = (A_\tau/2) + (B_\tau/2) + \frac{1}{2}\sqrt{(B_\tau - A_\tau)^2 + C_\tau^2}, \tag{4.96}$$

where $\{\tau_{en}\}$ is the endurance limit for the material's torsion with the symmetric loading cycle; $A_\tau = 0.125(\sigma_{Q1}^2 - \sigma_{Q1}\sigma_{Q2})$; $B_\tau = 0.125(\tau_{Q1}^2 - \tau_{Q1}\tau_{Q2})$;

$C_\tau = 0.25(2\sigma_{Q1}\tau_{Q1} - \tau_{Q1}\tau_{Q2} - \sigma_{Q1}\tau_{Q2})$, σ_{Q1} and τ_{Q1} are the largest working stresses of extension and rotation, respectively; σ_{Q2} and τ_{Q2} are the corresponding smallest stresses. For the helicoidal (pretwisted) strip with an elliptical section, we approximately have:

$$\sigma_{Q1} = (1 + 0.308\chi_{kp})4G\lambda\theta_1/(\kappa_p L) = U_\sigma\theta_1; \sigma_{Q2} = U_\sigma\theta_2;$$
$$\tau_{Q1} = 2Gh_1\theta_1/L = U_\tau\theta_1; \tau_{Q2} = U_\tau\theta_2, \text{and } \theta_1 = \theta_0 + \theta_d; \theta_2 = \theta_0.$$

Hence,

$$A_\tau = 0.125U_\sigma^2(\theta_1^2 - \theta_1\theta_2) = 0.125U_\sigma^2(\theta_d^2 + \theta_0\theta_d); \tag{4.97}$$

$$B_\tau = 0.125U_\tau^2(\theta_1^2 - \theta_1\theta_2) = 0.125U_\tau^2(\theta_d^2 + \theta_0\theta_d); \tag{4.98}$$

$$C_\tau = 0.5U_\sigma U_\tau(\theta_1^2 - \theta_1\theta_2) = 0.5U_\sigma U_\tau(\theta_d^2 + \theta_0\theta_d). \tag{4.99}$$

Substituting the expressions of these coefficients into (4.96), we finally obtain

$$\{\tau_{en}\}^2 = 0.063(\theta_d^2 + \theta_0\theta_d)(U_\sigma^2 + U_\tau^2 + \sqrt{U_\sigma^4 + U_\tau^4}). \tag{4.100}$$

Angle $\theta_{eq} = \sqrt{\theta_d^2 + \theta_0\theta_d}$ may be called the equivalent fatigue angle of the strip's crosspiece untwist. In the real sensor, the condition $\theta_{lim} > \theta_{eq}$ is required. The torsion stress endurance $\{\tau_{en}\}$ depends on the used material and the technology of the strip's treatment. Any torsion endurance value $\{\tau_{en}\}$ for the pretwisted strip from bronze is connected to the duration of its finite lifetime. This process is quasi-accidental because of the random relationships between the endurance stress (σ_{en} or τ_{en}) and the number N_c of cycles, random values of angles θ_d and θ_0, and may be classified in accordance with the "relaxation model." The frequency functions φ (τ_{en}, N_c) of the conditional distribution for the fatigue curve section at N_c = constant are usually considered as Weibull's distribution. The value of the equivalent stress τ_{eq} in the pretwisted strip at the

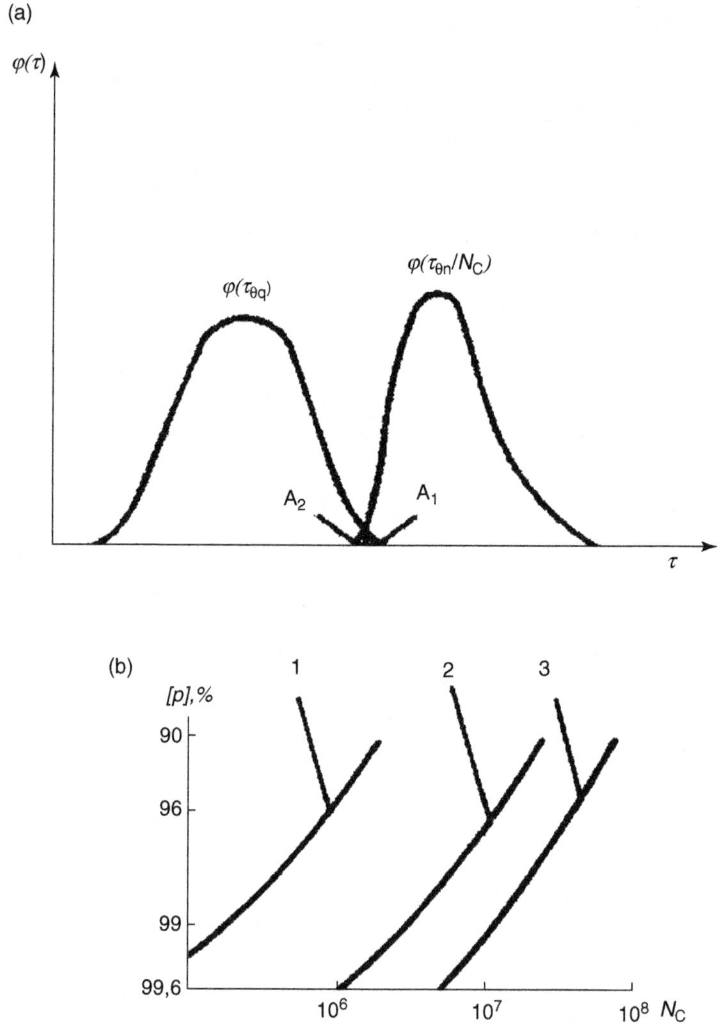

FIGURE 4.8 Helicoid's fatigue probability (a), and strain-stress diagram for its blank (b). *(After Tseytlin, 2006, reprinted with permission.)*

cyclic loading is a random one, because it depends on the variable random extension stress τ_u in each cycle, and the random stress value τ_0 from the preliminary tension in the sensor's strip. Therefore, with an equation analogous to the equivalent fatigue angle, we assume that

$$\tau_{eq} = \sqrt{\tau_u^2 + \tau_u \tau_0}. \tag{4.101}$$

Values τ_u and τ_0 depend on the adjustment of the structure, and its working regime. If the frequency function of the equivalent stress distribution equals $\varphi(\tau_{eq})$, then the probability of an absence of the strip's break $P(N_c)$ at the specified number N_c of loading cycles is approximately equal to:

$$P(N_c) = 1 - A_1 A_2, \tag{4.102}$$

where A_1 and A_2 are the hatched areas shown in Figure 4.8a.

Table 4.9 shows the results of the elastic limit's estimation for strips with different parameters.

We can find the values for the limit of the strip's elastic deformation close to experimental data, if we evaluate the reach of plasticity in a whole cross-section of the strip under extension by the approximate formula

$$\theta^{\circ}_{\lim} \approx \{\sigma_e\} A_c / j_\theta. \tag{4.103}$$

For responsible technical calculations of the elastic limit in the strip's work, one should certainly use expressions (4.87) and (4.100).

The helicoidal strip in the sensor always works in the asymmetric cyclic regime, therefore the author of this book and other researchers have conducted a series of experiments with helicoidal strips installed in the special driving cycling fixtures. Figure 4.8b shows the endurance probability $[p]$ experimental data for the microstrip of beryl bronze BrB2 at the stress level 1 with $\sigma_{en} = 450$ MPa, 2 with $\sigma_{en} = 290$ MPa, 3 with $\sigma_{en} = 245$ MPa. Moreover, the author of the book has found (Tseytlin, 2006) from the mass production practice, that the pretwisted strips have a break after 60×10^3 cycles, or even less, if the equivalent number θ_{eq} of the crosspiece turns exceeds the limiting calculated number $\theta_{\lim}$ for the strip's crosspiece rotation. It was found, on this basis, that the torsion endurance limit for the strip from zinc-bronze is equal to $\tau_{en} = 147$–176 MPa, and for beryl-bronze $\tau_{en} = 167$–245.

REFERENCES

Ayrton, W.E., Perry, J., 1884. The new form of spring for electric and other measuring. Proc. Roy. Soc. A 36, 297–319.

Biderman, V.L., 1963. Helical spring of rectangular cross-section as a cylindrical shell. Izv. AN USSR – Mekhanika i Mashinostroenie N4, 67–75 (in Russian).

Biderman, V.L., Shitikov, V.N., 1970. Tension and torsion on cylindrical ribbon spring in the presence of large displacements. Izv. AN SSSR – Mekhanika Tverdogo Tela 5 (1), 137–141, (121–125 in Mechanics of solids, English translation).

Boal, D., 2002. Mechanics of the Cell. Cambridge University Press, London, UK.

Cumpson, P.J., Hedley, J., Zhdan, P., 2003. Accurate force measurement in the atomic force microscope: a microfabricated array of reference springs for easy cantilever calibration. Nanotechnology 14, 918–924.

Filon, L.N.G., 1900. On the resistance to torsion of certain forms of shafting. Phil. Trans. Roy. Soc. Lond. A 193, 309–352.

McLachlan, N.W., 1964. Theory and Application of Mathieu Functions. Dover, New York, NY.
Perry, J., 1888. Twisted strips. Proc. Phys. Soc. Lond. 10, 343–349.
Quinn, T.J., Speake, C.C., Davis, R.S., 1997. Novel torsion balance for the measurement of the Newtonian gravitational constant. Metrologia 34, 245–249.
Quinn, T.J., Speake, C.C., Davis, R.S., 1992. Material problems. Phil. Mag. A 65, 261–276.
Ratajczyk, E., 1968. Výpočět převodů mikrokátorů. Merova Technika 7 (N6), 81–86 (in Czech).
Shorr, B.F., 1968. Naturally twisted beams. In: Birger, I.A., Panovko, Y.G. (Eds.), Strength, Stability, Vibrations. Handbook in 3 Volumes, 1, Mashinostroenie, Moscow, pp. 440–465 (in Russian).
Tseytlin, Y.M., 1965. Computation of spring-loaded instrument mechanism. Meas. Tech. 8, 223.
Tseytlin, Y.M., 2006. Structural Synthesis in Precision Elasticity. Springer, New York, NY.
Tseytlin, Y.M., 2012. Flexible helicoids, atomic force microscopy (AFM) cantilevers in high mode vibration, and concave notch hinges in precision measurements and research. Micromachines 3, 480.

Chapter 5

Dynamics of the Elastic Systems with Helicoids

5.1 DYNAMIC FUNCTIONS

Dynamic characteristics. The linear system's transient response $\theta_e(t)$ can be used for the estimation of motion s_{out} of the driven output element by convolution, if the motion of the input element s_{in} is specified: (i) in the operational imaging form

$$S_{out}(\lambda) = \lambda\theta_e(\lambda)S_{in}(\lambda), \tag{5.1}$$

where $S_{out}(\lambda)$, $S_{in}(\lambda)$, and $\theta_e(\lambda)$ are the output, input, and the transient function's Laplace transforms with argument λ, respectively; and (ii) in the regular form of Duhamel's superposition integral,

$$S_{out}(t) = \int_0^t \dot{\theta}_e(t - t_{st})\, s_{in}(t_{st})\, dt_{st}. \tag{5.2}$$

Periodic forcing function. The bowed beam corrector is similar to the bridge mechanism. Its transmission ratio increases with the decrease of beam's sag. However, the decrease of beam's sag has some limitations. This follows with the increase of the driving force, larger compression strain, and possible bifurcation and loss of the motion stability if the driving force is a periodical one. The latter phenomenon is similar to Mathieu's model (Sec. 5.3.3, McLachlan, 1964) shown in Figure 5.1a.

The dynamic differential equation for this beam (strip) structure,

$$m\ddot{y} + C_v\dot{y} + jy = (F_0 \cos 2\omega t)y,$$

reflects an option for instability at certain parameters of the beam-strip spring rigidity j, viscose resistance C_v per unit velocity of the transverse beam's motion, circular frequency ω, and amplitude F_0 of the driving force. This equation corresponds to the standard Mathieu's equation with stable and unstable regions:

$$\ddot{y} + 2k_z\dot{y} + (\bar{a}_z - 2q_z \cos 2z)y = 0,$$

where $k_z = C_v/2\omega t$, $\bar{a}_z = j/\omega^2 m$, $z = \omega t$ is the modulation, and $2q_z = F_0/\omega^2 m$.

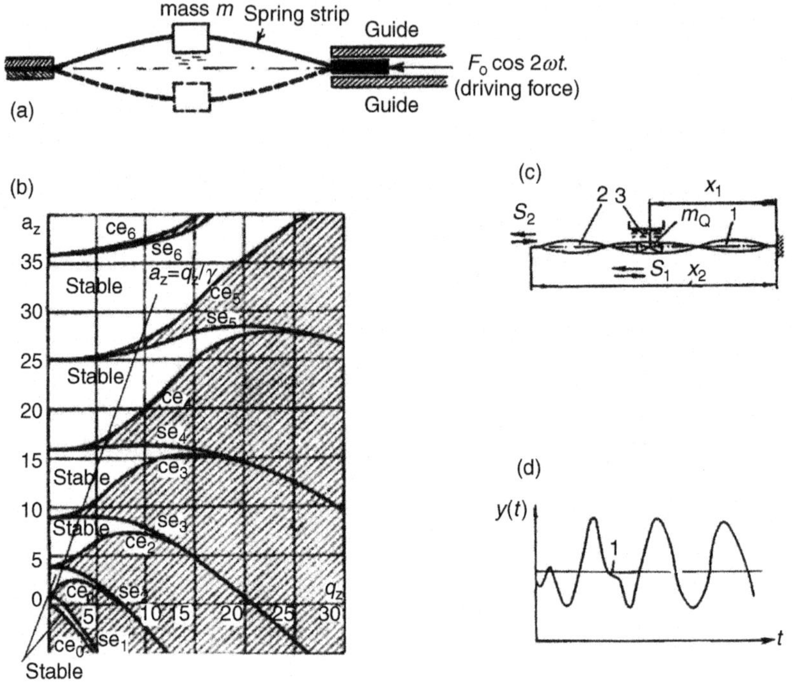

FIGURE 5.1 Models of periodical motion (a) with stable and unstable zones (b) and transients (c, d) in dynamics. *(After Tseytlin, 2006, reprinted with permission.)*

Figure 5.1b shows an example of the stability Strutt diagram (Strutt, 1932) at $C_v = 0$ with Mathieu's functions ce_n and se_n of the nth order (McLachlan, 1964). The hatched zones in Figure 5.1b represent the unstable regions for the equation with $q_z = \bar{a}_z \gamma_z$, where $\gamma_z = F_0/2j$. We assume that viscose damping k_z in the bowed beam application is very small.

5.2 HELICOIDS DYNAMICS

The most sensitive helicoidal strip mechanisms have been used in transducers for the measurement of length, temperature, force, and other parameters. The connection between the deformation and developing forces and moments in the helicoidal strips under extension is as follows:

$$F_1 = j_{kk} \frac{s_1}{L} - j_{\tau k} \frac{\theta}{L} \tag{5.3a}$$

$$F_2 = j_{kk} \frac{s_2 - s_1}{L} - j_{\tau k} \frac{\theta}{L} \tag{5.3b}$$

$$M_1 = j_{k\tau} \frac{s_1}{L} - j_{\tau\tau} \frac{\theta}{L} \tag{5.3c}$$

$$M_2 = j_{\tau k}\frac{s_2 - s_1}{L} - j_{\tau\tau}\frac{\theta}{L}, \tag{5.3d}$$

where F_1 and F_2 are the longitudinal forces; M_1 and M_2 are the torsion moments developing in segments 1 and 2 (Fig. 5.1c) of the strip, respectively; s_1 is the longitudinal displacement of mass m_Q on the crosspiece that is equal to the elastic deformation of the first strip's segment; and s_2 is the similar deformation for the second strip's segment connected to the driver. The rigidities of both segments are considered equal. The strip's mass m_s is practically small, in comparison with mass m_Q clamped on it. Therefore, we neglect the influence of mass m_s in this discussion. This mass is effectively used in research of the multivibrator's functions.

Let us express the differential equations for longitudinal-torsion displacements of mass m_Q in the following form:

$$m_Q\ddot{s}_1 = j_{kk}\frac{s_2}{L} - 2j_{kk}\frac{s_1}{L} - C_s\dot{s}_1, \tag{5.4a}$$

$$J_Q\ddot{\theta} = 2j_{\tau\tau}\frac{\theta}{L} + j_{k\tau}\frac{s_2}{L} - C_\theta\dot{\theta}, \tag{5.4b}$$

where J_Q is the moment of inertia of the movable mass; C_s is the coefficient of viscose resistance to the translational motion in damper 3; C_θ is a similar coefficient for the rotational motion in the same damper. These are the linear differential equations of the second order with separated variables.

Transient function. This separation of variables allows us to find the transient function $\theta_e(t)$ by the solution of differential Eq. (5.4b) at $s_2(t) = sU(t)$, where unit-step $U(t) = \int_{-\infty}^{+\infty}\delta(t)dt = 1$. Therefore,

$$J_Q\ddot{\theta} + \frac{2j_{\tau\tau}}{L}\theta + C_\theta\dot{\theta} = \frac{j_{k\tau}}{L}sU(t). \tag{5.5}$$

The solution of this equation can be derived by the Laplace-transform method with initial zero values $\theta(0) = 0; \dot{\theta}(0) = 0$, as follows:

$$£\theta_e(t) = \theta_e(\lambda) = \frac{j_{k\tau}}{L}s\left[\lambda(J_Q\lambda^2 + C_\theta\lambda + \frac{2j_{\tau\tau}}{L})\right]^{-1}. \tag{5.6}$$

The roots of this equation are correspondingly equal to:

$$\lambda_1 = 0;\quad \lambda_{2;3} = -[C_\theta/2J_Q] \pm \sqrt{\left(\frac{C_\theta}{2J_Q}\right)^2 - \frac{2j_{\tau\tau}}{J_QL}}.$$

The following partial cases influence the system's transfer function that depends on the relationship between the values of mass moment of inertia J_Q, specific torsion stiffness ($j_{\tau\tau}/L$) and the coefficient C_θ of the viscose resistance:

1. $(C_\theta/2J_Q)^2 > 2j_{\tau\tau}/J_Q L$ where the transfer function is with all real roots;
2. $(C_\theta/2J_Q)^2 < 2j_{\tau\tau}/J_Q L$ with two imaginary roots;
3. $(C_\theta/2J_Q)^2 = 2j_{\tau\tau}/J_Q L$ with two equal real roots.

Frequency response. When we look for the solution of a differential equation similar to Eq. (5.5), but with frequency forcing function

$$A_\Omega \ddot{y} + B_\Omega \dot{y} + C_\Omega y = F_m \cos(\Omega_1 t + \psi_\Omega), \tag{5.7a}$$

we find that this solution has two parts:

$$y(t) = y_{\text{st}}(t) + y_{\text{fr}}(t), \tag{5.7b}$$

where $y_{\text{st}}(t)$ is a stable part of the solution; $y_{\text{fr}}(t)$ is a free transient part of the solution. Chart $y(t)$ corresponding to Eq. (5.7a), and (5.7b) is shown in Figure 5.1d. One can see step 1 on the descending segment of the first semi-wave. This was noted in the experiments with the impulse sensor based on the helicoidal strip.

5.2.1 Conditional Radius

The stiffness of the double-coiled strips and their stability under dynamic loading depends on the relationship between the angle of crosspiece rotation and the torque induced on it. We (see Tseytlin, 2006; p.219-220) used a conditional radius for the comparison of different strips' performance in dynamics. This radius is equal to the ratio between the induced torque and the applied extension longitudinal force, when the crosspiece rotates on the specified angular value. The conditional radius ρ_c can be used to compare pretwisted and coiled strips in their ability to convert certain parts of the longitudinal force to the torque. The conditional radius for the pretwisted (helicoidal) strips can be calculated by the following formula:

$$\rho_c = \frac{j_{k\tau} - j_{\tau\tau} i_{\theta 1}}{j_{kk} - j_{\tau k} i_{\theta 1}}, \tag{5.8}$$

where $i_{\theta 1} = \theta_\gamma / s_\gamma$.

We obtained (5.8) by dividing the corresponding expressions of torque M and stretching force F_Q presented through the three partial rigidity parameters $j_{k\tau} = j_{\tau k}$, $j_{\tau\tau}$, and j_{kk} on the basis of Eq. (5.3a) and Eq. (5.3c), as follows:

$$M/F_Q = [j_{k\tau} s_\gamma - j_{\tau\tau}\theta_\gamma]/[j_{kk} s_\gamma - j_{\tau k}\theta_\gamma].$$

Let us also recall that for double helicoids $i_\theta = \theta_\gamma/2s_\gamma$ representing the strip's transmission ratio in sensors with right and left consequent helicoids. Samples of the helicoidal strips regularly used in precision instrumentation have ρ_c in the order of 0.001–0.004 mm. The double-coiled strips have ρ_c within the range of $\frac{M^t}{F_{pM}} \approx R = 0.05$ to 0.1 mm, as shown by their tests. This means that the pretwisted strips under extension transform the longitudinal force into torque, on the crosspiece, ten times less than the double-coiled strips under the same tension. But the pretwisted microstrips have a larger transmission ratio than the coiled strips, which is important for the precision sensors.

5.3 VIBRATION OF A THIN HELICOID

The author has developed a string system emitting a frequency signal in which the helicoidal thin strip, in addition to its usual transformation function of a spring multiplier, also functions as a string-type frequency sensor. As a result, this system, called a helicoidal multivibrator (multiplier vibrating), has new specific features.

The helicoidal multivibrator schematic is shown in Figure 5.2a. This sensor (Tseytlin, 1978) is a bifurcating one because it has two output signals at one input of strip 1 tension: (i) an amplitude-type signal through the angle of crosspiece 4 rotation, and (ii) a frequency-type signal from the helicoidal strip transverse vibration in the magnetic field 2 (permanent *N-S* or *ac* type). The bifurcating sensor with amplitude and frequency outputs automatically satisfies the well-known Abbe's principle for alignment of the compared arrangements, in order to avoid an additional uncertainty in the measuring chain.

When the electrical pulses from a self-excited oscillator are applied to the helicoidal string lying within the magnetic field, the Lorentz force acts on the string in a transverse direction, as a result of which the string moves out of equilibrium and starts to vibrate. These vibrations are maintained by the self-oscillator in the positive feedback chain, and can be measured with the standard digital or analog frequency (period) meters that are connected to the oscillator on one side, and to any intelligent calculating or servo system on the other side. The frequency output is useful for the large distance transmission of the signal, without disturbances. But the period output is faster because the frequency (period) meter has its own quartz oscillator with a natural frequency in MHz that is larger than the output frequency of the helicoidal multivibrator.

The helicoidal strip-string in the transverse vibration regime can be considered a string with the specific longitudinal stiffness at the quasi-static tension, $j_\theta\theta = j_\theta\, i_\theta\, s = j_s$. However, in the dynamics, its longitudinal stiffness is approximately equal to EA_c/L, as for regular round or flat strings with Young's modulus E, cross-section area A_c, and length L. Figure 5.2b shows the mechanical model of the multivibrator system. The double-helicoid multivibrator's system has indeed a set of interconnected oscillators (Fig. 5.2a,b): (I) of the helicoidal strip's segment L located in the transverse magnetic field; (II) the symmetrical part of the helicoidal strip that is not connected directly with the magnetic field; (III) of a mass M_{L1} on the elastic support; and (IV) of a mass M_{L2} with the mass moment

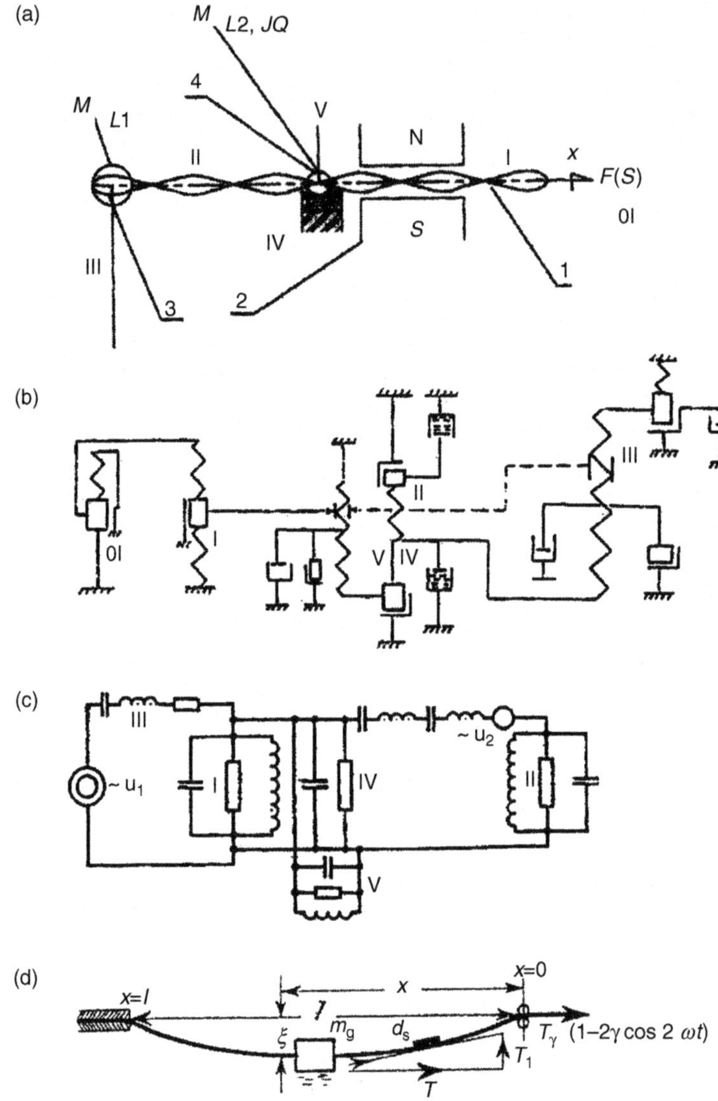

FIGURE 5.2 Schematic of mechanical (a) and electrical (b) wave band filter models of a helicoidal multivibrator, and model for longitudinal-transverse motion stability of the helicoidal string (d). *(After Tseytlin, 2006, reprinted with permission.)*

of inertia J_Q on the strip's crosspiece. Oscillators I–III have air damping, but oscillator IV may have liquid-type viscose damping. The natural frequencies of oscillators I and II are equal ($f_{\mathrm{I}} = f_{\mathrm{II}}$), and depend on the helicoidal strip's tension. Oscillator III usually has a higher natural frequency ($f_{\mathrm{III}} > f_{\mathrm{I,II}}$) that is practically independent from the strip tension. Oscillator IV has a comparatively low natural frequency ($f_{IV} \ll f_{I,II}$) that increases with the extension of the strip. The parametric resonance occurs at the frequency of the strip's vibration that is equal to the multiple of the natural frequency for the transverse vibration of oscillator IV. As a result of the action of the liquid viscose damping, oscillator IV can be in the nonperiodic vibration regime, especially at the weak tension of the helicoidal strip. In this case, the system of oscillators I, IV, and II is similar to the electrical wave band filter (Fig. 5.2c). In addition to the above mentioned oscillators, in certain realizations, input oscillator 0I can be found with the natural frequency f_{0I} that is equal to the natural frequency of the input sensing element (measuring spindle), and the output oscillator V with natural frequency f_V. The analog output oscillator V can be in the form of an arrowlike pointer, a light mirror, Hall's transducer chip, a small thin plastic or metallic nontransparent shield on the double-helicoid crosspiece. The analog output may be performed even without adding anything to the crosspiece, by indicating its angle of rotation with the laser beam diffraction.

5.3.1 The General Relations

Mechanical oscillators. The typical differential equations for the oscillators 0I, III, IV, and V (Fig. 5.2b) at the small natural vibrations can be represented as follows:

for oscillator 0I $$m_1\ddot{x} + j_1 x = 0, m_1\ddot{x} + j_1 x = F\delta(x-a), \tag{5.9a}$$

for oscillator III $$M_{L1}\ddot{y} + j_3\ y = 0, \tag{5.9b}$$

for oscillator IV $$M_{L2}\ddot{z} + C_x\dot{z} = 2L^{-1}F(z-z_1) + 2j_\theta i_\theta L^{-2}(z-z_1)^3, \tag{5.9c}$$

$$J_Q\ddot{\theta} + C_\theta\dot{\theta} = -2(j_{\tau\tau}\theta/L) + j_{k\tau}z^2/L^2, \tag{5.9d}$$

and for the oscillator V with cantilever pointer, whose two segments are supported in one intermediate point:

$$(EJ_{1,2}m_{1,2}^{-1}\,\partial^4 v/\partial x^4) + \partial^2 v/\partial t^2 = 0. \tag{5.9e}$$

In this section, we will study oscillators I, II, and IV, all of which are essential for the helicoidal multivibrators. Let us discuss this problem in more detail.

5.3.2 Helicoidal String

The helicoidal (initially pretwisted) beam-strip with fixed ends has flexural (in two planes), longitudinal, and torsional vibrations that are mutually related (Tseytlin, 2006). In the linear problem for quasi-constant relations at the vibration of the pretwisted beam-strip with double-symmetrical profile under longitudinal tension, this system splits into two independent systems: the flexural and the longitudinal-torsional vibrations. Let us note that the inductance voltage in the helicoidal multivibrator is generated only by its transverse motion that is perpendicular to the magnetic lines, that is, by the flexural vibration in one plane. Hence, in the linear problem with the elimination of nonused expressions for torsional vibrations, the transverse vibration of the helicoidal strip-string can be described by the partial differential equation with the periodic coefficients, such as follows:

$$F_0(\partial^2 v/\partial x^2) - EA\,\partial^2/\partial x^2[\kappa_j^2(x)\partial^2 v/\partial x^2] = \rho_\gamma A_c(\partial^2 v/\partial t^2), \quad (5.10)$$

where $\kappa_j^2(x) = (1\,/\,A_c)\,[J_\eta \sin^2\varphi(x) + J_\xi \cos^2\varphi(x)]$ is the equivalent of radius of gyration for the helicoidal strip's cross section, F_0 is the longitudinal pretension of the strip, ρ_γ is the density of the material, A_c is an area of the strip's cross section, J_η, J_ξ are the section's moments of inertia, $\varphi = (2\pi/S_0)\,x$, and S_0 is the pitch of the helix. Note that function $\kappa_j^2(x)$ is a periodical one. Let us evaluate the mean value $\bar{k}_j^2(x)$ in the limits of one helix pitch S_0 on the pretwisted (helicoidal) strip, equal to

$$\begin{aligned}\bar{k}_j^2(x) &= [1/(S_0 A)]\int_0^{S_0} dx[J_\eta \sin^2(2\pi/S_0)\,x + J_\xi \cos^2(2\pi/S_0)\,x] \\ &= \tfrac{1}{2}(J_\xi/A) + \tfrac{1}{2}(J_\eta/A_c) = \tfrac{1}{2}\,J_p/A_c. \end{aligned} \quad (5.11)$$

Furthermore, if the radius of gyration is constant and equal to the found $\bar{\kappa}_j(X)$, we have from (5.11) a solution similar to that for the rigid flat string with $\kappa_j^2(x) = \bar{k}_j^2(x)$, and at the fixed ends $Y(0) = Y(L) = Y'(0) = Y'(L) = 0$ that yields

$$f_n \approx (n_f/2L)\sqrt{F/\rho_\gamma A_c}\,[1 + (2/L)\sqrt{EJ_p/2F} + \left(4 + \frac{n_f^2\pi^2}{2}\right)EJ_p/(2FL^2)], \quad (5.12)$$

where f_n is the natural frequency of vibration for the elastic helicoidal multivibrator at $n_f \ll L/S_o$. As such, the number of harmonics n_f should be not more than 3, because in the typical multivibrator $L/S_o = 9$. The difference between Eq. (5.12) and the known analogous relationship for the flat string is in the substitution of the radius of gyration for the parameter $\sqrt{(1/2)J_p/A_c}$. At $J_p \to 0$, expression (5.12) corresponds to the formula for the vibration of the regular flat string under tension F. By substituting the expression $v = Y(x)\,e^{-2\pi i f t}$ into Eq. (5.10), we obtain the ordinary quartic differential equation of the fourth order with periodic coefficients:

$$-FY'' + [f(x)Y'']'' + \chi g_\chi Y = 0, \quad (5.13)$$

where

$$f(x) = \kappa_j^2[(2\pi/S_0)x],\ g(x) = 4\pi^2\rho_\gamma A_c;\quad \chi = f^2. \tag{5.14}$$

This conjugate equation with the conjugate boundary-value conditions $Y(0) = Y(L) = Y'(0) = Y'(L) = 0$ has eigenvalues with multiplicity not more than two. Eq. (5.13) fully corresponds to the problem for vibration of the rigid string with the section radius of gyration equal to the mean value of the radius of gyration for the initially pretwisted (helicoidal) string

$$k_j = \bar{k}_j = \sqrt{J_p/2A_c}. \tag{5.15}$$

Thus, we again come to the same expression (5.12) for the natural frequency of the helicoidal multivibrator, established earlier by the physical approach to the problem.

Uncertainty of the derived equation. The approximate expression similar to (5.12) for the rigid uniform flat or round string is an effective one at $EA_c\kappa_j^2 < L^2F_0$. But the differential equation for the vibration of the rigid string, in general, can be solved only by numerical methods. In Table 5.1, the uncertainty of the solution for (5.10) is shown in comparison with (5.12) of the numerical solution to a similar flat string (Tseytlin, 2006).

The uncertainty of natural frequency calculation in the case of the helicoidal multivibrator with preliminary untwist $\theta = 90°$ and $F_0L^2/(A_cE\bar{k}_j^2) = 243.2$ is less than 0.4%, and an uncertainty is less than 0.16% for the more typical adjustment with $\theta \geq 200°$ and $FL^2/(A_cE\bar{k}_j^2) \geq 564.3$. However, the influence of the section rigidity on the frequency value ω for the helicoidal multivibrator is significant, in accordance with the formula

$$\delta\left(\bar{k}_j^2, F_0\right) = 2\left(EA\bar{k}_j^2/F_0L^2\right)^{1/2} + (4+\pi^2/2)EA\bar{k}_j^2/\left(F_0L^2\right), \tag{5.16}$$

that predicts the following values of this influence, in percent,

$$\delta\left(F_0L^2/EA\bar{k}_j^2\right) = 27.77(106.5), 16.50(243.2), 10.00(564.3), 6.08(1358.0)$$

in comparison to the stretched string with negligible flexural rigidity. One can use a thinner strip with a smaller sectional polar moment J_p in order to decrease the influence of the section flexural rigidity at an equal preliminary tension of the helicoid.

5.3.3 Function of Impedance

The compliance of the end of the helicoid with the pointer's mass $M_{L2} = M_Q$ on the crosspiece (contour IV) is reflected in the differential equation of its transverse vibrations:

$$M_Q\ddot{\eta}_y + C_y\dot{\eta}_y - j_y\eta_y = 2F(\partial y/\partial x)_{x=L}\cos(\lambda_a at - \varepsilon_\lambda), \tag{5.17}$$

TABLE 5.1 Estimation of an Uncertainty in the Application of Eq. (5.12)

$\frac{F_0 L^2}{A_c E \bar{k}_l^2}$	$\sqrt{\frac{\rho_\gamma L^4}{E k_j^2}} \omega_0$ for(10)	$\omega \sqrt{\frac{\rho_\gamma L^4}{E \bar{k}_j^2}}$ from (5.12)	$\frac{\omega - \omega_0}{\omega_0}$ %	(i)		(ii)		(iii)	
				F_0 cN	$\theta°$	F_0 cN	$\theta°$	F_0 cN	$\theta°$
106.5	41.9	41.42	–1.10	2.17	48.4	–	–	–	–
243.2	57.3	57.08	–0.39	4.97	110.5	2.15	86.3	0.74	73.7
564.3	82.2	82.07	–0.16	11.54	256.4	5.00	200	1.71	170.9
1358.0	122.9	122.8	–0.08	27.76	617.0	12.05	500	4.11	411.3

Note: $\omega_0 = 2\pi f_0$; $\omega = 2\pi f$. Helicoidal strip of zinc bronze with dimensions in mm: (i) $0.008 \times 0.12 \times \pi \times 18$; (ii) $0.006 \times 0.10 \times \pi \times 18$; and (iii) $0.004 \times 0.08 \times \pi \times 18$.
Source: After Tseytlin (2006), reprinted with permission.

where C_y is a coefficient of the resistance to a motion in the liquid damper; $j_y = 2F/L$ is an equivalent rigidity of the transverse motion for the mass M_Q, and $y = \varphi(x)$ is the eigen form of the helicoid's segment natural vibration.

The natural frequency of vibration for the pointer (indicator) on the strip is approximately equal to:

$$f_{IV} \approx (1/2\pi)\sqrt{(2F/LM_Q)-(C_y/2)^2}. \tag{5.18}$$

One can neglect the transverse motion of the pointer if it is less than the amplitude of the string's vibration, $\eta_y \ll y_{x=L}$. However, if the frequency of helicoid vibration is close to $f = Nf_{IV}$ with N as an even integer, the parametric resonant influence effect can emerge. Let us discuss this in detail. Assume that the function of motion for the string is

$$y(x) = (\alpha_\lambda \sin\lambda_a x + \beta_\lambda \cos\lambda_a x)\cos(\lambda_a at - \varepsilon_\lambda)$$

with the following conditions at the ends of the helicoid:

(a) $x = 0 \Rightarrow y = 0$; (b) $x = L \Rightarrow M_Q\ddot{y} + C_y\ddot{y} + j_y y = -2F(\partial y/\partial x)_{x=L}$.

Condition (a) yields $\beta_\lambda = 0$, $y = \alpha\lambda \, sin \, \lambda_a x \cos(\lambda_a at - \varepsilon_\lambda)$.

Condition (b) after substitutions for y, y, $\ddot{y}$, and $(\partial y/\partial x)_{x=L}$ yields

$$\tan(\lambda_a L) = 2F\lambda_a/(M_Q a^2\lambda_a^2 - j_y) = 2/v_a \approx 2v_a/(1-v_a^2), \tag{5.19}$$

where $v_a = (M_Q\, a^2\lambda_a^{\,2} - j_y)\, F^{-1}\lambda_a$.

An analogous relationship can be obtained from the fundamental form for the elastic line of a string with the fixed ends and the vibrating middle point of mass M_Q that leads to the same result as in (5.19). Note that (5.19) is analogous to Rayleigh's relationship derived from investigation of vibrations for the string on the symmetrical compliant supports.

An influence of the mass on the crosspiece transverse displacements can also be evaluated through the impedance of its motion.

Impedance. The corresponding boundary-value conditions for the left-hand and right-hand end sections of the vibrating elastic helicoid are as follows:

$$Y(0) = 0;\, y(L) = (F/i\omega Z_L)\partial y/\partial x\,|_L.$$

The application of the fundamental functions in the form

$\psi_n\,(x,\omega) = \sin\,[(\omega_n/\rho)(x - a_n)]$ or $\psi_n = \sinh\,(i\omega_n/C_\rho)\,(x - a_n)/\sinh\,(i\omega_n/C_\rho)$ yields

$a_n = \pi\, n_f\, C_\rho/\omega_n$, $C_\rho = \sqrt{F/\rho_\gamma A_c}$, and

$$\tan[(\omega_n/C_\rho)(L - \pi n_f C_\rho/\omega_n)] = (F/i\omega Z_L)\omega_n/C_\rho, \tag{5.20}$$

where Z_L is an impedance.

Small ratio $F/(\omega Z_L)$ with large impedance. At this assumption, we have an expression for the circular frequency ω_n of vibration equal to

$$\omega_n = (n_f \pi / L)\sqrt{F/\rho_\gamma A_c}\left(1 + \frac{F}{i\omega_n L Z_L}\right), \qquad (5.21)$$

where impedance Z_L is equal to:

$$Z_L = C_y - i(\omega_n M_{L2} - j_y/\omega_n) \text{ with } j_y = 2F^{-1}N^2 \text{ and } N = 3,4,5,6\ldots$$

Substituting the impedance expression into (5.21) and assuming $C_y = 0$, we derive, after elementary transformations, the cubic equation in relation to the circular frequency

$$(LM_{L2})\omega_n^3 - \pi n_f M_{L2}\sqrt{F/\rho_\gamma A_c}\,\omega_n^2$$
$$- 2F\omega_n N^2 + 2\pi n_f L^{-1}\sqrt{F/\rho_\gamma A_c}\,F\left(N^2 - \tfrac{1}{2}\right) = 0 \text{ with } n = 1,2\ldots (5.22)$$

The cubic equation in relation to the linear frequency is a similar one:

$$4(LM_{L2})\pi^2 f_n^3 - 2\pi^2 n_f M_{L2}\sqrt{F/\rho_\gamma A_c}\,f_n^2$$
$$- 2Ff_n N^2 + n_f L^{-1}\sqrt{F/\rho_\gamma A_c}\,F\left(N^2 - \tfrac{1}{2}\right) = 0, \qquad (5.23)$$

where $N = 3,4,5,6\ldots$

If $M_{L2} = 0$, both parts of the double-helicoid are vibrating together, and

$$f_n = (n_f/2)(1/2L)\sqrt{F/\rho_\gamma A_c}\,.$$

Because the decrement Δ_M of (5.23) is less than zero

$$\Delta_M = -24(LM_{L2})\,F\pi^2N^2 - 4\pi^4 n_f^2 M_{L2}^2 F/(\rho_\gamma A_c) < 0.$$

Eq. (5.23) corresponds to a cubical parabola with three real roots. Its analysis shows that the two roots with plus and minus signs have an approximately equal module, and the third root corresponds to the natural vibrations of the string with fixed ends, which is equal to: $f_n \approx (1/2L)\sqrt{F/\rho_\gamma A_c}$.

Let us recall that (5.23) was derived with the assumption of a small ratio $F/\omega Z_L$, or a large enough impedance Z_L. But the system under consideration can have small impedance that, at resonance of mass M_{L2} (oscillator IV), can be close to zero. In this case, the discussed ratio cannot be considered a small one.

Large ratio $F/\omega Z_L$ with small impedance. It should be noted that the problem of string vibration in the whole field of the impedance values was not investigated before 2006. As such, we tried to make up a deficiency in the solution to the problem (Tseytlin, 2006) with only one assumption: that an influence of the liquid damper C_y is small. Now let us transform the transcendent Eq. (5.22) as follows:

$$\tan[(\omega_{vn}/C_\rho)(L - \pi n_f C_\rho/\omega_{vn})] = F\omega_{vn}/[C_\rho(\omega_{vn}^2 M_{L2} - 2FN^2/L)].$$

Then, we transform its right-hand side with the assumption that

$$\left(\sqrt{\rho_\gamma AL}/N\sqrt{2M_{L2}}\right)[\kappa_\omega/(\kappa_\omega^2-1)]=q_N(\kappa_\omega/(\kappa_\omega^2-1)\approx(\Delta\kappa_\omega)^{-1}q_N/2,$$

where $0 \leq \kappa_\omega \leq 1$; $\Delta\kappa_\omega = 1 - \kappa_\omega$. As a result, we obtain the following simplified expression:

$$\tan[(\omega_{vn}/C_\rho)(L-\pi n_f C_\rho/\omega_{vn})]=q_N(\kappa_\omega/(\kappa_\omega^2-1)\approx(\Delta\kappa_\omega)^{-1}q_N/2 \qquad (5.24)$$

that at $n_f = 1$ corresponds to $\omega_1 = (1/2)\,\pi C_\rho/L$. Therefore, we find from (5.24) that the circular frequency of oscillator I is

$$\omega_{vn}=(\pi nC_\rho/L)\{1-[(1/(\pi n)\,|\tan^{-1}q_N(\kappa_\omega/(\kappa_\omega^2-1)\,|]\},$$
$$\omega_{vn}=(N\pm k_\omega\mp 1)\sqrt{2Q_g/(LM_{L2})},\quad \text{and } f_1=\frac{\omega_{vn}}{2\pi}. \qquad (5.25)$$

The chart of this function (Fig. 5.3a) has typical zigzag deflections (deviations) near the frequencies $f_I = Nf_{IV}$.

There is a parametric resonance at the even values of integer N, with a break in the transformation function (shown by dashed lines), at uneven N. Following from (5.25), the sensitivity of a force to the frequency transformation and also a displacement to the frequency transformation is lower at the frequency value that is a multiple of the natural frequency for the vibration of mass M_{L2}, that is, when $\Delta\kappa_\omega \to 0$. The analysis of (5.25) shows that the influence of the motion of mass M_{L2} is small, when $\Delta\kappa_\omega > 0.05$, and the frequency characteristic approaches the usual function for a string sensor. Consequently, the operating range of the characteristic of a multivibrator without special scaling or correcting system does not exceed $\Delta f_1 = 0.9\,f_{IV}$. The operating range is selected during adjustment of the measuring sensor by varying the longitudinal tension F in the strip.

Effective parameters. Tension F_0, corresponding to $F_0L^2/AE\overline{k}_j^{\,2} > 560$, yields rotation of the crosspiece by an angle more than 170°. The relationship between the increment in frequency of oscillation and the tension in the pretwisted strip $\Delta f = ½\,\Delta F(f_0/F_0)$ under these conditions is a quasi-linear, within the limits of the operating range. If we substitute the elastic and geometrical parameters of the strip into the latter expression, then

$$\Delta f=1.99E(4+v_{if}h^2\kappa_p^2/\lambda^2)^{-1}\times$$
$$(1+\tfrac{1}{4}\,v_{jf}h^2\kappa_p^2/\lambda^2)h(F_o\rho_\gamma)^{-1/2}(4L^2\lambda^{1/4})^{-1}\Delta s, \qquad (5.26)$$

where Δs is the corresponding variation in the distance between the ends of the strip.

When choosing the parameters of the helicoidal strip, we should attempt to achieve a good frequency conversion $A_f = \Delta f/\Delta s = ½\, j_\theta\, i_\theta\, (f_0/F_0)$ and visual

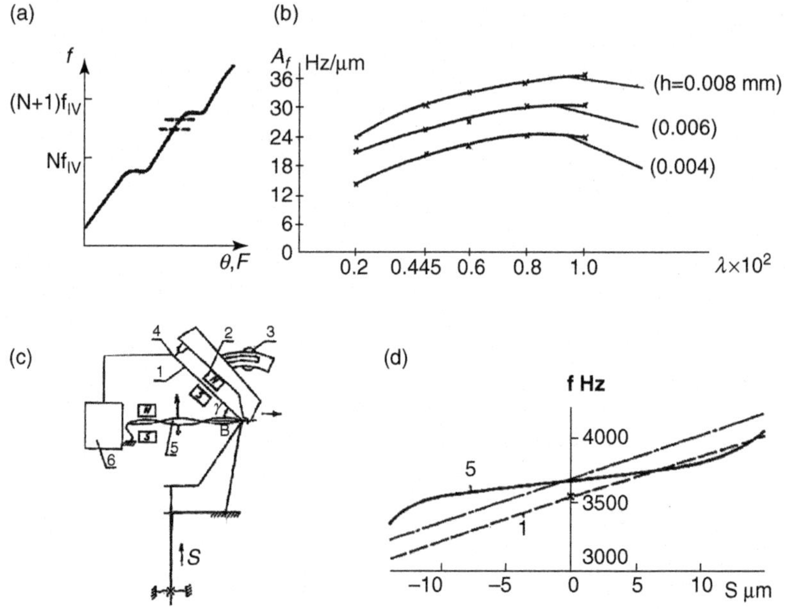

FIGURE 5.3 Frequency f_1 functions of the double helicoidal multivibrator with parametric resonance-break (a), frequency conversion A_f (b), differential multivibrator schematic (c), and its frequency functions (d).

amplitude conversion ($B_f = i_\theta = \Delta\theta/\Delta s$) at the same time, that is, to ensure that the ratio of A_f/B_f is a representative enough, or to achieve the greatest frequency conversion A_f with an adequate amplitude conversion B_f. Figure 5.3b shows charts of $A_f = \sqrt[4]{\lambda h^{-1}} i_\theta j_\theta (F_0 \rho_\gamma)^{-1/2}/4L$ for helicoidal strips with the rectangular cross section, $\rho_\gamma = 8.8$ g/cm^3, $k_p = \pi$/mm^{-1}, $L = 18$ mm, and preliminary tension $F_0 = 7.2$ cN. It is clear that in order to obtain a larger value for A_f, we should use a strip with larger thickness h. However, we need to have a thinner strip that is optimized with respect to λ ratio, in order to obtain a higher value $B_f = i_\theta$ of the amplitude transformation.

Optimization. Function (5.12) for the transformation of the longitudinal displacements into the frequency is a nonlinear one. There is a necessity of its linearization for the linear measurement systems. This can be executed by the following methods:

1. Connection of the calculator with preadjusted polynomial R_n(Z) coefficients to the output of the helicoidal multivibrator, for the appropriate approximation of the transformation function;
2. Installation of the differential-type multivibrator structure with the optimized ratio of the frequency outputs (Tseytlin et al., 1989, p. 145–147; Tseytlin, 2006, p. 239–244). Figure 5.3c shows a differential multivibrator's schematic (Patent USSR, 721658, 15.03.80) with two strings (5 with

pretwisted strip, and 1 with auxiliary flat strip or round wire) and additional magnet 2. Parts 3 and 4 are installed for the adjustment of strings' kinematics ($f_{5,1}$/s) and frequency ratio f_5/f_1. Both strings are electrically isolated from one end, and connected to the self-excited oscillator 6. The transformation characteristics f_5 for basic and f_1 for auxiliary strings are shown in Figure 5.3d;

3. Certain adjustments of the transducer's parameters (Tseytlin, 2006).

5.3.4 The Elastic Longitudinal Links

The dynamically stable fields of the helicoidal multivibrator at the longitudinal-transverse vibrations can be estimated with the help of Mathieu's model [McLachlan (1964), Fig. 5.1b and Fig. 5.2d with the exciting force T_γ $(1 - 2\gamma \cos 2\omega t)$] because our corresponding model has similar periodic coefficients in the differential equation

$$\ddot{S}_u + 2n_u \dot{S}_u + \omega_{v\Omega}^2 [1 - h_v \cos 2\omega_v t]\, S_u = 0, \tag{5.27}$$

where $S_u = \xi$ is the possible transverse displacement of the mass $M_{L2} = m_g$ that is clamped to the string's crosspiece, $h_v = 2\gamma$ is the depth of the periodical longitudinal modulation of the string's tension $F_{Q0} = T_\gamma$, $\omega_{v\Omega} = (2F_0/LM_{L2})^{0.5}$, and n_u is a coefficient of the liquid damper resistance to the transverse vibrations at the middle point of the string. The transverse vibrations are accompanied by the secondary longitudinal deformation of the string and by the corresponding longitudinal vibrations. It is proved (Tseytlin, 2006) that for the main region of instability, in the case of model similar to (5.27), the boundary value is

$$\omega_{v*} = 2\omega_{v\Omega}\sqrt{1 \pm \sqrt{\gamma^2 - (\Delta_v/\pi)^2}},$$

where $\Delta_v = 2\pi n_u / \omega_{v\Omega}$.

We can see from this expression that instability in the region near $\omega_v = \omega_{v\Omega}$ is possible, if $\gamma > \gamma_*$, where $\gamma_* = (\Delta_v/\pi)$ is a critical value of the excitement's coefficient. For nth region of the parametric excitement, the critical value of $\gamma_* \sim (\Delta_v/\pi)^{1/n}$. If the clamped mass and viscose damping are small, $m_g \to 0$, $n_u \to 0$, and the string's mass per unit length is equal m, then the corresponding differential equation for transverse motion $\xi = f_\xi(t)\sin_\xi x$ of the string may be expressed as standard Mathieu's equation, with possible stable and unstable zones

$$\frac{d^2 f_\xi}{dz^2} + (a_z - 2q_z \cos 2z) f_\xi = 0$$

where $a_z = b_\xi^2 T_\gamma / \omega_v^2 m$, $q_z = \gamma a_z$, and $z = \omega_v t$.

Critical values of a_z and q_z can be found on the Strutt diagram (Fig. 5.1b). In practice, owing to damping, the motion once started could die away with the increase in time.

5.3.5 Nonlinear Elasticity and Inertia

We recall that the differential equation for the natural vibration of a thin string has the following form:

$$\partial^2 v/\partial t^2 - a_f^2 \partial^2 v/\partial x^2 = 0, \tag{5.28}$$

where $a_f^2 = F/\rho_{\gamma L}$ and $\rho_{\gamma L}$ is the string's mass per unit length.

The longitudinal motion w and tension force F in the system with a helicoidal string that has longitudinal stiffness j_1, masses M_{L1}, M_{L2}, and the coefficient C_ε of the liquid damper resistance corresponds in vibration to the differential equation that describes the influence of the longitudinal w vibration

$$\begin{aligned} F = F_0 + j_\theta \Delta\theta + (EA_c/2L)w[1-(EA_c/L)(j_1+EA_c/L)^{-1}] \\ +j_1 w(EA_c/2L)(j_1+EA_c/L)^{-1} - (EA_c/L)M_{L1}\ddot{w}(j_1-EA_c/L)^{-1} \\ -M_{L2}\ddot{w}[1-(EA_c/L)(j_1+EA_c/L)^{-1}] - C_\varepsilon \dot{w}, \end{aligned} \tag{5.29}$$

where F_0 is the longitudinal pretension and $\Delta\theta$ is an angle of the crosspiece's untwist corresponding to the strip's extension. One can assume that, as a result of the negligible flexural rigidity of the strip, its transverse motion can be described by a sinusoidal function:

$$v = \varphi_t(t)\sin(\pi x/L) \quad \text{and} \quad w = L - \int_0^L \sqrt{1-(dv/dx)^2}\,dx = \frac{\pi^2\varphi_t^2}{4L} + \ldots, \dot{w} = \pi^2\varphi_t\dot{\varphi}_t/2L,$$

$$\ddot{w} = \frac{\pi^2}{2L}[\varphi_t\ddot{\varphi}_t + (\dot{\varphi}_t)^2] + \ldots$$

Substitution of these constituents into (5.29) leads to the differential equation with the functions $\varphi_t, \dot{\varphi}_t, \ddot{\varphi}_t$,

$$\ddot{\varphi}_t + \omega_v^2\varphi_t + \psi(\varphi_t, \dot{\varphi}_t, \ddot{\varphi}_t) = 0, \tag{5.30}$$

where $\psi(\varphi_t, \dot{\varphi}_t, \ddot{\varphi}_t) = \gamma_\varphi\varphi_t^3 + 2\varepsilon_L\varphi_t^2\dot{\varphi}_t + 2\aleph_\varphi\varphi_t[\varphi_t\ddot{\varphi}_t + (\dot{\varphi}_t)^2$,

$$\omega = (\pi/L)\sqrt{\frac{F_0 + j_\theta\Delta\theta}{\rho_{\gamma L}}}, \quad 2\gamma_\varphi = \omega^2\pi^4[(EA_c/L) + q_\pi]/[4L^3(F_0 + j_\theta\Delta\theta)],$$

$$q_\pi = (EA_c/L)(j_1 - EA_c/L)/(j_1 + EA_c/L), \quad \aleph_\varphi = \Phi(M_{L1}, M_{L2}, \rho_{\gamma L}, EA, j_1, L),$$
$$\varepsilon_L = \pi^4 C_\varepsilon/(2\rho_{\gamma L}L^3).$$

It is assumed that $EA_c/(Lj_1) \ll 1$.

The type of differential equation similar to Eq. (5.30) with members of the nonlinear elasticity $\gamma_\varphi\varphi_t^3$ and nonlinear inertia of the rod $2\aleph_\varphi\varphi_t[\ddot{\varphi}_t\varphi_t + (\dot{\varphi}_t)^2]$ are carefully studied, with the van der Pol method of slowly changing amplitudes, by Bolotin (1956), who found the relationship between the amplitude a_ω and the frequency ϖ of oscillations for the corresponding nonlinear system as follows:

$$\bar{\omega} = \omega\sqrt{\left[1 + \frac{3}{4}(\gamma_\varphi/\omega^2)a_\omega^2\right]/(1 + \aleph_\varphi a_\omega^2)}$$

or with respect to the negligibility of nonlinear corrections,

$$\bar{\omega} = \omega\{1+(a_{\omega}^2/2)[3/4(\gamma_{\varphi}/\omega^2)-\aleph_{\varphi}]\}. \tag{5.31}$$

It is important that, in the case of the helicoidal multivibrator, the value of $\gamma_{\varphi}/\omega^2$ decreases with the increase of untwist, that is, with the increase of the tension force $(F_0 + j_\theta\,\Delta\theta)$. Hence, the initial elastic nonlinearity decreases with the increase of the stretch and untwist. Then, after passing through the state of isochronic vibrations, we automatically come to the system with predominant nonlinear inertia. Let us analyze function $\varpi(\theta)$.

If the initial value of the argument's origin $\theta = \theta_\kappa$ at $\aleph_{\varphi} = ¾\,(\gamma_{\varphi}/\omega^2)$, then

$$\bar{\omega} = \omega\{1 \mp (a_{\omega}^2/2)(3\pi^4 j_\theta \theta_\kappa/32L^3)[(EA_c/L)+q_\pi]\}$$

and the nonlinear change of frequency in relation to the new origin is equal to:

$$|\bar{\omega}| = \omega(a_{\omega}^2/2)(3\pi^4 j_\theta \theta_\kappa/32L^3)[(EA_c/L)+q_\pi]. \tag{5.32}$$

It is clear that this nonlinear function has convexity in the preisochronic state of vibration, and concavity in the post-isochronic state of vibration. This conclusion corresponds to the zigzag frequency function (Fig. 5.3a), shown above, for the helicoidal multivibrator at the small impedance.

5.4 INTERNAL FRICTION (DAMPING) IN A HELICOID

Experimental data. The damping variations of the longitudinal displacements s of the ends in the helicoidal multivibrator corresponding to the certain crosspiece rotation θ in quasi-static deformations does not exceed $K_{\theta e} = 0.13\%$ for the tin-zinc bronze, or $K_{\theta e} = 0.05\%$ for the platinum-silver alloy. However, it has been found, from our experiments executed in the 1960s, that there is a stable difference $\delta\theta_\kappa$ in the angular displacements of the crosspiece with its indicator (pointer), when equal longitudinal distances between the ends of the naturally pretwisted strip have been established by fast and slow displacements. We can assume that $\delta\theta_\kappa = \psi_{\Delta\kappa}\Delta\theta_d$, where $\psi_{\Delta\kappa}$ is the coefficient of the elastic energy absorption due to internal friction, $\Delta\theta_d$ is the dynamic lag of crosspiece rotation behind the quasi-static transmission ratio, and κ is an index that characterizes the direction of the longitudinal deformation: $\kappa = t$ for tension and $\kappa = r$ for relaxation.

According to experimental data $\psi_{\Delta t} > \psi_{\Delta r}$ and $\psi_{\Delta t}/\psi_{\Delta r} = \delta_{tr}^2 \gg 1$. It has been established experimentally that similar effects also arise in difference for slow longitudinal deformations of an elastic helicoid with and without forcible immobilization (temporary arrest) of the crosspiece rotation (Table 5.2).

In Table 5.2, helicoid No 1 has dimensions $h \times b \times k_p \times L = 0.006 \times 0.1 \times \pi \times 18$; helicoid No 2 has dimensions $0.008 \times 0.12 \times \pi \times 18$, with the quasi-static transmission ratio 5.2 and 3.6 °/μm, respectively. The material is 4tin-3zinc bronze in both cases.

TABLE 5.2 Transmission Uncertainty in Helicoid Strips at a Different Working Regime

Helicoid No.	Nominal untwist angle $\theta°$	$(\theta°_{ts}-\theta°_{tf})$	$(\theta°_{rs}-\theta°_{rf})$	$(\theta°_{ts}-\theta°_{tf\Phi})$	$(\theta°_{rs}-\theta°_{tf\Phi})$	$(\theta°_{ts\Phi}-\theta°_{ts})$
1	180	8.2–11.6	–	–	–	–
	360	17.5	–	–	–	–
	720	13.9–19.7	0.7	13.9–27.3	0.6	–
2	180	5.4	0.0	6.6	0.6	–
	360	12.8	0.08	13.9	1.1	–
	440	–	0.4	16	1.9	12.6–15.9
	720	14.2	1.2	23	4.2	–

Note: $\theta°_{ts}$ is an angle position of the crosspiece relative to its regular angular position $\theta°$after a slow tension of the helicoid; $\theta°_{tf}$ is the same after a fast tension of the helicoid; $\theta°_{rs}$ and $\theta°_{rf}$ are the same after a slow and fast relaxation of the helicoid's tension. Subscript Φ marks the angular position of the crosspiece after the corresponding longitudinal deformation of the helicoid with forcible immobilization of the crosspiece rotation.
Source: After Tseytlin (2006), reprinted with permission.

Experimental data of these transmission uncertainties $\Delta\Theta°$, at slow and fast deformation in the double-helicoid sensor, can be presented as a regression (Fig. 5.4a), where chart N1 ($\Delta\theta_t$) is for tension, and chart N2 ($\Delta\theta_r$) is for relaxation. The experimental points are shown by the crosses for extension, and by dots for relaxation.

$$\Delta\theta(\bar{\theta})=\bar{\Delta\theta}\pm t_{q,n-2}(n-2)^{-1}\sqrt{1+n(\theta-\bar{\theta})^2/[\theta\theta]},$$

where n = 128 is the number of experimental data, $t_{q,n-2}$ = 2.241 is the 2.5% point of Student's distribution at the level of significance q = 2.5%; $\bar{\Delta\theta}$, and $\bar{\theta}$ are the corresponding mean values for uncertainty (deflections) and rotation angles of the pointer. Calculations show that for tension

$$\bar{\Delta\theta}_t(180°)=[5.8\pm0.9]°,\text{for relaxation}\,\bar{\Delta\theta}_r(180°)=[1.0\pm0.2]°,$$

$$\bar{\Delta\theta}_t(720°)=[20.0\pm1.3]°,\text{and}\,\bar{\Delta\theta}_r(720°)=[1.5\pm0.5]°.$$

The described phenomenon in some aspects reminds, largely increased by the transmission ratio in the pretwisted strip, the so-called micro-Baushinger effect of microplasticity, found by Carnahan in 1964, which showed that the stress required to operate the dislocation in materials depends on the shear modulus, Burgers' vector (denotes the amount and direction of atomic displacement under the motion of dislocation), and the length of dislocation.

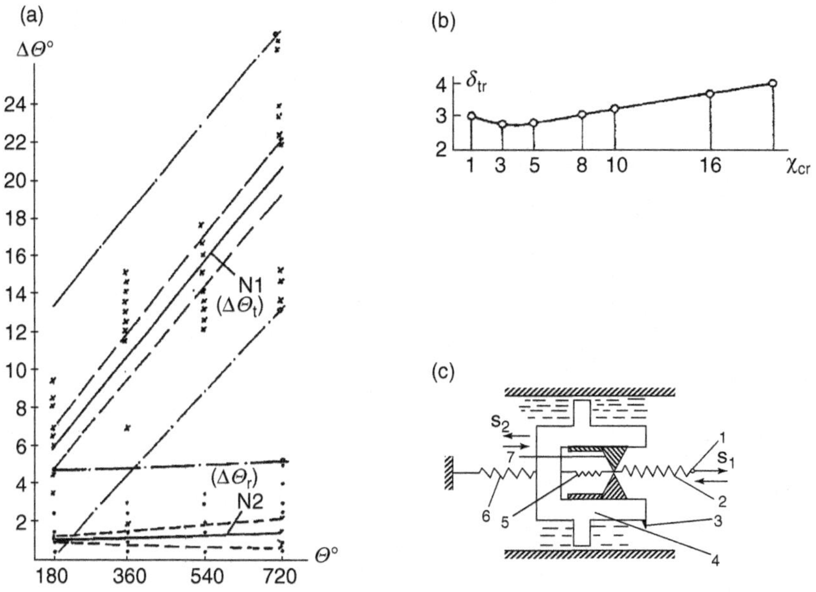

FIGURE 5.4 Internal friction asymmetry in helicoidal microstrip (a, b) and its macroscopic model (c). *(After Tseytlin, 2006, reprinted with permission.)*

Besides, the shear strain produced by the motion of a dislocation depends on the density of moving dislocations, and their pinning.

Let us look in the specifics of state of stress in the helicoidal strip under dynamic deformation.

State of stress. In this study, relationships from elasticity and plasticity theory are used to derive the theoretical ratio $\delta_{tr} = (\psi_{\Delta t}/\psi_{\Delta r})^{0.5}$ as a function of the elastic and geometric parameters of a helicoidal sensor (with multivibrator or amplitude version). Generally, the coefficient of absorption $\psi_{\Delta K}$ in the elastic helicoids depends on the conditions, limits, and type of deformation, as well as the inelastic properties of the material.

Chart N1 corresponds in Fig. 5.4a to $\Delta\Theta$ distribution at extension with different speed, and Chart N2 corresponds to distribution of $\Delta\Theta$ at the relaxation of the helicoid at different speed. The boundaries of the untwist angle distribution of $\Delta\Theta$ correspond to dash-dot lines, and correlation limits are shown by dashed lines. These latter limits are defined with the known statistical formula.

The stressed state of a naturally pretwisted (helicoidal) strip under elastic linear longitudinal-torsional deformations is described by a tensor $\hat{\sigma}_{\xi\eta}$ with the components

$$\hat{\sigma}_{\xi\eta} = \sigma_F^{\xi\eta} S_\gamma + \sigma_M^{\xi\eta} \theta_\gamma, \tag{5.33}$$

where $\sigma_F^{\xi\eta}$, $\sigma_M^{\xi\eta}$ are the components of the stress tensor in the respective cases: (i) longitudinal tension in the absence of the torque ($M = 0$; $F \neq 0$), and (ii) with torque M in the absence of tensile force F ($F = 0$; $M \neq 0$), S_γ is the longitudinal deformation per unit length, θ_γ is the torsional deformation per unit length of the helicoid, and ξ, η are the following coordinates of the cross section (Chapter 4). Since the transverse-vibration amplitude A_{tv} of a helicoidal multivibrator crosspiece is a small one in the normal operation ($A_{tv}/L < 2.10^{-3}$), these vibrations practically have a small influence on the longitudinal force in the helicoid. This is evident because the ends of the helicoidal multivibrator are mounted to an elastic support that provides a quasi-permanent tension force. During slow quasi-static longitudinal elastic deformation of the helicoid with $M = 0$, $F \neq 0$, the stress tensor has a following form at the most heavily loaded point of the strip's cross section:

$$\hat{\sigma}_{1T} = \begin{bmatrix} \sigma_Q^{11} & 0 & \sigma_Q^{13} \\ 0 & \sigma_Q^{22} & 0 \\ \sigma_Q^{31} & 0 & \sigma_Q^{33} \end{bmatrix}, \tag{5.34}$$

where $\sigma_Q^{11} \approx \sigma_Q^{22} = -(18/224)\ \chi_{kp}\ C_\sigma^{-1}\sigma_0^{33}$, $\sigma_Q^{13} = h_1\kappa_p C_\sigma^{-1}\sigma_0^{33}/(2\lambda)$, $\sigma_Q^{33} = (1 + 69\chi_{kp}/224)\ C_\sigma^{-1}\sigma_0^{33}$; $C_\sigma = 1 + \nu_{jf}\chi_{kp}$, ν_{jf} is the force factor for the certain form of the strip's cross section, $\sigma_0^{33} = F/A_c$; $\chi_{kp} = b_1^2\kappa_p^2\lambda^{-1}$, $\lambda = (h/b)^2 \in [0.16 \times 10^{-2} - 0.45 \times 10^{-2}]$ characterizes the relative thickness of the strip's cross section, $h = 2h_1 \in [0.002; 0.02]$ is the thickness (in millimeters), $b = 2b_1 \in [0.08; 0.2]$ is the width of the cross section (in millimeters), and $\kappa_p = 2\pi/S_0$ is the initial helix factor with a helix pitch S_0. When tension is quickly applied to the pretwisted strip, the longitudinal stress σ^{33} temporarily increases more than at slow tension because M ≠ 0, and

$$\sigma^{33} \approx (j_\theta\theta/A_c) + [I_Q\,\partial^2[\theta_e]_t/\partial t^2 + C_\theta\,\partial[\theta_e]_t/\partial t]/\rho_c A_c, \tag{5.35}$$

where I_Q is the mass moment of inertia of the driven pointer (indicator) on the crosspiece between the helicoids with left-handed and right-handed helices; $\partial[\theta_e]_t/\partial t$, $\partial^2[\theta_e]_t/\partial t^2$ are the angular velocity and acceleration of its motion, respectively, C_θ is the resistance of the liquid damper, $t = t_{max}$ is the time corresponding to the largest value of the derivatives of the displacement sensor's transfer function $[\theta_e]$. Parameters ρ_c, j_{kk}, $j_{k\tau}$, and $j_{\tau\tau}$ (Chapter 4).

With a sudden sharp extension of the preloaded helicoid, $\sigma^{33} = (j_\theta\ \theta_0/A_c) + ES_\gamma$ where θ_0 is the initial untwist angle of the crosspiece. On the other hand, if the tension in a pretwisted strip is relaxed quickly, the longitudinal tension is initially lower than under equivalent slow relaxation.

The state of stress in the most critical point of the cross section of the helicoidal strip can be approximately described under quasi-dynamic longitudinal tension by a tensor $\hat{\sigma}_{2T}$ with the components

$$\sigma_Q^{11} \approx \sigma_Q^{22} = -(18/224)\upsilon_\kappa^{11}\chi_{cr}C_\sigma^{-1}\sigma_0^{33};\quad \sigma_Q^{13} = \upsilon_\kappa^{13}h_1\kappa_p C_\sigma^{-1}\sigma_0^{33}/(2\lambda);$$
$$\sigma_Q^{33} = \upsilon_\kappa^{33}(1+0.308\chi_{cr})C_\sigma^{-1}\sigma_0^{33};\quad \chi_{cr} = \chi_{kp}.$$

Coefficients υ_κ^{ij} have different values under slow and fast longitudinal deformation. Under slow quasi-static deformation, $\upsilon_\kappa^{11} = \upsilon_\kappa^{13} = \upsilon_\kappa^{33} = 1$, and the tensor $\hat{\sigma}_{2T}$ becomes equal to tensor $\hat{\sigma}_{1T}$. Initially, with a fast tensile loading of the helicoid, the stress coefficients are

$$\upsilon_t^{11} \le 1, \upsilon_t^{13} = 0; 1 < \upsilon_t^{33} \le EA_c L^{-1} i\theta^{-1} j\theta^{-1}(1+0.308\chi_{cr})^{-1}C_\sigma,$$

and with a fast unloading (relaxation) of the helicoid their initial values are

$$\upsilon_r^{11} \le 1; \upsilon_r^{13} = 1; 0 \le \upsilon_r^{33} < 1.$$

5.4.1 Microplastic Shifts

The microplastic shifts e_s in the continuum media are proportional to the stress intensity σ_s that, for a fast relaxation of the helicoid, has the value

$$\sigma_{\text{sfr}}(\hat{\sigma}_{2T}, \upsilon_r^{33} = 0) = [I_2 \operatorname{dev} \hat{\sigma}_{2T}(\upsilon_r^{33} = 0)]^{0.5} \approx \sigma_Q^{13}(\upsilon_r^{13} = 1), \quad (5.36)$$

where I_2 dev $\hat{\sigma}_{2T}$ is the second invariant of the deviator of tensor $\hat{\sigma}_{2T}$. The angle factor of the stress tensor at the fast relaxation of the strip equals

$$\mu_{\text{fr}} = (\sqrt{3})\cot[0.33\cos^{-1}(3/2)(\sqrt{3})\sigma_{\text{fr}}^{-3} I_3 \operatorname{dev}\hat{\sigma}_{2T}(\upsilon_r^{33} = 0)] = 0, \quad (5.37)$$

that corresponds to a pure torsion.

Here I_3 dev $\hat{\sigma}_{2T}$ $(\upsilon_r^{33} = 0)$ is the third invariant of the deviator of tensor $\hat{\sigma}_{2T}$. The stress intensity during a fast stretching of the helicoid is initially equal to:

$$\begin{aligned}\sigma_{\text{sft}}[\hat{\sigma}_{2T}, 1 < \upsilon_t^{33} \ &\le EAL^{-1} j_\theta{}^{-1} i_\theta{}^{-1}(1+0.308\chi_{cr})^{-1}C_\sigma] \\ &= EA\sigma_0^{33}L^{-1} i_\theta{}^{-1} j_\theta{}^{-1}/\sqrt{3}.\end{aligned} \quad (5.38)$$

Then the angle factor of the stress tensor $\mu_{\text{ft}} = -1$ that corresponds to a pure tension.

Dissipation. The loss of the energy as a result of inelastic properties of the material with the coefficient of absorption $\psi_{\Delta\kappa}$ can be expressed by the formula

$$R_{V\kappa} = (1/L)\Delta\theta_{\max}^2 \psi_{\Delta\kappa} j_{\tau\tau}, \quad (5.39)$$

but $\partial R_{vt}/\partial\Delta\theta_{max} = M_t = 2\Delta\theta_{max}\psi_{\Delta t}\, j_{\tau\tau}/L$. Hence, the uncertainty of the angular position for the helicoid's crosspiece is

$$\delta\theta = M_t L/2 j_{\tau\tau} = \psi_{\Delta\kappa}\Delta\theta_{max}. \tag{5.40}$$

Assuming that the absorption coefficients are proportional to the work $R_{v\kappa}$ of the microplastic shifts that equals for tension $R_{vt} = \sigma_{st}\, e_{st}$ and for relaxation $R_{vr} = \sigma_{sr}\, e_{sr}$, respectively, we derive the sought relationship for the absorption coefficients ratio δ_{tr}

$$\delta_{tr}^2 = \psi_{\Delta t}/\psi_{\Delta r} = R_{vt}/R_{vr} = \sigma_{st}e_{st}/(\sigma_{sr}e_{sr}) \approx \sigma_{st}^2/\sigma_{sr}^2, \tag{5.41}$$

where e_{st} and e_{sr} are the microplastic shifts under the fast tension and the fast relaxation, respectively, which are proportional to the corresponding stress intensities σ_{st} under tension, and σ_{sr} under relaxation. Then, after substitutions and simple rearrangements, we finally obtain the following formula for the absorption coefficients ratio

$$\delta_{tr} = (\psi_{\Delta t}/\psi_{\Delta r})^{0.5} = \sigma_{st}/\sigma_{sr} = 2.32(\chi_{cr}^{-0.5} + v_{if}\chi_{cr}^{0.5}), \tag{5.42}$$

where v_{if} is the kinematic form factor for the cross section of the elastic helicoidal strip.

Analysis of δ_{tr} *curve.* It is seen from the plot of internal friction asymmetry δ_{tr} (Fig. 5.4b) that this function has a weak minimum δ_{trmin} at $\chi_{cr} = v_{if}^{-1}$ since the first derivative of the function (5.42) is equal to zero, and the second one is greater than zero, such that

$$\frac{d\delta_{tr}}{d\chi_{cr}}(\chi_{cr} = v_{if}^{-1}) = 0 \quad \text{and} \quad \frac{d^2\delta_{tr}}{d\chi_{cr}^2}(\chi_{cr} = v_{if}^{-1}) > 0.$$

The typical kinematic cross-sectional form factor of the initially pretwisted thin helicoids is in the range of $v_{if} = 0.32$–0.41, and the parameter χ_{cr} is in the range of $\chi_{cr} \in [3;12]$. Let us recall that for the elliptic cross section $v_{if} = 0.345$. Hence, $\delta_{tr} = \sigma_{st}/\sigma_{sr} = 2.32\ (\chi_{cr}^{-0.5} + v_{if}\ \chi_{cr}^{0.5}) = 2.6 - 3.5 \in \chi_{cr}$ [1; 12], $\delta_{trmin}(v_{if} = 0.345) = 2.72$, and the mean value of the absorption coefficients ratio $<\psi_{\Delta t}/\psi_{\Delta r}> = <\delta_{tr}^2> = 9$. The average value of $<\delta_{rr}> = 3$ corresponds to the ratio $\psi_{\Delta t}/\psi_{\Delta} \approx 6 - 12$ observed in the experiments with the average value of the absorption coefficients ratio $<\psi_{\Delta t}/\psi_{\Delta r}> = 9$.

Macromodel. The described effects can also be expressed through the macroscopic model with the elastic-viscose-plastic elements (Fig. 5.4c). The displacement of point 1 corresponds to the input displacement S_1 (longitudinal deformation), and the displacement of index 3 clamped to the inertial body 4 corresponds to the output (driven) displacement S_2. The ideal elastic links are represented by elements 2, 5, and 6. The plastic deformations are modeled by the dry friction link between driving body 7, and driven body 4. In addition, the viscose resistance acts on body 4. At the slow displacement of point 1 in direction S_1, the motion through elastic link 2 transmits to body 7, and connected to it body 4 which, at slow speeds and acceleration, meets a small viscose resistance and has negligible inertia forces. At the sharp fast displacement of point 1 in direction S_1, inertial body 4 lags behind body 7, and the sliding in the friction pair 7–4 occurs. In this case, at the same fast displacement S_1, the value of the output displacement S_2 will be less than at the corresponding slow displacement S_1. The sequential relaxation of the helicoid's tension causes the removal of the microplastic displacements by the backward forces (the backward displacement of the dislocation). And fast and slow relaxations will give a quasi-identical output.

The inertial body for the double-helicoid system corresponds to an element clamped to the crosspiece. The less the mass and the moment of inertia of this element the less the uncertainty of the transmission from the instability of the speed in longitudinal extension of the helicoid. If the moment of inertia for the body on the crosspiece is more than the critical one, $J_Q > (C_\theta)^2 L/8j_{\tau\tau}$, then the system of the helicoid can become sensitive to the instability of the relaxation speed, as well.

5.5 SHELLS AND QUASI-HELICOIDS

Helicoidal multivibrators can be built not only on the basis of the pretwisted thin solid or perforated strips, but also on the basis of coiled ribbons and pretwisted shells, as well as elastic systems with quasi-helicoid motion. We have two suitable types of shells: (i) coiled ribbon, and (ii) pretwisted tube. Type (i) is indeed a bar, but from a theoretical standpoint it should be considered a shell. Type (ii) is a pretwisted tube that can be simultaneously used as a helicoidal multivibrator and the pressure sensor with the effective internal kinematic transmission. These helicoidal multivibrators have special features, and can be used appropriately in different ways than the thin helicoidal multivibrators studied in Sec. 5.2.

5.5.1 The Coiled Ribbon Helicoids

A helicoidal multivibrator can be built on the basis of the double-coiled and combined coiled strips. In the first case, at least one coiled part of the double-coiled spring is located in the *N/S* magnetic field (Fig. 5.5a). In the second case, the initially flat strip 1 (Fig. 5.5b) or the coiled part 2 (Fig. 5.5c) may be located in the magnetic field *N/S*. However, electrical current in the thin-coiled nanospring may cause its compression, and bending by the generated magnetic force.

Natural frequency. Let us evaluate the natural frequency of the coiled strip-string. If we suppose that the thin coiled strip with a small diameter $D = 2R$ and a strip thickness h has negligible flexural rigidity, its natural frequency may approximately be calculated on the basis of the formula for a thin flexible string:

$$f_{nc} = (n_f/2L)\sqrt{F_p/\rho_\gamma A_c} = (n_f/2L)\sqrt{F_p \cos\alpha/\rho_\gamma bh} \tag{5.43}$$

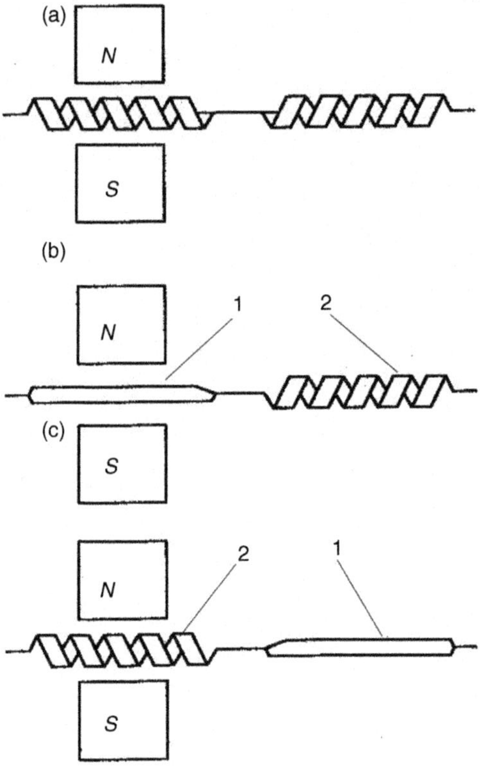

FIGURE 5.5 Quasi-helicoidal multivibrators with the coiled ribbon in magnetic field (a), initially flat strip in the magnetic field (b), and initially flat strip outside the magnetic field (c).

because a cross-section area of the coiled ribbon is equal to $A_c = bh/\cos\alpha$, L is the length of a coiled part, F_p is the force of longitudinal tension, and ρ_γ is the density of the strip's material. If the flexural moment M^f in the double-coiled strip equals zero, the corresponding longitudinal force is equal to: $F_{p\,M^f=0} = M^t_{M^f=0}/R$, which, after substituting $\frac{1}{R} = \left(\frac{\cos\alpha_0}{R_0} - \frac{\theta}{L_r}\right)/\cos\alpha$, yields the following expression:

$$F_{p\,M^f=0} = 2(1-\nu_p)(Db/R_0)\left(\frac{\cos\alpha_0}{R_0} - \frac{\theta}{L_r}\right)\times$$
$$[\sin(\alpha-\alpha_0)/\cos(\alpha-\alpha_0)^2]\times\{\cos(\alpha-\alpha_0) - K_\xi(\alpha-\alpha_0)\}$$
$$-2(1-\nu_p)\cos^2\alpha\sin\alpha$$
$$+(1-\nu_p)^2\cos^2\alpha(2\alpha-\alpha_0)]\}/[1-K_\xi(\sin^2\alpha+\nu_p\cos^2\alpha)^2]. \quad (5.44)$$

For small displacements, we assume that $(\alpha - \alpha_0) = \Delta\alpha$ is small, and $K_\xi \to 0$.

$$\text{Then } F_{\text{p}M\,f=0} = \frac{2(1-\nu_{\text{p}})Db\Delta\alpha}{R_0^2 \cos\alpha_0}. \tag{5.45}$$

Substituting the expression for flexural cylindrical rigidity $D = \frac{Eh^3}{12(1-\nu_{\text{p}}^2)}$ and $\Delta\alpha = \frac{sS_{av}}{2\pi R_0 L / \cos\alpha_0}$, we finally have a linear expression for the tension force at small displacements s of the double-coiled strip's ends, as follows:

$$F_{\text{p}M\,f=0} = \frac{Ebh^3 S_{av}}{12\pi R_0^3 L} s. \tag{5.46}$$

If we recall from Chapter 4 that $s = \frac{\pi D_{r0}^2}{0.0573 S_{av}} \theta$, and $D_{r0} = 2R_0$, then the rigidity of the double-coiled strip's extension related to the angle θ of its cross-piece rotation is equal to:

$$j_\theta = \frac{F_{\text{p}M\,f=0}}{\theta} = C_Q \frac{Ebh^3}{D_{r0} L}, \tag{5.47}$$

where coefficient $C_Q \approx 5.8/(1+\nu_{\text{p}})$, if j_θ is expressed in cN/(angular degree).

5.5.2 Combined Spring Calibration

In the version of a combined coiled spring with the initially flat strip in a magnetic field, the natural frequency of the latter strip at zero twist is given by the known expression

$$f_0 = 0.5L\sqrt{1/(\rho_\gamma A)}[F_{\text{p}M} + (4Ej_1/L^2)]^{1/2}, \tag{5.48}$$

and the natural frequency of a twisted strip is equal to:

$$\begin{aligned} f_{0t} &= f_0(1 + K_f A_{f\theta})^{1/2} \\ &= f_0\{1 + K_f[1 - (\sin\theta/\theta) + (\theta\sin\theta/(4\pi^2 - \theta^2))]\}^{1/2}, \end{aligned} \tag{5.49}$$

where $K_f = (J_2 - J_1)/[(L^2 F_{\text{pM}}/2\pi^2 E) + 2J_1]$, J_2 and J_1 are the cross-section moments of inertia at zero twist ($\theta = 0$) with respect to cross-section coordinate axes, $F_{p\text{M}}$ is the axial tension, and the untwist angle at the input displacement s is equal to $\theta = i_{\theta\text{m}}\, s$. Axial tension F_{pM} in accordance with (5.46) is a variable that makes expressions (5.48) and (5.49) more complicated for linearization. If the strength conditions and attainable frequencies allow, one can substitute the initially flat strip by a round string with equal torsional rigidity, a calibration formula that is simpler, as for a regular string where the small stretch s and the change of natural frequency Δf will have the following relationship:

$$\Delta f = ½\Delta F_{pM}(f_0/Q_{pM0}) = \frac{1}{2}\frac{Ebh^3 S_{\text{av}}}{12\pi R_0^3 L}(f_0/F_{pM0})S. \tag{5.50}$$

Here, the ratio between initial frequency and tension force (f_0/F_{pM0}), as well as other parameters, are assumed to be constant.

5.5.3 The Pretwisted Helicoidal Tube

These tubes find an effective use as sensitive elements in the manometers for high-pressure measurement. They can also be used as double-helicoids in kinematic systems. These helicoids have a smaller longitudinal rigidity $j_{\theta tu}$ than the solid helicoids with the same outer parameters – fact that is clear from the following expression

$$j_{\theta tu} = 69.6(G\lambda\pi/L\kappa_p)\begin{bmatrix} a_{1(1)}b_{1(1)}(1+vj_f b_{1(1)}^2\kappa_p^2/\lambda) \\ -a_{1(2)}b_{1(2)}(1+v_{jf} b_{1(2)}^2\kappa_p^2/\lambda) \end{bmatrix}, \tag{5.51}$$

where $a_{1(1)}$, $b_{1(1)}$ and $a_{1(2)}$, $b_{1(2)}$ are the larger and smaller semi-axes of an outer surface with subscript 1 and an inner surface with subscript 2, $\kappa_{p1} = \kappa_{p2} = \kappa_p = 2\pi/S_o$ if S_o is the helix pitch, and $\lambda_1 \approx \lambda_2 \approx \lambda = (b_{1(i)}/a_{1(i)})^2$. The pretwisted double-helicoid tube has a significantly smaller transmission ratio than the thin double-helicoid from a pretwisted strip.

5.6 DYNAMIC RATE IN HELICOIDAL SENSORS

We assume that helicoidal sensors may be functionally considered as experimental installations for testing EHM DNA's and protein (Shi et al., 2015) dynamic functions.

5.6.1 Photoelectric Sensors

Several types of photoelectric sensors with the elastic helicoids were developed: (i) sorting multicommand with up to 50 dimensional groups; (ii) signaling with adjustable limits; (iii) functional with continuous amplitude output; (iv) photo-pulse with irreversible count; and (v) photo-pulse with reversible count (Tseytlin, 2006, Sec. 8.1).

Photo-pulse sensors. As shown in Figure 5.6a, these sensors with reversible count have measuring spindle 1 on the elastic guides, spring transmission 2 and 3 (elastic helicoid) with rotatable mirror 4. The direction of the light rays between the source 5 and mirror 4 is constant, and it depends on the angular motion of mirror 4 in the space between this mirror and the annular raster grids 9 and 10 with radius R_{gr} and the raster's pitch intervals S_p. Moving the light spot 11 on these reflecting segment grids produces rays' amplitude modulation with the shifted phase. Their phase shift depends on the mutual angular position γ_s of grids 9 and 10. Each of the modulated rays is focused on a certain light sensing element 12 and 12'. Changing the phase shift of the raster grids by pinions 6 and 7, we can change the value of pulse from $c_i = S_p/i_{om}$ to $c_i = 0.5\ S_p/i_{om}$, where i_{om}

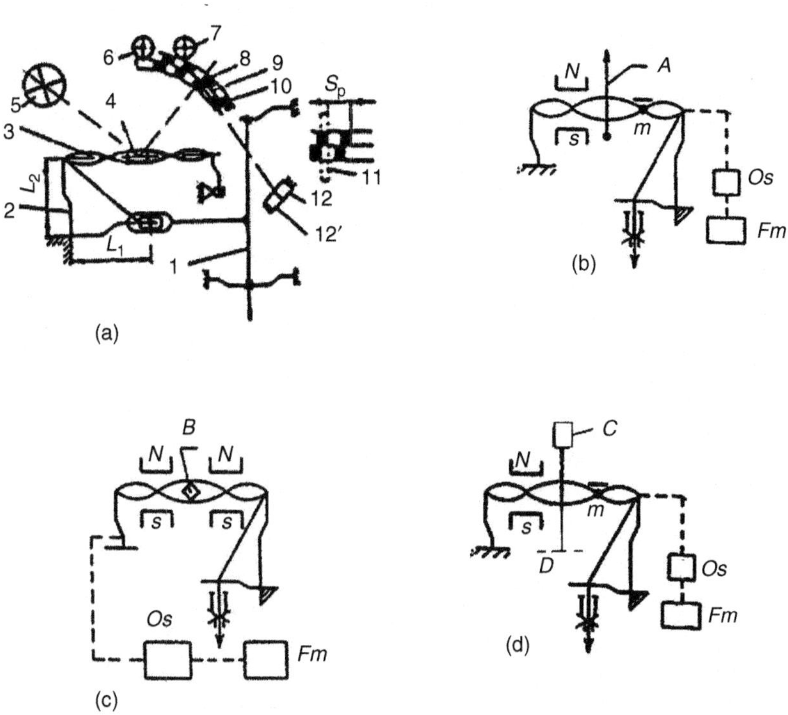

FIGURE 5.6 Helicoidal sensors with photopulse output (a), multivibrators with arrow-pointer A (b), mirror B (c), and laser C (d); *Os* is an self-excited oscillator, *Fm* is a frequency (period) meter.

corresponds to the transmission ratio of the installed opto-mechanical system. A random reverse or parasitic vibration of the light spot on the grids does not cause additional uncertainty. However, in photo-pulse transducers with only one raster grid and irreversible pulse count, a random reverse or parasitic vibration of the light spot on the grid will cause a large uncertainty in the measurement.

The approximate transient functions and dynamic working rate for photoelectric sorting transducer spring-optical sensors are presented in Table 5.3.

The sensor 1 in Table 5.3 has a pretwisted strip with the following parameters: thickness $h \times$ width $b \times$ length L (mm) $\times$ twist parameter k_{p0} (mm^{-1}) = $0.006 \times 0.12 \times 18 \times \pi$. Sensor 2 has the pretwisted strip with the following parameters: thickness $h \times$ width $b \times$ length L (mm) $\times$ twist parameter k_{p0} (mm^{-1}) = $0.016 \times 0.17 \times 18 \times \pi$.

Problem I. It is necessary to estimate a transmission's time lag in sensor 1. We assume that the depth of the sensor's spindle retraction is equal to 0.02 mm. The measuring voltage on the light sensing elements can be switched on only after the light index comes to a new position corresponding to the dimension of a measurand. When spindle returns to a measuring position after retraction,

TABLE 5.3 Transient Function–Step Response θ_e [U (t)], and Dynamic Rate of Microspring-Optical Sensor

No.	Type of sensor	Resolution r_r, μm	η_s	Approximate expression	Rate, Hz
1	Helicoidal strip for case 1 in Eq. (5.6)	0.5 1 2	1.8 0.9 0.45	$\theta_e[U(t)] = \eta_s s\ [67–74 \exp(-100t) + 6.7 \exp(-1100t)]$ 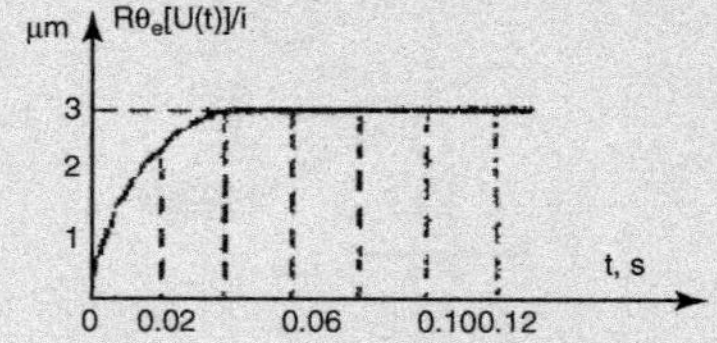 	10 2.2
2	Helicoidal strip for case 3 in Eq. (5.6)	–	–	$\theta_e[U(t)] = s\left[28 - \frac{338 \cdot 10^5}{z}\exp(-n_\theta t)\frac{z\cos zt - n_{\theta \sin zt}}{n_\theta^2+z^2}\right];$ $n_\theta = \frac{\varsigma_\theta}{2J_{mr}} = \frac{1}{\tau}\ln\frac{A_{i+1}}{A_i};\ z = \sqrt{1.2 \cdot 10^6 - n_\theta^2}$ 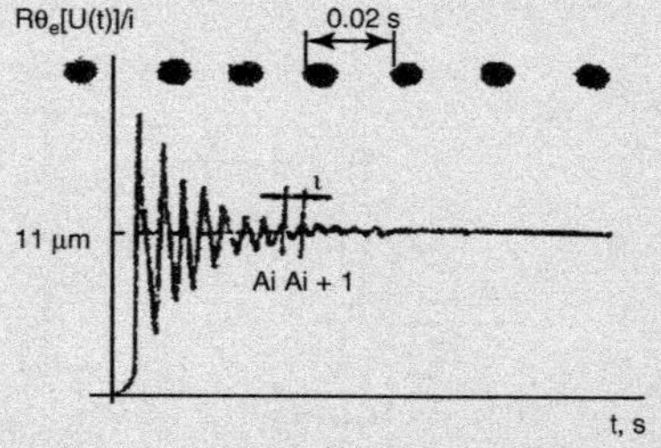 	2 0.87

Note: U (t) is a unit step; ζ_θ is a coefficient of viscose resistance in the damper; J_{mr} is the mass moment of inertia for the rotatable mirror and shellac bead; A_i and A_{i+1} are sequential amplitudes of the sensor's index vibration on the dial; $\alpha_{T\theta}$ is the temperature coefficient for the crosspiece of the pretwisted strip untwist in angular degree per 1°C; s is the displacement of the spindle, i is the transmission ratio.

Source: After Tseytlin (2006), reprinted with permission.

the switching light index should pass an angular distance equal to: $\theta = \eta_s \cdot 0.02 \cdot 67 = 0.9 \cdot 0.02 \cdot 67 = 1.2$ rad, and the time of its motion, in accordance with the transient function of Table 5.3, equals $t_M \approx 0.06$ s. This should be reflected in the cyclogram of the automatic measuring system. The process of the light index displacement is considered finished after an angle of the mirror rotation corresponds to 0.98θ.

Problem II. The accurate wedge moves under the measuring tip of spindle with the constant speed V_{wed} in the direction that is perpendicular to the spindle of sensor 1. Therefore, we can represent the input motion of the spindle with the following expression $s_{in} = V_{wed}\, t\, tan\,\psi_{wed}$, where ψ_{wed} is the slope angle of the wedge plane. It is necessary to estimate the difference Δs in the position of the light pointer of the sensor on the dial, and the set of photoresistors corresponding to the quasi-statical and dynamical measurement of the same wedge's section dimensions. Substituting into Duhamel's integral (5.2) the spindle's motion presentation and the derivative of the sensor's transient function (Table 5.3, sensor 1), we obtain the following expression for the sensor's pointer output motion:

$$\theta_{out}(t) = \int_0^t \eta_s[7400e^{-100(t-\tau)} + 7330^{e-1100(t-\tau)}] \times$$
$$V_{wed}\tau \tan\psi_{wed} d\tau \approx \eta_s V_{wed} \tan\psi_{wed} \times (67t - 0.74 + 0.74e^{-100t}).$$

The actual dimension of the wedge section, above which the measurement tip passes in the moment t, is equal to $V_{wed}\, t\, tan\,\psi_{wed}$, and the position of the light pointer on the dial that corresponds to this dimension in the statical state is

$$\theta_{out.st} = 67\eta_s t V_{wed} \tan\psi_{wed}.$$

Let us note that number 67 represents the quasi-statical transmission ratio of the sensor's mechanism. As a result, the sought difference is equal to: $\Delta\theta = \theta_{out}(t) - \theta_{out.st} \approx -\eta_s V_{wed}(0.74 - 0.74e^{-100t})\tan\psi_{wed}$.

It is clear from this equation that the dynamic uncertainty of the sensor will increase with an increase of the wedge's motion velocity V_{wed}, or with the increase of its slope angle ψ_{wed}. These conclusions were, of course, expected.

5.6.2 Versions of String Helicoidal Multivibrators

The structure of the sensors with bifurcating helicoidal multivibrators are: (i) with pointer A (Fig. 5.6b); (ii) with mirror B (Fig. 5.6c); (iii) by indicating the crosspiece angle of rotation with the laser C (Fig. 5.6d) ray diffraction picture D. The transmission in which the crosspiece has no additional mass is effective for a faster measuring rate, than that with a mirror. Therefore, the amplitude output in the amplitude and frequency sensors with the helicoidal strip may be performed even without adding anything to the crosspiece.

5.7 FLEXURAL WAVE PROPAGATION

Recent studies of nanostructures' dynamics, such as dispersion of phase velocity v_{pc} in carbon nanotubes at high frequency vibration, are based on the different beam models: the traditional Euler-Bernoulli and Timoshenko elastic beams, as well as their nonlocal elastic versions. All four models are effective at wavelengths larger than 6.3×10^{-9} m. Predictions of flexural waves' dispersion at very high frequencies (in the THz region) with wavelengths less than 3×10^{-9} m are more effective on the basis of the Timoshenko nonlocal elastic beam vibration (Wang and Hu, 2005). The Timoshenko nonlocal elastic beam takes not only the rotary inertia (Rayleigh correction) and shear deformation, but also the second-order gradient of strain that characterizes the microstructure of the system under consideration. These considerations are important for high frequencies, when a vibrating beam is subdivided by nodal cross sections into comparatively short portions. For example, the influence of shear and rotation inertia on the simply supported short rectangular beam's frequency of oscillations may be represented (Panovko, 1968) by the following formula:

$$f_k^* = f_1 \frac{k_w \lambda_{\rho L}}{\pi} \sqrt{\frac{1}{2}\left(B_{kL} \pm \sqrt{B_{kL}^2 - 4\beta_{GE}^2}\right)},$$

where $f_1 = \frac{\pi}{2L^2}\sqrt{EI/m}$ is the lowest resonance frequency calculated without account for the shear and rotation inertia, m is the mass per unit length of the beam, $B_{kL} = 1 + \beta_{GE}^2 + \frac{\beta_{GE}^2 \lambda_{\rho L}^2}{\pi^2 k_w^2}$; β_{GE} is the factor depending on the cross section geometry ($\beta_{GE} = 1.2$ for rectangular cross-section); $\lambda_{\rho L} = L/\rho_{gr}$ is the beam's flexibility, ρ_{gr} is the radius of cross-section gyration, and k_w is the number of half-waves in the beam at oscillations. Another useful formula based on the Rayleigh –Timoshenko model is shown in Weaver et al. (1990), where one can see an influence of shear and rotation inertia on the frequency of a short simply supported beam. Let us recall that the shear deformation in static contributes under uniform load only few percents to the comparatively short prismatic cantilever's free end total deflection. In turn, this latter depends on the square of the ratio of depth h of the beam to its length L (Oden and Ripperger, 1981).

5.8 LIMITS OF THE TRANSMISSION ENTROPIC FLUCTUATIONS

The expected limits of statistical fluctuations for helicoidal strip sensor and coiled ribbon are shown in Table 5.4. These fluctuations correspond to the normal statistical distribution, and are in agreement with the equipartition

TABLE 5.4 Limits of Thermal Statistical Fluctuations in Helicoidal Sensors

Type of model	Fluctuation limits $\Delta_{\varepsilon M}$	Example of estimation at normal temperature 293.15 K
Pretwisted strip	$\Delta_{\varepsilon M} = \pm 3\sqrt{\dfrac{2k_B TL(\lambda + \nu_{if} b_1^2 k_p^2)}{EA(\lambda + 0.18 b_1^2 k_p^2)}}$	$h \times b \times k_p \times L = 0.004 \times 0.06 \times \pi \times 20$ $E = 112.8$ GPa $\Delta_{\varepsilon M} \approx \pm 0.8 \times 10^{-8}$ mm
Coiled ribbon	$\Delta_{\varepsilon M} = \pm 3\sqrt{\dfrac{k_B T \pi D_r^3 L}{0.36 S_{av} b h^3 E}}$	$D_r = 0.2$ mm; $S_{av} = 1$ mm; $b = 0.5$ mm; $h = 0.01$ mm; $E = 200$ GPa; $L = 76$ mm $\Delta_{\varepsilon M} = \pm 4.4 \times 10^{-8}$ mm

theorem that states that the object's fluctuation variance is proportional to Boltzmann's constant, absolute temperature, and inversely proportional to the object stiffness.

Medium with shape memory. Most elements of precision elastic systems proportionally change their dimensions with the change of temperature. The hysteretic damping of this process in elastic systems is usually small, if the change of temperature is not very large and fast. However, there are some metal alloys that have a so-called shape memory (e.g., titanium-nickel or nitinol (Ti-Ni), Ti-Pd, Cu-Al, Au-Cu, Fe-Mn alloys). In these cases, small changes of an ambient temperature relative to the specific level may cause significant changes in the shape of the elastic element made of the material with shape memory. This process is not accurate enough for precision systems, and is more applicable to servo-motion systems, grippers, and medical injection applications.

Speed break options. Fast deformations of elastic elements are followed by the elastic and plastic waves in them. The simplest differential equation for elastic and plastic wave propagation along the beams is as follows:

$$\rho_\gamma \, \partial^2 s / \partial t^2 = E_s \, \partial^2 s / \partial x^2, \tag{5.52}$$

where $E_s = \partial\sigma_\varepsilon / \partial\varepsilon_s$ is the modulus of the longitudinal deformation from the diagram of the material extension (Fig. 5.7); ρ_γ is the density of the material, s is the elastic deformation, and x is a current coordinate for points in the beam's length, along with the beam's axis. It is natural that for elastic deformations of material with small internal friction $E_s = E$, where the latter is the Young's modulus. With the fast displacement of one of the beam's ends, while its other end is fixed and immovable, the deformations and stresses in the material can be much larger than that for the equal quasi-static displacement. Speed V_1 of the

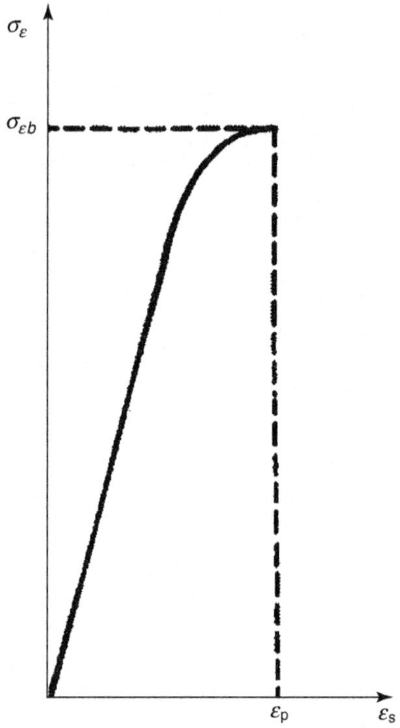

FIGURE 5.7 Material extension diagram with strain ε_p corresponding to break stress $\sigma_{\varepsilon b}$.

movable end displacement and the strain ε_s of the material are connected by the following relation:

$$V_1 = \int_0^{\varepsilon} s(E_s/\rho_\gamma)^{0.5}\, d\varepsilon. \tag{5.53}$$

Stress $\sigma_{\varepsilon b}$ and end displacement ε_p correspond to the break of the beam.

REFERENCES

Bolotin, V.V., 1956. The Dynamic Stability of Elastic Systems. Gostekhteorisdat, Moscow (in Russian). English transl. V.I. Weingarten et.al., Holden-Day, 1964.

Carnahan, R.D., 1964. Microplasticity – its measurement and application to guidance and control components. J. Metals 16, 990–994.

McLachlan, N.W., 1964. Theory and Application of Mathieu Functions. Dover, New York.

Oden, J.T., Ripperger, E.A., 1981. Mechanics of Elastic Structures, second ed. McGraw-Hill, New York.

Panovko, Y.G., 1968. Natural and forced vibrations of beams and beam systems. In: Birger, I.A., Panovko, Y.G. (Eds.), Strength, Stability, Vibrations. Handbook in Three Volumes, 3, Mashinostroenie, Moscow, pp. 285–346 (in Russian).

Shi, F., Zhang, Q., Wang, P., et al., 2015. Single-protein spin resonance spectroscopy under ambient conditions. Science 347, 1135–1138.

Strutt, M.J.O., 1932. Laméshe-Mathieushe-und vermandte funktionen in physik und technik. Verlag J. Springer, Berlin, (in German).

Tseytlin, Y.M., 1978. Automation of spring measuring heads. Meas. Tech. 21, 34–36.

Tseytlin, Y.M., Skachko, Iu.V., Kapirin,V.V., 1989. Modified String Transducers for Measuring of the Geometric Quantities. Standartov, Moscow (in Russian).

Tseytlin, Y.M., 2006. Structural Synthesis in Precision Elasticity. Springer, New York, NY.

Wang, L., Hu, H., 2005. Flexural wave propagation in single-walled carbon nanotubes. Phys. Rev. B 71, 195412.

Weaver, W., Timoshenko, S.P., Young, D.H., 1990. Vibration Problems in Engineering, fifth ed. Wiley, New York.

Chapter 6

Double-Stranded DNA Elasticity

6.1 HELIX CHAIN

The elasticity of the DNA molecule with double-stranded (dsDNA) and single-stranded (ssDNA) helical structure has been extensively and carefully studied. We should mention that the DNA molecule's model in the shape of a thin cylindrical elastic rod and elastic cylindrical core, wrapped with wires (Gore et al., 2006; Swigon, 2009), used by some authors, seems to be not relevant to the physical structure of DNA – that resembles an elastic helix; notwithstanding that, estimates based on elastic rod or cylindrical core models agree with some experimental data.

From a theoretical stand point, a DNA molecule is modeled as a worm-like helix chain (Schellman, 1974) with a simple relationship between the local distortion of helix regularity in the deviation of direction angles θ_i and a free energy $E_g(\theta)$ temperature dependent variation similar to the statistical fluctuation in the macro-model of precision elastic systems (Chapters 2, 4, 5, 7) where the standard deviation σ_i (θ_i) in accordance with the equipartition theorem equals: $\sigma_i(\theta_i) = \sqrt{\dfrac{K_B T\theta_i^2}{2E_g(\theta)}}$.

When tension in a nicked DNA molecule is increased to 65 pN, it displays a reversible, cooperative transition to an extended form that is approximately 70% longer than normal B-DNA (Chapter 7), and with a substantially reduced twist. Twist-stretch coupling has also been found in these molecules. There exists evidence that overstreched DNA (polynucleotide) may present one of the following structures: (i) base-paired (dubbed *S* form), or (ii) two independent strands of ssDNA that are similar in structure (Bhushan, 2010) to polypeptide single helix.

6.2 PROPERTIES OF EXPLICIT HELICOIDAL MODEL (EHM)

We assume that the explicit functional helicoidal models (EHM) of dsDNA (Mod 16 a,b,c,d) can rotate as a result of nicks and kinks in the DNA strands (Vologodskii and Frank-Kamenetskii, 2013) especially for short dsDNA molecules.

Experimental data and theoretical analysis suggest that kinks may represent openings of isolated base pairs that have been experimentally detected in linear

DNA molecules. The most natural way to strongly bend DNA is to convert it into a circular form. The calculation suggests that, although the probability of kinks opening in unstressed DNA is close to 10^{-5}, it increases sharply in small DNA circles reaching 1 open bp per circle of 70 bp.

Spatial mathematical rod models of closed molecules of DNA were studied by Iliykhin and Timoshenko (2008).

The efficiency of the cyclization process is quantitatively described (Vologodskii and Frank-Kamenetskii, 2013) by the so-called *j*-factor that is equal to

$$j = 2M_0 \lim_{t\to 0} \frac{C(t)}{D(t)},$$

where M_0 is the initial concentration of the fragments in solution, $C(t)$ and $D(t)$ are the concentration of ligated monomeric circles and dimers, respectively, at the time moment t.

On this basis, we assume that our rotatable model (Fig. 6.1a) is good for a DNA molecule with short length of at least up to 150 bp. We can stop the

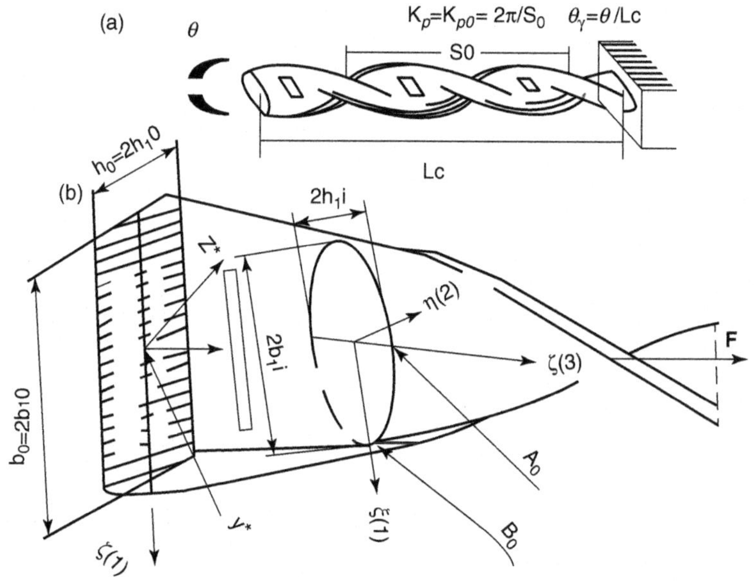

FIGURE 6.1 Pretwisted hollow nanostrip model. (a) Schematic of the instrumental model; (b) the calculation fragment, $\xi(1), \eta(2), \zeta(3)$ is the helix coordinates'system, coordinate axis $\zeta(3)$ corresponds to the strip's length direction, Z_* and Y_* are the following coordinates, A_0 and B_0 are the most critical stress points of the cross section, $h_o = 2\ h_{1o}$ and $b_o = 2\ b_{1o}$ are the thickness and the width of the strip's cross section outer contour, $2h_{1i} = h_i$ and $2b_{1i} = b_i$ are the thickness and the width of the strip's cross section inner contour. *(After Tseytlin, 2013).*

rotation by the application of certain set of forces and moments (torques), if necessary.

Subsequently, we developed a more complete model with nonlinear relations between the helicoidal punched pretwisted nanostrip's length L_c and the variance of thermomechanical fluctuations that are similar to experimentally found (Mathew-Fenn et al., 2008a,b; 2009) length fluctuations in the DNA molecule.

The parameters of the developed EHM mechanical model are close to the mechanical features of DNA molecules; these are now in use for different purposes, including precision instruments in physical research, nanotechnology manipulators, nanoscissors handles, etc.

Any mechanical model requires knowledge of its constituents' mechanical properties, particularly Young's modulus and Poisson's ratio. For DNA, some researchers equate Young's modulus E_e with the stretch modulus E_{stm} = 1,000…1,100 pN that has units of force. However, Young's modulus should be expressed as a stress value: $E_e = F_{stm}/A_{cDNA}$, where A_{cDNA} is the cross-sectional area of the DNA model. For example, $A_{cDNA\ 1} = \pi\, d_D{}^2/4$ for a model with a circular cross section (Smith et al., 1996) of diameter d_D, while $A_{cDNA\ 2} = (b_o \times h_o - \pi b_{1i} \times h_{1i})$ for our EHM model with the pretwisted hollow nanostrip (Fig. 6.1b), where $b_{o,i} = 2b_{1o,i}$ is the width and $h_{oi} = 2h_{1o,i}$ is the thickness of the outer rectangular (index o) and inner (index i) elliptical cross sections, respectively. Therefore, for a given stretch modulus, the corresponding Young's modulus E_e can have different values depending on the effective model cross section. For the case of a circular cross-section, the molecular diameter d_D is usually taken to be 2.0…2.2 nm that corresponds to E_{e1} = 346–350 pN/nm^2 = 3.46–3.5 × 10^8 Pa. For the nanostrip model (EHM), we assume that $b_o = 2\, b_1o = 2.1$ nm or that it is approximately equal to the DNA cell circumference diameter d_D, $h_o = 2h_{1o}$ = 0. 295 nm, $(h_o/b_o)^2 = \lambda = 0.0197$; $b_i = 2\, b_{1i} = 1.75$ nm, $h_i = 2\, h_1 i = 0.22$ nm (Fig. 6.1b) and A_{cDNA2} = 0.317 nm^2. Therefore E_{e2} = 3470 pN/nm^2 = 3.47 GPa. Note that the product $E_e A_{cDNA} = F_{stm} \approx$ 1,000–1,100 pN remains the same for both models. The average published value (Ikai, 2008) for DNA Young's modulus is equal to 1 GPa. Poisson's ratio v_p is usually taken as equal to 0.3…0.5 as for polymers. The initial twist parameter of the pretwisted strip is given by $k_{p0} = 2\,\pi/S_0 = 1.847$ nm^{-1}, where S_0= 3.4 nm is the typical pitch of the B-DNA molecule helix. The dimensionless parameter of the strip's relative pretwist is $\chi_{kp} = k_{p0}^2\,(b_o/2)^2\lambda^{-1} = 190.9$, the ratio between the section torsional stiffness T_0 and area A_c value (Shorr, 1968) equals

$$v_{T_0/A_c} = T_0/A_c = \{0.3333\, b_o^4\lambda^{3/2} - [(\pi/16)b_i^4\lambda^{3/2}/(1+\lambda)]\}/A_{cDNA2} = 0.0408\ \text{nm}^2. \tag{6.1}$$

Calculation shows that the persistence length for the nanostrip model with these parameters is equal to 50 nm, and the twist-stretch coupling is 90 pN/nm, which are in agreement with the published data for DNA molecule properties.

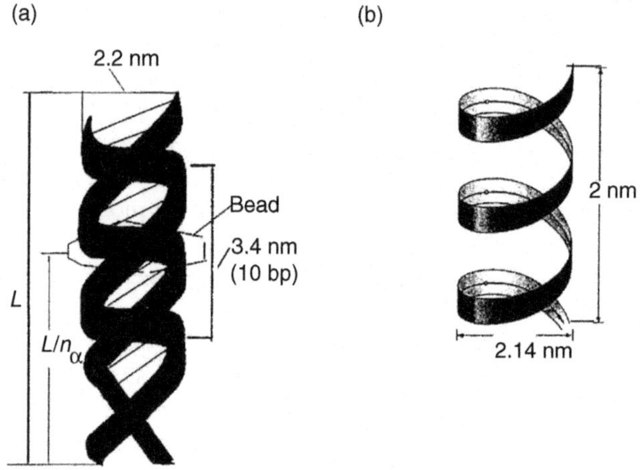

FIGURE 6.2 Helico. *(After Tseytlin, 2006, reprinted with permission.)*

6.3 TWIST-STRETCH COUPLING

We assume now that the stretched dsDNA model (i) is similar in its mechanical aspects to the elastic helicoid (Fig. 6.2a) made of a punched pretwisted strip, and model (ii) of ssDNA is similar (Fig. 6.2b) to a coiled ribbon structure (see Chapter 4). Nanohelicoidal models may be synthesized by chemical methods with templates of DNA-nanoparticle systems. Let us consider (Tseytlin, 2011) the features of the base-paired structure (i).

For twist-stretch coupling g_p estimation, we can use the modified conditional radius formula (5.8) of elastic pretwisted strip:

$$\rho_{dc} = \frac{M}{F} = \frac{j_{k\tau} - j_{\tau\tau} i_{\theta m}}{j_{kk} - j_{\tau k} i_{\theta m}}, \tag{6.2}$$

where $j_{\tau\tau} = GT_0(1 + \nu_{ipf} b_1^2 k_p^2 / \lambda)$; $j_{kk} = EA_c$; $i_{\theta m} = \dfrac{Ek_p}{4n_a G\lambda}(1 + \nu_{ipf} b_1^2 k_p^2 / \lambda)^{-1}$, and $j_{k\tau} = j_{\tau k} = EA_c b_1^2 k_p / 4$.

Here $j_{k\tau} = j_{\tau k}$ is the tension stiffness of a pretwisted strip (see Chapter 4) under its tension with a simultaneous twist, or the stiffness of this strip under its twist with a simultaneous tension, $j_{\tau\tau}$ is the twist stiffness of a pretwisted strip under twist with an applied torque, but without tension force, j_{kk} is the tension stiffness of the pretwisted strip under tension without twist, $i_{\theta m} = |\theta/(sn_a)|$ is the modified ratio in rad/nm between the untwist bead angle θ and corresponding stretch s of the pretwisted strip, $n_a = 1\ldots2$ is a factor of the experimental submicrometer rotor bead location on the DNA molecule with $1/n_a$ as a length fraction (see Fig. 6.2a and Gore et al., 2006), and $T_0 = A_c \times \nu_{T_0/A_c} = 0.0408 \times 0.317 = 0.0129$ nm^4 in accordance with Eq. (6.1).

Calculations with Eq. (6.2) correspond well to DNA experimental data. For example, $\rho_{dc} = -0.087$ nm and $E_{stm}\rho_{dc} = g_p \approx -95$ pN·nm at $n_a = 1.65$ and accepted parameters of our DNA model.

We see a similar value of $g_p = -90 \pm 20$ pN Nm in Gore et al. (2006), because in linear transmission $x^*F = \theta M$ and

$$\frac{\partial x^*}{\partial \theta} = \frac{M}{F} = \rho_{dc} = -\frac{g_p}{E_{str}},$$

where x^* is an extension of the molecule.

We estimate the *stretch force-untwist rotation* angle $(F-\theta)$ function of the pretwisted nanostrip by the following formula:

$$\theta = Fj_{\theta}^{-1}. \tag{6.3}$$

We can estimate the spring stiffness k_h of DNA hydrogen bonds as a product of the translational s and rotational θ motion transformation $i_\theta = \theta/s$ and the torsional rigidity at the pretwisted strip extension $j_\theta = F/\theta$ that yields

$$k_h = \frac{EA_c(1+v_{jf}\chi_{kp})}{L_b(1+v_{if}\chi_{kp})} = 1{,}131.6 \text{ pN/nm}, \tag{6.4}$$

where $v_{jf} = (1/2)[0.1052(1 + v_p) + (1 + v_p)/8]$ is the rigidity factor for the hollow strip with $v_{jf} = 0.14$ at $v_p = 0.2$; $v_{if} = 0.41$ is the transformation factor for the perforated pretwisted strip, $\chi_{kp} = k_{p0}{}^2(b_o/2)^{2-1} = 190.9$, and $L_b = 0.34$ nm is the length of one DNA bones' base pair.

6.4 STATISTICAL FLUCTUATION

The expected limits of statistical fluctuations for different elastic systems correspond to a normal statistical distribution and are in agreement with the equipartition theorem. The limit of statistical fluctuations' standard deviation for the pretwisted strip structure (see Table 5.4) is equal to:

$$\sigma_{\varepsilon M} = \sqrt{\frac{k_B T L_c(\lambda + v_{ipf}\, b_1^2 k_{p0}^2)}{EA_c(\lambda + 0.18\, b_1^2 k_{p0}^2)}}, \tag{6.5}$$

where L_c is a length of the molecule.

Therefore the variance of these fluctuations equals

$$\sigma_{cM}^2 = \frac{k_B T L_c(\lambda + v_{ipf}\, b_1^2 k_{p0}^2)}{EA(\lambda + 0.18\, b_1^2 k_{p0}^2)}. \tag{6.6}$$

Let us recall that $k_B T = 4$ pN nm at the reference temperature $T = 293$K. We can also assume that, for double stranded DNA, the expression $EA_c(\lambda + 0.18 b_1^2 k_p^2)/L_c(\lambda + v_{ifp} b_1^2 k_p^2) \approx F/s$ is a stretch stiffness, and may be equal to 65/(3.4 × 0.7) = 27 pN/nm for the nicked molecule of 10 base pairs (bp) in length (see Bustamante et al., 2003; Strunz et al., 1999), where F is an acting stretching force, and F/s is the molecule's elastic stretch rate. Therefore, from (6.6), we can obtain the limit of the statistical fluctuations' standard deviation within $\sigma_{\varepsilon M} = 0.4$ nm for a DNA molecule with the parameters shown. If the flexural persistence length of the stretched DNA under the force of 10 pN is equal to A_{bp} = 50 nm (as shown in many published sources), and the elongation of the DNA molecule under the force of up to 65 pN is linear and equal to s = 0.1 A_{bp} = 5 nm, when the force reaches 65 pN (see Sarkar and Alexandridis, 2015), then the elastic rate F/s = (65–10) /5 ≈ 0.14 pN/nm. In this case, the standard deviation of the fluctuations is equal to $\sigma_{\varepsilon M} = \sqrt{4/14} = 0.5\,\text{nm}$. If we assume that Young's modulus of the DNA molecule is equal to (3.46 ± 0.3) × 10^8 Pa (see Smith et al., 1996), then our prediction for a standard deviation of the fluctuations in a molecule with the flexural persistence length A_{bp} = 50 nm, on the basis of expression (6.5), is not less than

$$\sigma_{\varepsilon M} = \sqrt{\frac{4 \times 50 \times 1.35}{3.46 \times 100 \times 1.46 \times 0.63}} = 0.9\,\text{nm},$$

where 1.46 nm^2 = A_c is the estimation of the molecule's cross-section area, 1.35 and 0.63 are the numerical values of ($\lambda + 0.39 b_1^2 k_p^2$) and ($\lambda + 0.18 b_1^2 k_p^2$) members, respectively, with the above accepted parameters λ, b_1, and k_p of the DNA molecule. Let us recall also that, on the basis of the wormlike DNA model, Schellman (1974) and Bustamante et al. (2003) found the expression for Force (F_D) – Stretch S_D relationship (see **Mod 1** in Chapter 1), as follows:

$$F_D = \frac{K_B T}{A_{\text{bp}}}\left[\frac{S_D}{L_{\text{cont}}} + \frac{1}{4(1 - S_D/L_{\text{cont}})^2} - \frac{1}{4}\right], \tag{6.7}$$

where A_{bp} is the persistence length and $L_{\text{cont}} = S_0 N_t$ is the contour length of DNA molecule with N_t turns, and helix pitch S_0 = 10 bp or 3.4 nm. The expression $F_D - S_D$ has linear and nonlinear members. However, if $S_D/L_{\text{cont}} = 1$, the situation becomes unfeasible (division-by-zero) with infinite force $F_D = \infty$.

Let us note that the variance in Eq. (6.6) is a linear function of the molecule's length that does not correspond to recent experimental results (Mathew-Fenn et al., 2008a,b; 2009) that show the nonlinear fluctuation variance with respect to the molecule's length. But we have another estimate for variance of the ds-DNA molecule length fluctuation in the last section of this chapter, on the basis of nonlinear relations in the pretwisted strip model.

6.5 STRETCH STIFFNESS AND POISSON'S RATIO

The rigidity form coefficient for elastic pretwisted (helicoidal) strip

$$v_{if} = (1/2)[0.1052(1 + v_p) + (1 + v_p)/8] \tag{6.8}$$

is given here as the average of the corresponding coefficients for rectangular (outer) and elliptical (inner) contours. We recall that Poisson's ratio ν_p for different regular materials can be in the range 0.0–0.5. The latter value of 0.5 was assumed for our calculations (Tseytlin, 2011). However, the real value of ν_{if} and subsequently Poisson's ratio ν_p can be verified through the experimental measurement of the DNA molecule's stretch stiffness

$$j_{sp} = dF/ds,$$

where F (Fig. 6.1b) is a stretching force, and s is the strip's corresponding stretch. In this case, we can use the simplified linear relations for the transformation ratio i_θ and rotational stiffness $j_{\theta o}$ (see Chapter 4) that are equal to:

$$i_\theta = -\frac{0.0286\,Ek_{p0}}{4G\lambda(1+\nu_{if}\chi_{kp})} \tag{6.9}$$

and

$$j_{\theta 0} \approx 69.6GA_{cDNA}\lambda(1+\nu_{jf}\chi_{kp})/L_c k_{p0}, \tag{6.10}$$

where G is the shear modulus of material.

As a result, the ν_{jf} value can be deduced (after multiplication and certain cancellations) as follows:

$$\nu_{jf} = \left[\frac{j_{sp}(1+\nu_{if}\chi_{kp})L_c}{0.498E_{e2}A_{cDNA}} - 1\right]\chi_{kp}^{-1} \tag{6.11}$$

Here, we assume ν_{if} = 0.41 as for the perforated helicoidal strip (see Chapter 4, Table 4.8). If we assume from Bustamante et al. (2003) and Strunz et al. (1999) that for nicked DNA with released torsion in the strand j_{sp} = 27 pN/nm = 65/(0.7 × 3.4) and, for example, L_c = 6.93 nm (20 base pairs), then

$$\nu_{if} = \left[\frac{27(1+0.41\times 190.9)\times 6.93}{0.498\times 3479\times 0.317} - 1\right]\times 190.9^{-1} = 0.14$$

and from Eq. (6.8)

$$\nu_p = \frac{2\times 0.14}{0.2302} - 1 = 0.2.$$

6.6 PERSISTENCE LENGTH RELATIONS AND POISSON'S RATIO

The conditional Poisson's ratio of DNA may be related to its persistence length A_{bp} if we assume the following expression for the latter:

$$A_{bp} = E_{e2}^{*}J_\eta/[n_{bp}(k_B T)], \tag{6.12}$$

where $E^*_{e2} = E_{e2}/(1-v_p^2)$ is an appropriately modified elastic modulus for the strip bending, J_η is the moment of inertia for the hollow cross-section (Fig. 6.1b) of the model with pretwisted strip. We also assume that under different conditions (including temperature, type of solution), the value of factor n_{bp} should be $\sqrt{2}$ or 2 for the helix body with two surfaces: outer (*o*) and inner (*i*). This follows from the possible combination of thermomechanical statistical influence on the two surfaces as $\sqrt{(k_BT)^2+(k_BT)^2} = k_BT\sqrt{2}$, or from a necessary thermomechanical energy $2(k_BT)$ at larger material deformation, with larger v_p. Our experience with the published experimental data shows that the first value $n_{bp} = \sqrt{2}$ is more appropriate for a molecule with $v_p \le 0.2$, $v_{jf} = 0.14$ and the second value $n_{bp} = 2$ for a molecule with $v_p = 0.5$, $v_{jf} = 0.18$ in standard solution with 70 mM Tris-Hl, pH 8.0, and 10 mM ascorbic acid.

Eq. (6.12) yields persistence length values at $T = 293$ K within 49–53 nm, for both cases. In any case, the experimental value of j_{sp} and A_{bp} should preferably be measured or evaluated at the corresponding conditions, because the persistence length depends on them, as well as on the molecular length. A mean experimental value of the DNA persistence length in different cases and different solutions is equal to $A_{bp} = (30 + 80)/2 = 55$ nm. The value of v_{jf} is indeed a scale factor that can and should be verified on an experimental and theoretical basis for certain representative point of the function in evaluation.

In accordance with Eq. (6.12), the relevant Poisson's ratio may be found from the following expression:

$$v_p = \sqrt{1-\frac{E_{e2}J_\eta}{A_{bp}n_{bp}(k_BT)}}$$

where $E_{e2} = 3470$ pN/nm^2, $J_\eta = \frac{1}{2}(J_{\eta o} - J_{\eta i}) \approx (b_o^3 h_o/24) - (\pi b_{1i}^3 h_{1i}/128) = 0.085\ \text{nm}^4$ is the moment of inertia for the hollow cross-section of a helicoidal strip. Similar values of the DNA persistence length of 50–53 nm are shown in many published sources. The persistence length (see Boal, 2002) in our model is equal to

$$A_{bp} = E_{e2}J_\eta\left(1-v_p^2\right)^{-1}/\left[\sqrt{2}(k_BT)\right] = 51\ \text{nm at } v_p = 0, \text{ and at } v_p = 0.2 \quad A_{bp} = 53\ \text{nm}.$$

The twisting of stretched DNA can lead to other structural transitions. For example, after a critical amount of overwinding has been introduced into a molecule, it may buckle. Complex force-extension curves with multiple force plateaus are seen when single DNA molecules are twisted in either direction, and pulled to high forces. This allows us to assume possible jumps of hysteresis asymmetry at a fast and slow extension, and the relaxation of the DNA molecules similar to those in the pretwisted helicoids (see Chapters 5, 7).

The persistence length of dsDNA depends on the ionic strength of the solution (see Baumann et al., 1997) as

$$l_p = l_{po}^{ds} + l_{e1} = l_{po}^{ds} + \frac{1}{4k^2 l_B} = l_{po}^{ds} + 0.324 I^{-1}\ \text{Å},$$

where $l_{po}^{ds} = 500$ Å and l_{e1} are the nonelectrostatic and electrostatic contributions to persistence length l_p, $1/k$ is the Debye-Hückle screening length, l_B is the Bjerrum length (7.14 Å in water at 25°C for double-stranded DNA), and I is the ionic strength of the medium in molar units. In the case of ssDNA, $l_{p0}^{ss} = 1.481\,\text{nm}$ and $l_p = l_{p0}^{ss} + \dfrac{0.03977}{I}$ nm (Ambia-Garrido et al., 2011).

6.7 OVERWINDING OPTIONS

The overwinding of DNA upon the beginning of stretching was discovered by a group of researchers in 2006 (Gore et al., 2006). Simple physical intuition predicts that DNA should unwind under tension, as pulled toward a denatured structure. However, for small distortion, contrary to intuition, DNA overwinds under tension, reaching a maximum twist at a tension of approximately 30 pN. As tension is increased above this critical value, the DNA begins to unwind. The observed twist-stretch coupling predicts that DNA should also lengthen when overwound under constant tension.

6.7.1 "Toy" Model

For the initial research of the overwinding phenomenon, a simple **Mod 13** model (Type3, Table 1.3) was presented that explains these unusual properties. These results have implications for the action of DNA-binding proteins that must stretch and twist DNA to compensate for variability in the lengths of their biding sites. The energetics of these distortions is governed by the mechanical properties of DNA. For small deformations, DNA in physiological buffer was usually modeled as an isotropic rod with bending rigidity $B_p = 230 \pm 20$ pN nm^2, twist rigidity $C_p = 460 \pm 20$ pN nm^2, and stretch modulus $E_{stm} = 1100 \pm 200$ pN. A fourth mechanical parameter is allowed in the linear theory of a deformable rod: the twist-stretch coupling g_p that specifies how the twist of the helix changes when the molecule is stretched. At forces sufficient to suppress bending fluctuations, the energy of a stretched and twisted DNA molecule is written (Gore et al., 2006) as

$$\epsilon_{DNA} = \frac{C_p}{2L}\theta^2 + g_p\theta\tfrac{x}{L} + \frac{1}{2}\frac{E_{str}}{L}x^2 - xF,$$

where L is the contour length at zero force, x is the distance that the DNA is stretched beyond its contour length, and θ is the angle through which the DNA is twisted from its unperturbed equilibrium value.

Interpolation between the B-form helix and denatured or overstreched forms of DNA suggests that g_p should be positive, so that DNA unwinds as it is stretched. However, for small deformations, DNA indeed overwinds when stretched. It was found that the twist of DNA increases until the tension reaches a critical value $F_c = 30$ pN, beyond which the DNA begins to unwind. The negative twist-stretched coupling g_p observed at lower tensions implies that DNA should lengthen when overwound.

To explain these experimentally found DNA features, Gore et al. (2006) proposed the artificial model **Mod 13** (Type 3, Table 1.3), a model rod with shrinking radius as it is stretched. In simulation of DNA stretching, using all-atom potentials, a reduction in the radius of the double helix has been seen, concurrent with stretching and overwinding. As a result, the artificial simple model ("toy") was constructed that consisted of an elastic cylindrical rod (core) with stiff wire affixed helically to the outside surface. The inner rod is constructed from material with Poisson's ratio $\nu_p = 0.5$, so it conserves volume under stress. As this system is stretched, the inner rod decreases in diameter. When the model is stretched without constraining twist, changes in helicity arise from the tendency of the stiff outer wire to resist changes in contour length. The inner rod with radius R_r has stretch modulus $E_{stm} = \pi R_r^2 E_r$, bending rigidity $B_r = \pi R_r^4 E_r / 4$, and torsional rigidity $C_r = B_r / (1 + \nu_p)$, where E_r is the Young's modulus of the rod material. The outer wire has a larger stretch modulus $E_{strw} > E_{strm}$ and helix angle $\alpha = \arctan(3.4 / 2\pi R_r)$.

The presence of the outer wire does not change the bending rigidity or stretch modulus appreciably, but it does stiffen this system to torsion, and also generates a nonzero twist-stretch coupling, g_p. Although construction of this "toy" DNA was motivated by the discovery of negative twist-stretch coupling in the beginning at the molecule tension, it also provides a possible explanation for the anomalously large torsional rigidity of DNA. For an isotropic rod, the torsional rigidity C_r must be smaller than the bending rigidity B_r unless the rod is constructed from a material with negative Poisson's ratio (having the unlikely property that its radius increases as the rod is stretched). It has, therefore, long been puzzling that most measurements of C_r have been larger than B_r for DNA (Gore et al., 2006; Baumann et al., 1997). However, unlike the isotropic rod model, this "toy" model displays the correct values of B_r and C_r without resorting to exotic material properties. The outer "helix" stiffens the system to torsion because any twisting of the DNA model requires stretching or compression of the outer "helically wrapped" wire. This model (Gore et al., 2006) included the following parameters: $R_r = 0.924$ nm, $E_h = 965$ pN is the stretch modulus of the outer wire, $E_r = 0.393$ GPa, $B_{reff} = 225$ pN·nm^2, $C_{reff} = 460$ pN·nm^2, $g_p = -90$ pN·nm, and $E_{str} = 1{,}081$ pN.

In the "toy" model, negative twist-stretch coupling occurs only for shallow helix angle below a critical angle $\alpha_c = \arctan\left(\sqrt{\nu_p}\right) \approx 0.62$ rad. The geometry of B-form DNA lies just in the regime of negative coupling with $\alpha = \arctan(3.4 / 2\pi R_r) \approx 0.53$ rad $< \alpha_c$. Although the "toy" model is only intended to capture the behavior of DNA for small distortions, it does predict a sign reversal of twist-stretch coupling upon stretching, reminiscent of the behavior observed at high tensions.

Anomalous twist-stretch coupling of DNA has important implications for plasticity in site recognition by DNA-binding proteins. Even if the change does not directly involve a DNA-protein contact, the protein must overcome geometric mismatch in extensions and twist, in order to bind the DNA. Proteins are able to recognize binding sites with variable sequence lengths; this can be achieved by simultaneously stretching and overwinding (or compressing and underwinding) the DNA (Gore et al., 2006).

Finally, we see that while the "toy" DNA model is useful for initial studies of overwinding phenomenon, the model does not answer such simple question as the limiting of the stretch force to 30 pN. Besides, between the outer wire and core rod real model will have outer friction at the core extension that is bad for the system with nanometer precision.

6.7.2 Explicit Helicoidal Model (EHM)

Therefore, now we turn to the explicit helicoidal model (**Mod 16a**, type 6a, Table 1.3).

The effect of overwinding in our model of DNA is caused (see Fig. 6.3 and Eq. 4.49) by the negative lateral stress $\sigma^{11} = -(9/112)\chi_{kp} C_{\sigma}^{-1}\ \sigma_0^{33}$ in the cross section of the pretwisted strip under applied tension force F_{str} with the average stress $\sigma_0^{33} = F_{\text{str}}/A_{\text{cDNA2}}$ and parameter $C_{\sigma}^{-1} = (1+0.179\chi_{kp})^{-1}$.

One can find that, in our case,

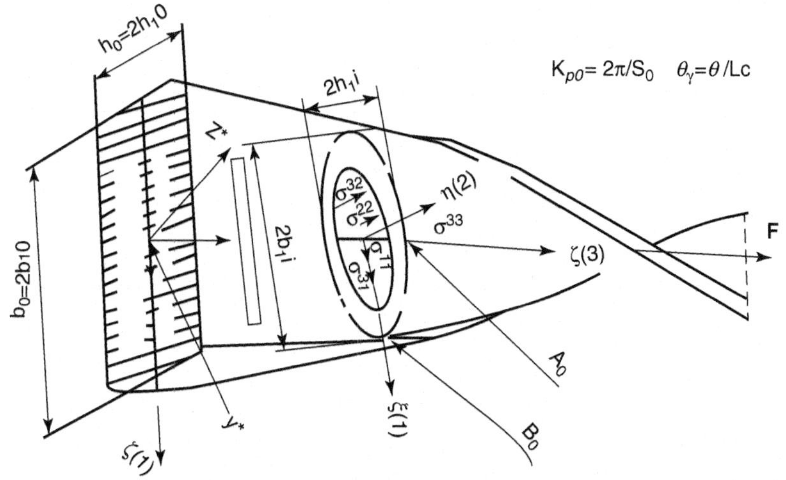

FIGURE 6.3 Pretwisted hollow nanostrip model with the same parameters as in Figure 6.1 and σ^{ii} as the stress tensor contravariant members.

$$\sigma^{11} = -0.436 F_{\text{str}} / A_{\text{cDNA}\,2} \tag{6.13}$$

and the corresponding decrease of the strip's width is equal to

$$2\Delta b_0 = -0.436 \frac{F_{\text{str}} b_o}{A_{\text{cDNA}\,2} E_{e2}}. \tag{6.14}$$

We can see that the decrease of the strip's width with negative displacement of its outer filament causes overwinding. The original DNA helix pitch $S_0 = 3.4$ nm corresponds to the outer filament's point path of $2\,\pi b_{1o}$ and the new helix pitch S_{01} after stretching corresponds to the point's path of $2\pi\,(b_{1o} - \Delta b_0)$. The difference of these two pitches equals $2\,\pi\Delta b_0$. The number of winds N_0 with the original pitch for the helix with the length L_c is equal to $N_0 = L_c/3.4$. Substitution of $F_{\text{str}} = 30$ pN and other known parameters into (6.14) yields

$$\Delta b_o = -0.436 \frac{2.12 \times 30}{2 \times 0.317 \times 3470} = 0.0125 \text{ nm}.$$

Therefore, $S_{01} = 3.4 - (2\,\pi \times 0.0125) = 3.32$ nm and the number of winds with the smaller pitch equals $N_{01} = L_c/3.32$. It is clear that $N_{01} > N_0$ and the helix overwinds. This overwinding can continue until the compression stress σ^{11} reaches the compressive yield stress for the model material. The overwinding stops at this point because any further extension of the nanostrip will not decrease its width. Substitution of $F_{\text{str}} = 30$ pN and $A_{\text{cDNA2}} = 0.317$ nm^2 into the equation for σ^{11} yields $\sigma^{11} = -34.1$ MPa, which value is close to the compressive yield strength of some polymers and DNA elements. Let us now define the critical point between the option of overwinding and unwinding the pretwisted strip under tension. The overwind option is expressed in radians, through the following relationship:

$$2\pi(N_{01} - N_0) = 2\pi\left[\frac{L_c}{S_0 - 2\pi\Delta b_o} - \frac{L_c}{S_0}\right] = \frac{k_{p0} L_c 2\pi\Delta b_o / S_0}{1 - 2\pi\Delta b_o / S_0}, \tag{6.15}$$

where $1\text{–}2\,\pi\Delta\, b_o/S_0 \approx 1$, because $2\,\pi\Delta b_o/S_0 \ll 1$, and Δb_o can be expressed through the pretwisted strip parameters and the average tension stress σ_0^{33} with the help of Eq. (6.13) and (6.14), as follows:

$$\Delta b_o = -(9/112)\frac{k_{p0}^2 R^3 \sigma_0^{33}}{(\lambda + 0.179 k_{p0}^2 R^2) E_{e2}} \tag{6.16}$$

where $R = b_o/2$.

Substitution of (6.16) into (6.15) yields the following expression for the possible overwinding:

$$2\pi\,(N_{01} - N_0) = \frac{k_{p0} L_c (9/112) k_{p0}^3 R^3 \sigma_0^{33}}{(\lambda + 0.179 k_{p0}^2 R^2) E_{e2}} \tag{6.17}$$

The unwinding option under tension in radians is expressed through the following formula:

$$F_{str}/j_{\theta} = \frac{2(1+\nu_p)L_c k_{p0}\sigma_0^{33}}{57.3E_{e2}(\lambda+\nu_{jf}k_{p0}^2R^2)}, \tag{6.18}$$

where j_θ is the torsion rigidity at the pretwisted strip extension (see Chapter 4), $Ee_2/[2(1 + \nu_p)] = G$ is the shear modulus, ν_p is Poisson's ratio of the strip's material, $\nu_{jf} = 0.179$ is the stiffness form factor for the strip's cross-section.

Therefore, the critical point for the equal options of overwinding and unwinding of the strip under tension corresponds to the equality of the right sides in (6.17) and (6.18) that, after some cancellations, yields

$$k_{p0}^3R^3 = \frac{224(1+\nu_p)}{57.3\times 9}. \tag{6.19}$$

Let us recall that $1/(k_{p0}R) = S_0/(2\pi R) = \tan\alpha$, where α is a helix angle to the cross-section's plane. Hence, we have found in (6.19) the relationship between the critical helix angle $\alpha_c = \tan^{-1}(1/(k_{p0}R))$ and the Poisson's ratio of the strip's material. The solution of (6.19) with $\nu_p = 0.5$, typical for polymers and DNA molecule, yields $\alpha_c = 0.856$ rad. A helix with angle $\alpha < \alpha_c$ has more probable options for overwinding, while a helix with angle $\alpha > \alpha_c$ has a good probability to unwind under the longitudinal tension. In the case of DNA molecule, $\alpha = \tan^{-1}(3.4/2\,\pi \times 1.05) = 0.476$ rad $< \alpha_c$. Hence, the DNA molecule can exhibit initial overwinding. Similar conclusions about the overwinding of DNA molecule and its critical helix angle as a function of Poisson's ratio are shown in Gore et al. (2006), but with some difference in the numerical value of $\alpha_{c1} = \tan^{-1}\left(\sqrt{\nu_p}\right) = 0.62$ rad. However, this difference is not very important for the DNA molecule, nor for our model with $\alpha < \alpha_{c1}$.

In the case of the pretwisted bronze microstrip (Tseytlin, 2006) with $\nu_p = 0.279$, for example, the critical helix angle is equal $\alpha_c = 0.822$ rad and for typical pretwisted strip with the cross-section's width $2R = 0.12$ mm and the helix pitch of $S_0 = 2$ mm $\alpha = \tan^{-1}(2/2\,\pi 0.06) = 1.38$ rad $> \alpha_c$ and α_{c1}. Therefore, the helicoidal sensor with the metallic pretwisted microstrip has good options for unwinding under longitudinal tension, and less or no options for overwinding.

Torsional rigidity of the helicoidal model at torsion is equal to $j_{\tau\tau} = GT_0(1+\nu_{if}\chi_{kp})$, where $\nu_{if} = 0.41$ as for the perforated strip, $\chi_{kp} = 190.9$ in this model, $G = 1445.8$ pN/nm^2, and $T_0 \approx (A_{cDNA})^4/(40\,J_p)$ in accordance with the approximate Saint-Venant's formula. Here, J_p is the polar moment of inertia that, for helicoidal model cross-sections, is approximately 0.086 nm^4. Therefore, the torsional rigidity of this model equals $C_r = j_{\tau\tau} = 1445.8\times[(0.317)^4/(40\times 0.086)]\times(1+0.41\times 190.9) = 336.42\ \text{pN}\cdot\text{nm}^2$.

Meanwhile, the bending rigidity of the same helicoidal model is equal to

$$B_r = E_{e2}\, J_\eta = 3470 \times 0.085 = 294.95\ \mathrm{pN \cdot nm^2}.$$

Therefore, $C_r > B_r$ corresponds to the claim made about this relation in Gore et al. (2006). However, the numerical value for C_r in our case is different from that of Gore, fact that may be due to our application of the approximate Saint-Venant's formula for the evaluation of T_0. We will receive an increased C_r value if we use Eq. (6.1) for the estimation of $T_0 = A_c \times \nu_{T_0/A_c}$. For a more accurate solution, we should evaluate the following integral:

$T_0 = \int_{Ac} (R^2 + \varphi'_\psi)\, dA_c$, where A_c is the cross-section area, R is the contour vector-radius, $\varphi'_\psi = \varphi'_\eta\,(\xi - \xi_j) - \varphi'_\xi\,(\eta - \eta_j)$, and $\varphi'_\eta = \partial\psi^o/\partial\eta$; $\varphi'_\xi = \partial\psi^o/\partial\xi$, where ψ^o and T_0 are the function of rotation and torsional constant for nontwisted beam with the same cross-section, respectively. We will leave this exercise for future study.

The transmission ratios $i_\theta, j_\theta, j_s = i_\theta\, j_\theta$ of the EHM model may be adjusted by choosing different secondary parameters, such as thickness of outer h_o and inner h_i contours, and using a model with two adjacent helicoidal strips (AEHM).

6.7.3 P-DNA Phase

The stress that appears when twisting DNA is characterized by a torque (M_z) applied to the molecule. As the torsional stress increases, it may alter the DNA structure. When the torque on the DNA reaches 9 pN·nm, an underwound DNA molecule locally denatures, releasing the torsional stress (by about 10 bp for every extra turn). If the torque reaches ~20 pN·nm (in low salt), an overwound DNA molecule adopts a new inside-out overtwisted structure called P-DNA. The stress at $M_z = 20$ pN · nm in our helicoidal model is equal to

$$\sigma_{pt} = \frac{M_z}{2A_{cD}(b_0/2)} = \frac{20}{2 \times 0.317 \times 1.07} = 29.5\ \mathrm{pN/nm^2} \approx 30\ \mathrm{MPa}.$$

This stress is close to some polymer materials' ultimate tensile stress.

6.8 NONLINEAR LENGTH FLUCTUATIONS

6.8.1 Remeasuring the Double Helix

A nonlinear thermomechanical length L_c fluctuation variance *Var* (L_c) of DNA molecules was found in the experimental studies at "remeasuring the double helix" (Mathew-Fenn et al., 2008a,b; 2009).

The experimental sample (Fig. 6.4) has the following elements: 1 and 5 are the gold nanocrystal probes, 2 and 4 are the linkers, and 3 is the double-stranded DNA molecule whose thermomechanical fluctuation is under consideration.

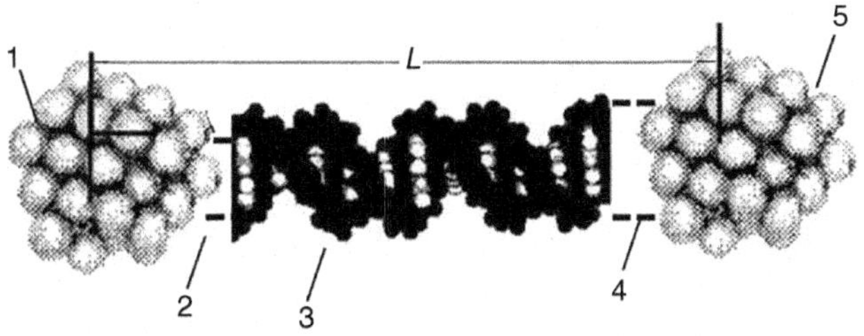

FIGURE 6.4 Schematic of the experimental sample of double stranded DNA with attached gold nano-crystals, used for remeasuring the double helix: gold nanocristal probes, 1 and 5; are the linkers, 2 and 4; double-stranded DNA, 3. *(Mathew-Fenn et al., 2008b; open access.)*

A geometric model (**Mod 21**, Chapter 1) of the double helix (Mathew-Fenn et al., 2008b) for measuring the distance L between two probes is a function of the number of intervening base steps N with a sum of axial and radial separations:

$$L(N) = \sqrt{\text{axial}^2 + \text{radial}^2} = \sqrt{(\text{axial}_0 + r_a N)^2 + d_1^2 + d_2^2 - 2d_1 d_2 \cos\left[\theta_0 + \frac{2\pi}{10.4}N\right]}. \quad (6.20)$$

The term r_a is the axial rise per base step. The terms d_1 and d_2 are the radial displacements of the two probes off the helical axis. The term axial_0 is the axial distance between two probes separated by zero base steps. The term θ_0 is the azimuthal angle between two probes separated by zero steps. In case with symmetrical probe geometry (specifically $d_1 = d_2 = d_s$), the radial separation distance simplifies to radial $= 2d_s \sin[\theta/2]$. The azimuthal angle as a function of base steps was calculated in Mathew-Fenn et al. (2008b) as $\theta = \theta_0 + 2\pi N/10.4$. The constant in the denominator of Eq. (6.20) derived from the fact that 10.4 base pairs of DNA in solution make one full turn around the helix axis. The values for θ_0 were determined by inspection of the Dickerson dodecamer structure. For probe attachment to 3′-phosphates, θ_0 was set to 1.34π. For probe attachment to exocyclic methyl groups of T bases, θ_0 was set to 1.58π. For probe attachment to 5′-phosphates, θ_0 was set to 0.95π. Two gold nanocrystals are attached at the terminal 3′-phosphate positions of an A:T base pair through three atom linkers. The axial separation between the nanocrystals corresponds to the parameter axial_0. For the scattering interference data, axial_0 optimized to 23.9 Å.

Results of these measurements are shown in Figure 6.5 and Table 6.1.

All data from Mathew-Fenn et al. (2008a,b; 2009) support their claim for nonlinear length variance function in tested DNA samples. Each sample has corresponding sequences and linkers, for example, they are:

5′-CGACTACGTACCGATGCATCACTACGCAGCGC-3′ at 4 base pairs.

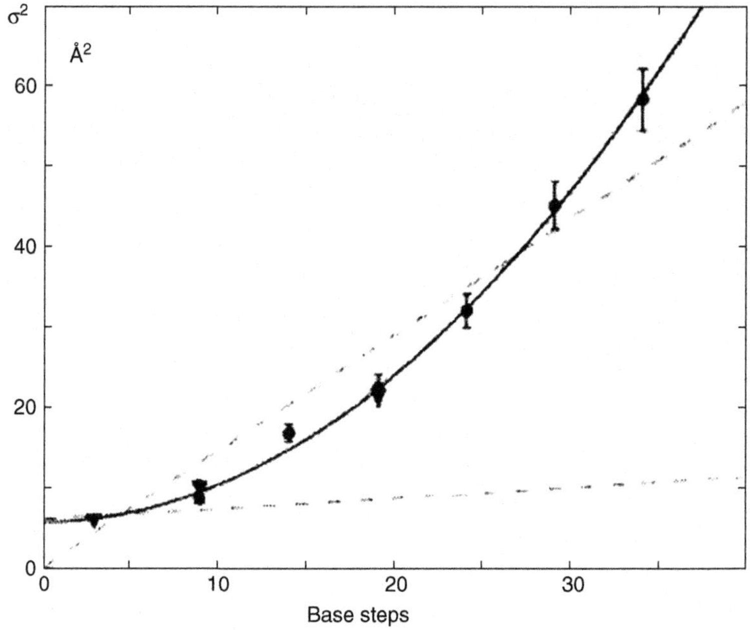

FIGURE 6.5 Variance σ^2 of dsDNA length thermomechanical fluctuations. *(Mathew-Fenn et al., 2008b; open access.)*

TABLE 6.1 Mean Nanocrystal Separation Distance and Distance Variance

Ruler	DNA length (base pairs)	Mean (Å)	Variance (Å^2)	Bending correction	Corrected mean
XSI	9	56.1	8.7	0.34	56.4
XSI	14	69.3	15.8	0.77	70.1
XSI	19	86.1	21.2	1.37	87.5
XSI	24	101.2	28.4	2.15	103.4
XSI	29	118.3	42.4	3.10	121.4
XSI	34	131.1	49.4	4.22	135.3

Source: After Mathew-Fenn et al. (2008b).

6.8.2 Negation of Nonlinear Fluctuation

However, Becker and Everaers (2009) insist that, when accounting for a subtle linker leverage effect, the data presented by Mathew-Fenn et al. (2008a,b) can be understood in the framework of standard linear and noncooperative DNA elasticity, when properly accounting for the geometry of the molecular linker between DNA and nanoparticles.

Becker and Everaers (2009) claim that the distance fluctuations measured by Mathew-Fenn et al. (2008a,b) are due to the superposition of various modes originating from DNA, DNA-nanoparticles linkers, and other experimental factors. They predict that the DNA contribution to the variance depends on its compressional variance per base pair n, helical rise per base pair l_m, and bending persistence length l_p. It is dominated by longitudinal fluctuations $\sigma_b^2 = l_m^4/(90 l_p^2) n^4$ take over. In the Gaussian long-chain limit, the distance variance approaches $\sigma_G^2 = [1 - 8/(3\pi)] 2 l_p l_m n$. Using standard literature values for n, l_m, and l_p determined in single-molecule experiments, these scaling regimes are reproduced by microscopically parameterized simulations. Therefore, they concluded that standard DNA elasticity alone does not account for the data reported in Mathew-Fenn et al. (2008a). As emphasized by Mathew-Fenn et al. (2008a,b), adding an n-independent correction to include isotropic linker fluctuations and other experimental noise sources is insufficient to reproduce the observed increase in variance. Therefore, Becker and Everaers (2009) supposed that the authors of "Remeasuring the double helix" disregarded lever-arm effects in the conversion of DNA bending fluctuations into the nanoparticles distance variation. They insist that, in the absence of this leverage, the DNA bending contribution scales as $\sigma_b^2 \propto L^2 \langle \alpha^2 \rangle^2$ in terms of the contour length $L = l_m n$ and the total mean square deflection angle $\langle \alpha^2 \rangle = 2L/l_p$. Including an axial lever arm length axial$_0$/2 per label yields

$$\sigma_b \approx (1 + \text{axial}_0)^2 \langle \alpha^2 \rangle^2 \propto (l_m n + \text{axial}_0)^2 n^2.$$

Similarly, a radial offset D of the nanoparticles from the helical axis leads to an additional oscillating term $\propto D^2(1+\cos\theta)\langle \alpha^2 \rangle \propto n$, where θ is the total azimuthal angle between the nanoparticles. These corrections cannot be absorbed into an n-independent constant offset. The small change in DNA end-to-end distance coincides with a large change in nanoparticle distance. The linkers are included in the simulation for testing this problem. Using rigid linkers with values reported in Mathew-Fenn et al. (2008a), at radial offset $D = 9$ Å, total axial offset axial$_0 = 24$ Å, and azimuthal angle $\theta_0 = 1.34\pi$, yields variances that have about 90% of the total variance increase. As a result of the radial offset, both the nonfitted and fitted variance curves exhibit helical oscillations.

Becker and Everaers (2009) concluded that the data presented in Mathew-Fenn et al. (2008a,b) publications are consistent with standard, noncooperative, linear DNA elasticity using "canonical" microscopic and mesoscopic elastic and structural parameters.

Mazur (2009a,b) also gave a negative estimation of "remeasuring the double helix" results on the nonlinear length variance.

6.8.3 EHM Estimate

Let us now check this problem of the double stranded DNA molecule nonlinear length fluctuation with our explicit nanohelicoidal model (EHM) that does not have linkers and attached nanocrystal probes.

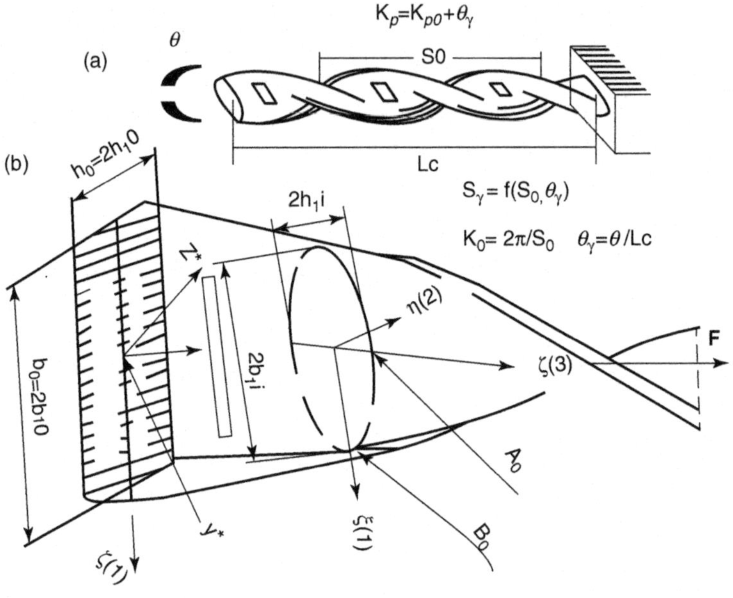

FIGURE 6.6 EHM nonlinear model (a) and its calculation schematic (b). *(After Tseytlin, 2013; open access.)*

Mod 16a. As we can show, our explicit nanohelicoidal model (Fig. 6.6) with the hollow pretwisted nanostrip sensor has nonlinear variance fluctuation features. The length fluctuations correspond to the normal statistical distribution, and are in agreement with the equipartition theorem.

Hence, the variance of the model length is equal to

$$\mathrm{Var}\,(L_c) = \aleph k_B T / j_s, \tag{6.20a}$$

where $\aleph$ is the units' conversion factor between the left and right sides of the formula in our calculations, k_B is Boltzmann's constant, T is the absolute temperature in Kelvin, and j_s is the object stiffness.

The nonlinear explicit helicoidal (pretwisted) nanostrip system should have the "following" Cartesian coordinates Z*, Y* rotating with variable speed, because its helix parameter $k_p = k_{p0} + \theta_\gamma$ will change at the strip section rotation, unlike that of the constant helix parameter k_{p0} in the linear system. It should also be remembered that, at the stretching of the pretwisted strip, the angle θ of untwist rotation, after a small overwinding zone, is negative, $\theta/L_c = \theta_\gamma < 0$.

As a result, we find the relationship (Eq. 4.24a, b): between components ε_{ik} of strain tensor for nonlinear body in the "following" system of coordinates, and the strain tensor components ε^0_{ik} for a linear body of the same pretwisted strip:

$$\varepsilon_{11} = \varepsilon^0_{11};\ \varepsilon_{22} = \varepsilon^0_{22};\ \varepsilon_{12} = \varepsilon^0_{12} = 0;\ \varepsilon_{33} = \varepsilon^0_{33} - \theta_\gamma \Gamma_a \partial\psi^0/\partial\vartheta;$$
$$\varepsilon_{13} = \varepsilon^0_{13} - 2\theta_\gamma \Gamma_b \xi^2 \eta;\ \varepsilon_{23} = \varepsilon^0_{23} - 2\theta_\gamma \Gamma_c \xi \eta^2,\ \Theta = \Theta^0 - \theta_\gamma \Gamma_a \partial\psi^0/\partial\vartheta,$$

where 1, 2, 3 are the coordinate axes numbers in Figure 6.6b.

The potential energy (4.25) at the stretching of the elementary ($L = 1$) nonlinear naturally pretwisted beam, without an external twist moment ($M_{xL} = 0$), is expressed as

$$R_v = G\iint_A \{[\varepsilon_{11}^{0\,2} + \varepsilon_{22}^{0\,2} + (\varepsilon_{33}^0 - \theta_\gamma \Gamma_a \partial\psi^0/\partial\vartheta)^2] + 2[(\varepsilon_{13}^0 - 2\theta_\gamma \Gamma_c \xi^2\eta)^2 + (\varepsilon_{23}^0 - 2\theta_\gamma \Gamma_c \xi\eta^2)^2] + (\Theta^0 - \theta_\gamma \Gamma_a \partial\psi^0/\partial\vartheta)^2\}\, dA - QS_\gamma = R_v^0 + R'_v,$$

where $G = E_c/[2(1+\nu_p)]$ is a material shear modulus; $A = A_{cDNA}$; ψ^0 is the function of the strip rotation, Θ^0 is the invariant of the tensor of deformation, $S\gamma = s/L_c$, L_c as a length of the structure under consideration, and s is the stretch value. $R_v^0 = U_v$ is the known expression (4.20) of the potential energy at a small deformation of the pretwisted strip, and absence of the twist moment $M_{xL} = 0$, R'_v is an additional member that shows the influence of the large degree of strip untwist. The constants Γ_a, Γ_b, Γ_c are defined with the principle of stationary potential energy when deriving the solution by the method of variations.

The solution of the simultaneous Eq. (4.32)…(4.36) in terms of the constants Γ_a, Γ_b, Γ_c and variable S_γ; θ_γ yields the nonlinear equation for the average transformation ratio $i_{\theta av}$ and the torsional rigidity at the extension $j_\theta = F_Q/\theta$ of the flexible helicoids similar to Eq. (4.37) and (4.42), but with the specified dimensional units:

$$i\theta\text{av} = \left|\theta_\gamma/2S_\gamma\right| = 57.3(\kappa_{p0}+\theta_\gamma)\frac{E[1+\theta_\gamma(\kappa_{p0}+\theta_\gamma)^{-1} - 3(\kappa_{p0}+\theta_\gamma)b_1^2\theta_\gamma]}{4G\lambda[1+\nu_{if}(\kappa_{p0}+\theta_\gamma)^2 b_1^2\lambda^{-1} + 3\theta_\gamma b_1^2\lambda^{-1}(\kappa_{p0}+\theta_\gamma)/8]},\quad (°/\text{nm}), \tag{6.20b}$$

$$j_{\theta o} = F/\theta \approx 0.0696 G A \lambda (1+\nu_{if}\chi_{kp})/L\kappa_{p0}, \tag{6.20c}$$

where F (pN) is the stretching force at the strip elongation with corresponding untwist angleθ; $j_{\theta 0}$ is in pN/(angle degree); G is the shift modulus in pN/nm^2. Note that the i_θ value does not depend on the structure's length, and force F should be less than 65 pN, if overstretching is excluded.

We can derive an equation for the nanostrip model stretch nonlinear stiffness j_s as the product of nonlinear average transformation ratio $i_{\theta av}$ and the torsional rigidity j_θ at the pretwisted strip extension: $j_s = i_{\theta av} j_{\theta o}$.

The inverse value of nonlinear stiffness j_s^{-1} is, therefore, equal to:

$$j_s^{-1} = \frac{L_c[1+\nu_{if}(1+\nu_H)^2\chi_{kp} + 3(1+\nu_H)\nu_H\chi_{kp}/8]}{E_{e2}[(1+\nu_H)+\nu_H - 3k_{p0}^2(1+\nu_H)^2\nu_H b_{1o}^2](1+\nu_{if}\chi_{kp})A_{cDNA2}}, \tag{6.21}$$

where L_c is the molecular length, $\nu_{if} = 0.39$–0.41 is the transformation form coefficient for the punched pretwisted nanostrip, $\chi_{kp} = k_{p0}^2(b_o/2)^2\lambda^{-1} = 190.9$ is the dimensionless parameter of the strip's relative pretwist, $\lambda = (h_o/b_o)^2 = 0.0197$ with height h_o and width b_o dimensions of the outer rectangular contour

(Fig. 6.6b), $k_{p0} = 2\pi/S_0 = 1.847\ \text{nm}^{-1}$ is the initial twist parameter with helical pitch $S_0 = 3.4$ nm, $E_{e2} = 3470\ \text{pN/nm}^2$ is Young's modulus of a DNA molecule in our model, and $A_{cDNA2} = 0.317\ \text{nm}^2$ is its cross-section area.

The parameters $\nu_H = -\theta/(k_{p0}L_c) < 0$ and θ are the stretched strip moving end's rotation relative and absolute angles, respectively. The values of the parameter ν_H are inversely proportional to the molecular length L_c if the angle θ = constant. Therefore,

$$\nu_{Hi}/\nu_{Hj} = L_{cj}/L_{ci}. \tag{6.22}$$

It is only one conditional restrain for the model that requires the stretch of pretwisted strip samples to the same angle of rotation on the moving end. In further calculations, we accept $\theta = 1$ rad/1.57 $\approx$ 36.5 angular degree (see also Table 1.1 with $\Omega_r = 36.1°$ as the rotational twist per residue typical for the B structure of dsDNA).

Note that ν_p can be dependent on the length of the DNA molecule in accord with the recent study (Eslami-Mossallam and Ejtehadi, 2011) of persistence length dependence on the molecular length.

Table 6.2 shows the results of our calculations using Eq. (6.20a), (6.21), and (6.22), and assuming ν_H (L_c = 131.1 Å) = –0.15, ν_{jf} (ν_p = 0.5) = 0.18 for version *a*. The difference, |δ|%, between the calculated and experimental values from Mathew-Fenn et al. (2008b) is less than 10%. These data are represented in Figure 6.7 where the solid red line (ν_{jf} = 0.18) denotes the calculated results, and the circles correspond to the experimental values. Another set of experimental data in Table 6.2 for version *b**with ν_{jf} = 0.14 at ν_p = 0.2 also corresponds well to the calculation using Eq. (6.20a) and shown in Figure 6.7. For example, a DNA molecule in the standard buffer solution plus 200 mM NaCl has (Mathew-Fenn et al., 2009) persistence length of 50 nm, while in the standard buffer solution with 0 mM NaCl the persistence length is A_{bp} = 55.2 nm.

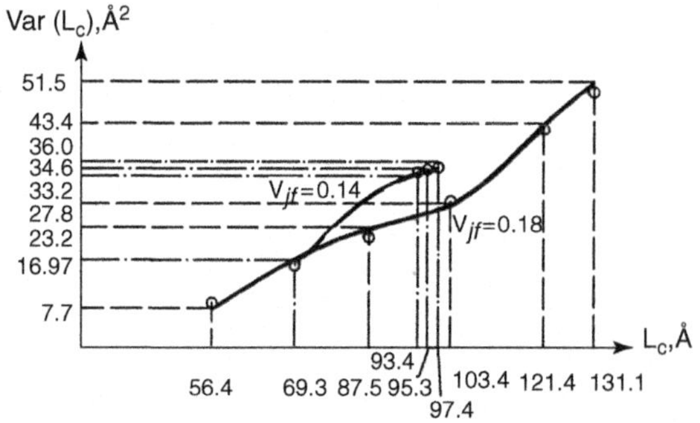

FIGURE 6.7 Thermomechanical variance of the molecular length with solid lines corresponding to our theory (Tseytlin, 2012; open access) for cases with ν_{jf} = 0.14 and 0.18, and experimental data (Mathew-Fenn et al., 2008a,b; 2009) denoted by circles.

TABLE 6.2 Thermomechanical Variance of DNA molecule's Length L_c at $T = 293$ K, $\theta = 1/1.57 = 0.64$ rad; $v_{if} = 0.39–0.41$

	Calculation						Experiment					
	v_H		L_c (Å)		*Var* (Å^2)		L_c (Å)		*Var* (Å^2)		|δ| (%)	
Nº	a	b*	a	b*	a	b*	ae	b*e	a e	b*e	a	b*
1	−0.35	−0.284	56.4	69.3	8.4	16.97	56.4	69.3	8.7	15.8	3.4	7.4
2	−0.228	−0.210	86.1	93.4	22.5	33.2	86.1	93.4	21.2	34.0	6.1	2.3
3	−0.194	−0.206	103.4	95.3	27.9	34.6	103.4	95.3	28.4	34.5	1.8	0.4
4	−0.167	−0.205	121.4	95.8	43.4	35.05	121.4	95.8	42.4	35.0	2.4	0.1
5	−0.15	−0.203	131.1	97.4	51.5	36.0	131.1	97.4	49.4	35.4	4.2	1.7

Version *a* $v_{if} = 0.18$, experimental *ae* data after Mathew-Fenn et al. (2008a,b); version $b^* v_{if} = 0.14$, experimental *b*e* data after Mathew-Fenn et al. (2009).
Source: After Tseytlin (2012), with permission.

Hence, the average A_{bp} = 52.6 nm. The experimental variance (Mathew-Fenn et al., 2009) is Var(L_c) = 35.0 Å^2 at the length L_c = 95.8 Å. Calculations with Eq. (6.20a) at L_c = 95.8 Å, $\nu_H = -0.205$, and $\nu_{jf} = 0.14$ show a variance value of 35.05 Å^2 that corresponds to the above experimental data at an uncertainty within 0.15%. Calculations with (6.20a) at L_c = 97.4 Å, $\nu_H = -0.203$, and $\nu_{jf} = 0.14$ show a variance value of 35.9997 Å^2 that corresponds to the experimental result (Mathew-Fenn et al., 2009) of 34.9 Å^2 with an uncertainty within 3.1% (see the solid line in Fig. 6.7 with $\nu_{jf} = 0.14$). Our calculations with (6.20a) for L_c = 69.3 Å at $\nu_H = -0.2838$ and $\nu_{jf} = 0.14$ show a variance of Var(L_c) = 16.97 that corresponds to Mathew-Fenn et al. (2009), experimental result within 7.4%.

The results of our study show that our DNA helicoidal model (EHM) works well for the evaluation of nonlinear variance of the molecular length thermomechanical fluctuations in different solutions, with an appropriate estimate of Poisson's ratio, persistence length, and stretch spring constant of the DNA molecule.

Here, we have effective relations between the model scale parameters and the functioning conditions of DNA molecules. This research shows closely related features of nanohelicoidal structures in the organic and inorganic fields.

Unfortunately, we don't know all of the stress conditions of experiments where the DNA molecules have been in the solution. Mazur (2009b) suggested the influence of the DNA molecule bending, and the linker to end clusters' rotation with azimuthal angles from $\theta_{0a} = 1.34\pi$ to 0.91π at radial offset D_r = 0.9 nm on its stress conditions, in this case. Deflection of the linker from the margin positions by $\pm 0.215\pi = \pm 0.68$ rad will cause a displacement of the pretwisted strip end by $\Delta = 0.68 D_r = 0.61$ nm. Similar small stretch numbers are shown in Becker and Everaers (2009) as well. If we now consider that the linear transformation ratio for the helicoidal strip sensor is equal to

$$i_\theta = 57.3 \frac{(1+\nu_p)k_{p0}}{2\lambda(1+\nu_{if}\chi_{kp})} \approx 51 \quad \text{(angular degree)/nm},$$

then the strip moving end's rotation is $\theta = i_\theta \Delta = 51 \times 0.61 = 31$ (angular degree) that is close to the value of angle 36.5 (0.64 rad) accepted in our calculations with EHM (see Table 6.2 and Table 1.1). In principle, the accepted angle θ should only be constant for the whole bunch of the molecules under fluctuation test.

Hence, we see that nonlinear thermomechanical variance is a feature of the pure mechanical sensor with the pretwisted nanostrip that may resemble the DNA molecule's geometric structure and mechanical properties. However, influence of the attached golden clusters, excluded volumes, ions, electric field, methods of measurement, and other conditions, can cause large differences and uncertainties in the experimental data. We see those differences in results of measurements by different methods in Mathew-Fenn et al. (2008b), where mean length and length variance of DNA duplexes were measured by three different methods: XSI – X-ray scattering interference; trsmFRET – time-resolved single-molecule fluorescence resonance energy transfer; and DEER – double electron–electron spin resonance. In order to fit a helical rise value, the mean distances were corrected for the apparent shortening caused by DNA bending. The calculation results with our EHM model correspond better to the XSI measurement method data.

Thus, our study shows that the experimentally discovered nonlinear stretching fluctuations of the DNA's double helix is an intrinsic structural feature that can be understood by modeling DNA as a purely mechanical sensor of the helicoidal pretwisted nanostrip with nonlinear functions. In this case, the appearance of a nonlinear variance of the molecular thermomechanical length fluctuations is independent of the specific experimental technique used.

Length dependence of variance in short DNA. Padinhateeri and Menon (2013), examine how the variance $\left(\langle L_R^2\rangle - \langle L_R\rangle^2\right)$ of the end-to-end distance L_R varies with the length (L) of the DNA filament. They show that the variance for short chain ($L < 150$ bp) fits to a power of 3.4. For longer chains ($L > 300$ bp), the variance grows linearly with the size of the chain. They concluded that such behavior is expected because once the chain lengths exceed the correlation lengths, the central limit theorem should accurately predict the statistics of the variance. A similar tendency is shown in Figure 6.7, where the chart for our **Mod 16** (EHM) is straightened at a longer length L_c.

6.9 SPEED FACTOR

Mod 22. The drag force is the sum of the drag from the transverse and longitudinal modes. The longitudinal displacement (i.e., extension along applied force) is determined by the dynamics at finite pulling speed, rather than by diffusive motion that has been considered elsewhere (Lee and Thirumalai, 2004). The drag force arising from the motion in the direction of the pulling is

$$F^{\ln} = \frac{1}{2}\eta_0 v_0 u_{\parallel} L$$

and the total drag force (Lee and Thirumalai, 2004) is

$$F = \frac{1}{2}\eta_0 v_0 u_{\parallel} L + l_p^{-1} u_{\parallel}(t) + \frac{l_p^2}{4D_\eta N^2}\frac{u_{\parallel}^2(t)}{1-u_{\parallel}^2(t)} v_0,$$

where η_0 is the solvent viscosity, v_0 is the pulling speed, $u_{\parallel}$ is the relative longitudinal displacement, L is the molecule contour length, l_p is the persistence length, $D_\eta = k_B T / \varsigma_0$ is the monomer diffusion constant, and t is time.

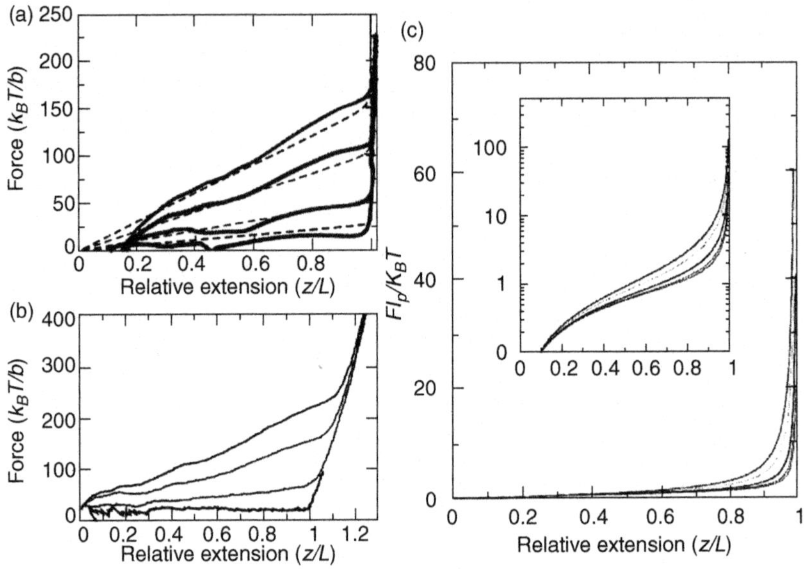

FIGURE 6.8 Comparison of force F-extension z/L curves. (a) the dashed lines are the theoretical results and the thick lines (6.8a and 6.8b) represent Langevin simulation results, (c) theoretically determined dynamical (F, z) curves for WLC at various pulling speeds. *(Lee and Thirumalai (2004), reprinted with permission.)*

The values of the pulling speeds in Figure 6.8 change in the order $V_0 = (0.5, 1, 2, 3)\ b/t_0$ from bottom to top. The persistence length $l_p = 20b$ at $N = 100$, the bond length $b = 2.5$–2.8 nm. The insert in Figure 6.8c is in the linear-log scale. The simulations were performed for eWLC chain with the bending spring constant $k_b = 10{,}000 k_B T/b^2$.

As described by Brouchard-Wyart (1995), the nonuniform profiles of the DNA molecule (trumpet and stem-flower profiles) manifest themselves when the pulling speed reaches $v^c(F) = k_B T/\eta_0 b_f^2 > v_0^c$, where b_f is the transverse size of the chain.

The chain relaxation time under tension is $\tau_z^f = \eta_0 b_f^3/k_B T$. Below the threshold pulling speed v_0^c, the chain confirmation can be approximated as a uniform cylinder. Let us recall the comment to **Mod 22** in Chapter 1: The value of v_0^c at water viscosity $\eta \sim 1$ cP for typical parameters of DNA is $v_0^c \sim 800\ \mu$m/s. The typical pulling speeds used in experiments are $v_0 = (1$–$10)\ \mu$m/s, they are at least two orders-of-magnitude smaller than v_0^c. Therefore, below the threshold pulling speed, the chain confirmation can be approximated as a uniform cylinder. However, the nonequilibrium response considered here may be observed in the force-extension profiles of longer DNA molecules.

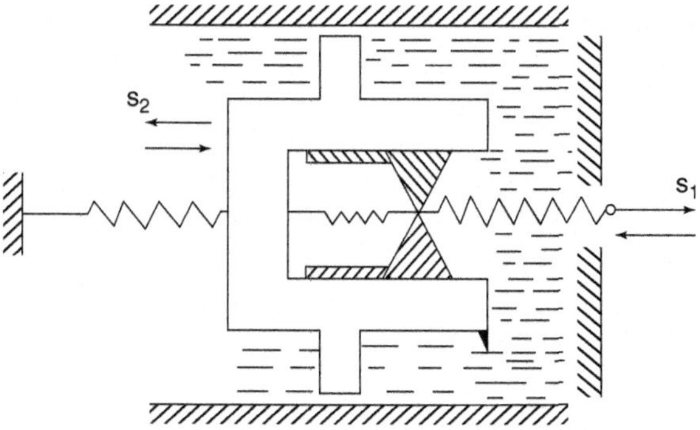

FIGURE 6.9 EHM-explicit nano-helicoidal model with external and internal damping (see also Fig. 5.4c).

Mod 16.

The differential equation for the EHM model rotation with damping may be presented as follows: $J_R\ddot{\theta}+\frac{2j_{\tau\tau}}{L}\theta+\varsigma_\theta\dot{\theta}=\frac{j_{k\tau}}{L}sU(t)$, where U (t) is a unit step, J_R is the dynamic moment of inertia on the strip crosspiece.

The transition functions for the pretwisted (helicoidal) microstrip-photo-optical sensor is shown in Table 5.3.

With our nano-helicoidal model (Fig. 6.9), we can represent the dragging force at the different pulling speed V_0 in the following form:

$$F=i_\theta j_\theta\frac{s}{L}+\varsigma_\theta i_\theta V_0;$$

$$m_s\ddot{S}=j_{kk}\frac{s_2}{L}-2j_{kk}\frac{s_1}{L}+\dot{s}_1\varsigma_L.$$

The sensor's resolution $r_r=0.5$ μm (Fig. 6.10) corresponds to the double speed extension on the step response of the pretwisted strip with thickness × width ×

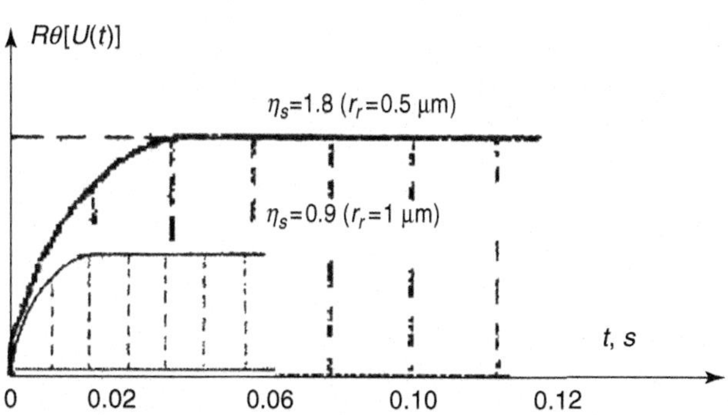

FIGURE 6.10 Transient functions-step response (no scale) at different transmission ratios for EHM.

length (mm) $\times$ twist parameter (mm^{-1}) = $0.006 \times 0.12 \times 18 \times \pi$ in comparison with the extension of the same strip in the sensor with the resolution $r_r = 1$ μm.

Note that the EHM model is a comprehensive physical model of the DNA molecule, unlike many other known models (WLC, FJC, DPC, etc.) that correspond only to the mathematical representation of a specific feature of a DNA molecule. This model may be studied in different size scales.

The next chapter is focused on the application of the EHM to dsDNA transition representations at higher stretching forces. The EHM model stability and reliability can be estimated by Lyapunov's method, evaluation of condition numbers of stress structural matrices (Tseytlin, 2006) and other appropriate methods.

REFERENCES

Ambia-Garrido, J., Vairub, A., Pettitt, B.M., 2011. A model for structure and thermodynamics of ssDNA and dsDNA near a surface: a coarse grained approach. Comput. Phys. Commun. 181 (12), 2001–2007.

Baumann, C.G., Smith, S.B., Bloomfield, V.A., Bustamante, C., 1997. Ionic effects on the elasticity of single DNA molecules. Proc. Natl. Acad. Sci. USA 94, 6185.

Becker, N.B., Everaers, R., 2009. Comment on Remeasuring the double helix. Science 325 (5940), 538.

Bhushan, B., 2010. Springer Handbook of Nanotechnology. Springer-Verlag, Berlin-New York.

Boal, D., 2002. Mechanics of the Cell. Cambridge University Press, London, UK.

Brouchard-Wyart, F., 1995. Polymer-chains under strong flows: stems and flowers. Europhys. Lett. 30(7), 387–392.

Bustamante, C., Bryant, Z., Smith, S.B., 2003. Ten years of tension: single-molecule DNA mechanics. Nature 421, 423–427.

Eslami-Mossallam, B., Ejtehadi, M.R., 2011. Contribution of nonlocal interactions to DNA elasticity. J. Chem. Phys. 134, 125106:1–1125106.

Gore, J., Briant, Z., Nollman, M., Le, M.U., Cozzarelli, N.R., Bustamante, C., 2006. DNA overwinds when stretched. Nature 442, 836.

Iliykhin, A.A., Timoshenko, D.V., 2008. Mathematical models of the closed molecules of DNA. Proc. Saratov Univ. 8 (5), 32–40, (in Russian).

Ikai, A., 2008. The World of Nano-Biomechanics. Elsevier, Amsterdam.

Lee, N.-K., Thirumalai, D., 2004. Pulling-speed-dependent force-extension profiles for semiflexible chains. Biophys. J. 86 (5), 2641–2649.

Mathew-Fenn, R.S., Das, R., Harbury, P.A.B., 2008a. Remeasuring the double helix. Science 322, 446–449.

Mathew-Fenn, R.S., Das, R., Silverman, J.A., Walker, P.A., Harbury, P.A.B., 2008b. A molecular ruler for measuring quantitative distance distribution. PLoS One 3, e3229, Supplementary materials section.

Mathew-Fenn, R.S., Das, R., Fenn, T.D., Schneiders, M., Harbury, P.A.B., 2009. Response to comment on remeasuring the double helix. Science 325, 538-c.

Mazur, A.K., 2009a. Comments on Remeasuring the Double helix, arXiv:0904.2678v1[q-bio.BM].

Mazur, A.K., 2009b. Analysis of accordion DNA stretching revealed by the gold cluster ruler. Phys. Rev. E 80, 010901, (R).

Padinhateeri, R., Menon, G.I., 2013. Stretching and bending fluctuations of short DNA molecules. Biophys. J. 104, 463–471.

Sarkar, B., Alexandridis, P., 2015. Block copolymer-nanoparticle composites: structure, functional properties, and processing. Prog. Polym. Sci. 40, 33–62.

Schellman, J.A., 1974. Flexibility of DNA. Biopolymers 13 (1), 217–226.

Shorr, B.F., 1968. Naturally twisted beams. In: Birger, I.A., Panovko, Y.G. (Eds.), Strength, Stability, Vibrations. Handbook in 3 Volumes, 1, Mashinostroenie, Moscow, pp. 440–465 (in Russian).

Smith, S.B., Cui, Y., Bustamante, C., 1996. Overstretching B-DNA: the elastic response of individual double-stranded and single-stranded DNA molecules. Science 271, 795.

Strunz, T., Oroslan, K., Schafer, R., Guntheodi, H.-J., 1999. Dynamic force spectroscopy of single DNA molecule. PNAS 96, 11277–11282.

Swigon, D., 2009. The mathematics of DNA structure, mechanics and dynamics. In: Benham, C.J., Harvey, S., Olson, W.K. (Eds.), Mathematics of DNA Structure, Function and Interactions. Springer, New York, NY.

Tseytlin, Y.M., 2006. Structural Synthesis in Precision Elasticity. Springer, New York, NY.

Tseytlin, Y.M., 2011. An effective model of DNA like structure: with length fluctuation nonlinearity. AIP Adv. 1, 012118.

Tseytlin, Y.M., 2012. Flexible helicoids, atomic force microscopy (AFM) cantilevers in high mode vibration, and concave notch hinges in precision measurements and research. Micromachines 3, 480.

Tseytlin, Y.M., 2013. Functional helicoidal model of DNA molecule with elastic nonlinearity. AIP Adv. 3, 062110.

Vologodskii, A., Frank-Kamenetskii, M.D., 2013. Strong bending of the DNA double helix. Nucleic Acids Res. 41 (14), 6785–6792.

Chapter 7

DNA Elasticity with Transition

7.1 DISCRETE PERSISTENCE CHAIN (DPC)

Cluzel et al. (1996) and Smith et al. (1996) first observed that the stretching of double-stranded DNA is quite different from that of ssDNA. The experiments showed that when a force around 65–70 pN is applied, the dsDNA sample suddenly snaps open (an "over-stretching transition"), extending to almost twice its original contour length before entering a second entropic regime. The second regime clearly represents a DNA configuration quite different from ordinary double-stranded or B-DNA, which has been dubbed S-DNA. The B to S transition is represented by two states in the DPC model.

Mod 8a, Type 8a, Table 1.2. This model (Fig. 1.3d) contains several specified chain parameters (Storm and Nelson, 2003). Each segment of the chain carries a discrete variable σ that takes the values ± 1. $\sigma = +1$ means the segment is in the B state and $\sigma = -1$ corresponds to the S state. The factor by which a segment elongates when going from B to S (Storm and Nelson, 2003) is called ς_s, that is, $b^{\mathrm{S}} = \zeta_s b^{\mathrm{B}}$ with $\zeta_s > 1$. A bend stiffness parameter A^{B} is assigned to B-DNA and $A^{\mathrm{S}} = \beta\zeta_s A^{\mathrm{B}}$ to S-DNA. β is a dimensionless parameter with $\beta\zeta_s < 1$. A bend stiffness ηA^{B} is assigned to a hinge joining B and S segments.

The force F-extension z function is found by the evaluation of $\left\langle \frac{z}{L} \right\rangle = k_{\mathrm{B}}T \frac{d\Omega}{dF}$ with $\Omega \equiv b^{-1} \times \max \ln Y(\omega_1, \omega_{-1})$ through numerically maximizing the expression $\ln Y(\omega_1, \omega_{-1}) = \frac{b}{2}(P + R + \sqrt{(P-R)^2 + 4Q^2})$, where $P = \bar{\alpha} + \left(\frac{F}{k_{\mathrm{B}}T} - \frac{\omega_1}{2A^{\mathrm{B}}}\right)\left(\coth(2\omega_1) - \frac{1}{2\omega_1}\right)$; $R = -\bar{\alpha} + \left(\frac{\varsigma F}{k_B T} - \frac{\omega_{-1}}{2A^B \beta}\right)\left(\coth(2\omega_{-1}) - \frac{1}{2\omega_{-1}}\right)$; and

$$Q = \frac{\bar{g}\sqrt{\beta}}{\eta}\left(\frac{\omega_1 \omega_{-1}}{\sinh(2\omega_1)\sinh(2\omega_{-1})}\right)^{1/2}\left(\frac{2\sinh(\omega_1 + \omega_{-1})}{\omega_1 + \omega_{-1}}\right),$$

$2\bar{\alpha}\mathrm{k_B T}$ is the free energy per unit length required to flip B-DNA into S state and is measured in J/nm, $\bar{g}$ and $\bar{\alpha}$ are fixed parameters, $\bar{\alpha} = 5.45\,\mathrm{nm}^{-1}$, $\omega_{1,-1}$ gives the degree of alignment of the monomers (how forward-peaked their probability is). Q measures the cooperativity of the transition and has units 1/nm. A^{B} is the bend stiffness parameter of B-DNA in units of nm. The dimensionless parameter β is the ratio of the B and S DNA bend stiffnesses. $E^{(\mathrm{B})}$ and $E^{(\mathrm{S})}$ are the stretch stiffness with moduli $F(1 + F / 2E^{(S,B)})$ of B and S DNA, respectively, and they are measured in pN. Data set fit parameters are given by: $\bar{\alpha}_0 = 4.82\,\text{nm}$, $\beta = 0.08$, $Q = 0.23$, $\zeta_s = 1.71$, $E^B = 7.3 \times 10^2$ pN, and $E^S = 3 \times 10^4$ pN. $\chi 2 = 2.15$ at $N = 339$. Points with $1.15 < \langle z/L \rangle < 1.5$ were excluded from the fit.

The discrete persistent chain (DPC) (Type 8a, Table 1.2) displays an overstretching transition of dsDNA to S-DNA.

Finally, the least-square fit of this DPC model gives the overstretching results that are similar to the typical chart presented in Figure 1.3e.

The accepted fit strategy was as follows: first, the part of the stretching curve well below 65 pN was fit to one state of the continuum model (i.e., to the eWLC), determining its effective spring constant and stretch modulus. The values thus obtained are used as initial guesses in a fit of the full curve to the DPC model. To improve convergence, two parameters were estimated as follows: $E^{(B)}$ was accurately estimated from the low forces data and it was held fixed to this value during the full fit; the effective spring constant k was obtained analytically for low-force data as function of the model parameters. Then the remaining five parameters ($\bar{\alpha}_0$, β, Q, E^S, and ζ_s) were fit to the dataset.

The number of parameters used, the complexity of their estimation, and their complex mathematical relations show that practical application of this model is not a simple solution to the transition of B-DNA to S-DNA.

7.2 REVERSE WLC MODELING

Mod 15a (Model 5a, Table 1.3). This model (Cizeau and Viovy, 1997) combines several approaches that describe the regime of extreme stretching, that is, the entropic elasticity of the DNA as a wormlike-chain (WLC), the chain elongation, and a statistical model for transition from the B state to S state. The DNA molecule with all of the base pairs (bp) is first considered in the same state (B or S). The molecule has a Kuhn length A_{Ku}, and the number of bp is N. Its curvilinear (contour) length is $L_0 = Nl_d$, where l_d is the distance between two bp in the absence of elastic deformation. Elasticity of the chemical links is characterized by a longitudinal elastic modulus k_l in the WLC model. Each of the bp can be in either of two states, the regular one for B, and the extended state for S, with rise per base pair l_b and l_s, respectively. The internal energy of a bp has an absolute minimum at l_b, and a relative minimum at l_s, with an energy difference $\ni_s$. These minima are locally quadratic (defining the elastic modulus k_l).

The chain is a sequence of groups of bp in states B or S. n_b denotes the fraction of bp in the B state. The stretching and bending elasticity of the chain depends only on n_b, and not on the details of the distribution of states. Then B and S sections have Kuhn length A_{kb} and A_{ks}, and a longitudinal moduli k_{lb} and k_{ls}, respectively. The spatial degrees of freedom (spatial configuration and longitudinal elasticity of chain) and the internal degree of freedom (state of each bp) are independent, but only for a given value of n_b. The total energy per bp, integrated over the spatial degree of freedom, and for a given composition n_b is

$$e(n_b) = n_b l_b \phi_b + (1 - n_b) l_s \phi_s + e_{comp},$$

where $\phi = -zF + \int F(z')\,dz' + \frac{F^2}{2k}$ is the total free energy per unit length for the chain and at fixed force, and e_{comp} is the internal energy per bp.

The two first terms of above equation are linear in n_b. Therefore, they appear as an external field for the model. Standard calculations lead to the following results for mean distance per bp $\varsigma(z)$ versus force F:

$$\varsigma = n_b l_b \left[z_b(F) + \frac{F}{k_{kb}} \right] + (1 - n_b) l_s \left[z_s(F) + \frac{F}{k_s} \right],$$

where $2n_b = 1 - [1 + \sinh^{-2}(h_\Phi / 2k_B T)\exp(-2 \in / k_B T)]^{-1/2}$,

$h_\Phi = \mu - l_b \phi_b + l_s \phi_s$, $z_b(F)$ and $z_s(F)$ are the inverse relation of WLC equation. The best fit parameters with respect to experimental data are as follows: $l_b = 3.5$, $l_s = 5.6$ Å, $k_b = 2.35\ \text{nm}^{-1}$, and $k_s = 8\ \text{nm}^{-1}$. They are in good qualitative agreement with earlier determinations. The Kuhn length of B state is 35 nm, consistent with what is expected at the high salt concentration used in the experiments.

This theoretical approach does not take into account important aspects of the problem, such as the possible dependence of the parameters on the sequence, the possibility of denaturation, and the effect of ionic strength of the medium. It also ignores the coupling of the chain elongation with torsion energy, that is, it assumes that the chain is free to rotate.

7.3 APPROXIMATION BY SEQUENCE OF LINEAR SPRINGS

Xiao (2013) presented DNA elastic nonlnearity with fourfold smooth transitions between soft and hard linear springs up to breaking, hyperbolic tangent functions, and matching test data for dsDNA and ssDNA.

Mod 15b (Type 5b, Table 1.3) models the nonlinear transition of dsDNA to S-DNA (ssDNA) through the sequence of connected linear springs functions. In the model Type 5b (Table 1.3), DNA elastic nonlinearity is explained (Xiao, 2013) as a multiple smooth combination of soft and hard linear springs (Fig. 7.1) with the constants: k_0 for the initial soft spring, k_1 for a hard spring with high rigidity, $k_{BS} = k_{02}$ for a soft overstretching spring with low rigidity, k_2 for a hard spring with high rigidity, and breaking at force F_m with negative slope k_3. Besides, the first and second hardening extensions x_1 and x_2 lie in the middle of the two hard spring functions.

At the midpoint of the overstretching soft spring part comes overstretching force F_{BS} and the maximum force F_m emerges at the peak. Approximate

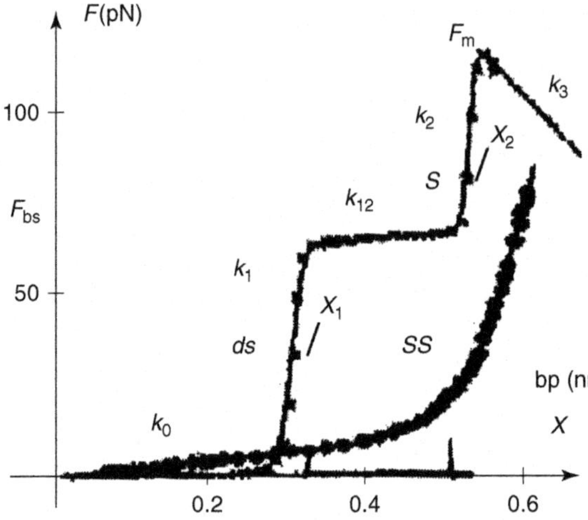

FIGURE 7.1 Stretching charts of multiple soft and hard linear springs dsDNA and ssDNA models. *(Xiao, 2013; reprinted with permission.)*

equations for the force F versus extension x contain k_0 as initial rigidity, and six constants: a_{1k}, b_{1k}, β_1 and a_{2k}, b_{2k}, β_2 that are presented as follows:

$$a_{1k} = F_{BS} - k_0 x_1 - 1/2(x_2 - x_1)k_{BS}; b_{1k} = k_{BS} - k_0; \beta_1 a_{1k} = 2k_1 - k_0 - k_{BS};$$
$$a_{2k} = F_m - F_{BS} - 1/2(x_2 - x_1)k_{BS}; b_{2k} = k_b - k_{BS}; \beta_2 a_{2k} = 2k_2 = k_{BS} - k_b.$$

In Figure 7.1, $k_{BS} = k_{12}$ and $k_b = k_3$.

This approximation has nine constants that are obtained from test data.

Xiao captured the strongly nonlinear features of the fourfold smooth transitions just by hyperbolic tangent functions, in conjunction with the linear functions of the characterizing springs. He introduced two hyperbolic tangent functions, as follows:

$$S_i = \frac{1}{2}(\tanh \beta_i (x - x_i) + \tanh \beta_i x_i), \quad i = 1, 2, \tag{7.1}$$

where (x_1, x_2) are first and second hardening extensions and $\beta_1 > 0$, $\beta_2 > 0$ are two dimensionless constants. For large β_1 and β_2, there are two very small extensions, $0 < \delta_1 \ll x_1$ and $0 < \delta_2 \ll x_2$ so that the two hyperbolic tangent functions S_1 and S_2 display the following sharp transition properties:

$$S_i \approx \begin{cases} 0, x - x_i < -\delta_i \\ 1, x - x_i > \delta_i \end{cases}. \tag{7.2}$$

Namely, the hyperbolic tangent functions S_i (here i is either 1 or 2) may be treated as either 0 or 1 outside a very narrow zone centered at the hardening extension x_i. Moreover, within the very narrow zone just indicated, S_i is linear, with a very steep slope. It may be clear that the sharp transition properties of S_1 and S_2 can exactly quantify and characterize the fourfold smooth transitions, together with their separate linear transition parts characterizing two hard linear springs with high rigidities. Two linear functions with very small slopes may be introduced in order to represent two soft linear springs with high compliances.

The force-extension relation of a DNA molecule, $y = f(x)$, is found to be of the form:

$$y = k_0 x + (a_1 + b_1(x - x_1))S_1 + (a_2 + b_2(x - x_2))S_2. \tag{7.3}$$

In the above equations, k_0 is the initial spring rigidity, a_1, b_1, β_1 and a_2, b_2, β_2 are six constants. From the sharp transition properties of S_1 and S_2, it may be clear that the two soft springs are given by Eq. (7.3) over three wide stretch ranges, separately, while the two hard springs are also given by Eq. (7.3) within two very narrow stretch ranges, separately. Therefore, it may be concluded that the fourfold smooth transitions and all intrinsic characteristics exhibited in the force-extension relation of DNA molecules are indeed incorporated into the explicit expression Eq. (7.3). Furthermore, the foregoing six constants in Eq. (7.3) may be given just by nine intrinsic constants:

$$\begin{cases} a_1 = f_{\mathrm{BS}} - k_0 x_1 - \dfrac{1}{2} k_{\mathrm{BS}}(x_2 - x_1), \\ b_1 = k_{\mathrm{BS}} - k_0, \\ \beta_1 a_1 = 2k_1 - k_0 - k_{\mathrm{BS};} \end{cases} \quad \text{and} \quad \begin{cases} a_2 = f_{\mathrm{m}} - f_{\mathrm{BS}} - \dfrac{1}{2} k_{\mathrm{BS}}(x_2 - x_1), \\ b_2 = k_{\mathrm{b}} - k_{\mathrm{BS}}, \\ \beta_2 a_2 = 2k_2 - k_{\mathrm{BS}} - k_{\mathrm{b}}. \end{cases}$$

Two hyperbolic functions S_1 and S_2 in conjunction with the constants β_1 and β_2 play an essential role in characterizing the fourfold smooth transitions between three pairs of soft and hard linear springs, and will be referred to as the first and second transition index of DNA molecules, respectively. The greater these two indices, the sharper the transitions among the three pairs of soft and hard linear springs. On the other hand, mild transitions may be expected with moderate values of β_1 and β_2.

Unlike dsDNA molecules, there is no overstretching transition, and only at breaking may a sharp transition emerge for an ssDNA molecule. Prior to breaking, a mild transition takes place, and is characterized by the transition index β_2. The latter is much smaller than in the case of dsDNA molecules. In this case, one soft spring with the initial stiffness k_0 and one hard spring with the hardening stiffness k_2 are involved, together with the hardening extension x_2 and maximum force F_{m}, as well as the breaking slope k_{b}.

As we see, this model is useful for the approximation of the averaged experimental data on dsDNA and ssDNA stretching, but does not reveal the real mechanism of the dsDNA and ssDNA molecules' elasticity and overstretching behavior. It uses an imaginary combination of "multiple soft and hard linear springs" that does not exist in a real DNA molecule structure. Besides, the number of required constants depends on the accuracy of experimental data, and may not be known *a priori*.

7.4 EXPLICIT HELICOIDAL MODEL (EHM) TRANSITION

Force-stretch relations.

The average nonlinear transformation ratio $i_{\theta av}$ of a pretwisted strip sensor has been expressed by us as follows (see Chapters 4, and 6, Fig. 6.6):

$$i_{\theta av=|\theta_\gamma/2S_\gamma|} = (\kappa_{p0}+\theta_\gamma)\frac{0.0286E[1+\theta_\gamma(\kappa_{p0}+\theta_\gamma)^{-1}-3(\kappa_{p0}+\theta_\gamma)b_{10}^2\theta_\gamma]}{4G\lambda[1+\nu_{if}(\kappa_{p0}+\theta_\gamma)^2 b_{10}^2\lambda^{-1}+3\theta_\gamma b_{10}^2\lambda^{-1}(\kappa_{p0}+\theta_\gamma)/8]}, \tag{7.4}$$

and its torsional rigidity at extension with one nonlinear helix (unlike regular mechanical sensors with two consequently connected helices with opposite directions) system $j_{\theta\eta} = F_q/\theta$ is

$$j_{\theta\eta} = 0.698\,GA_{c\sec}(1.4)^{n_j}\lambda(1+\nu_{jf}\chi_{kp})/L_c k_{p0}, \text{pN/(angular degree)}, \tag{7.5}$$

where $1.4 \sim\sqrt{2}$, $\theta_\gamma = \theta/L_c$. Here, θ is the negative untwist angle of the strip under stretching force F_q. The dimensionless stretch is denoted by $S_\gamma = s/L_c$ with the strip's corresponding stretch s in nanometers and the molecular length L_c in nanometers.

Power in (7.5)

$$n_j = \frac{\ln S_\gamma}{\ln 1.4} \mp 1 \tag{7.5a}$$

with -1 for $S_{\gamma 1} < 1$ and plus for $S_{\gamma 1} \geq 1$. As an exception, we assume $n_j = -2$ if its value calculated by the shown formula is $n_j \approx 0$ (i.e., at $S_\gamma = 0.68$, $S_{\gamma 1} = 1.36$). Note that $S_{\gamma 1} = 2S_\gamma$.

The factor $\nu_{if} = 0.39–0.41$ is the transformation form coefficient for the punched pretwisted nanostrip. The parameter $\chi_{kp} = k_{p0}^2(b_o/2)^2\lambda^{-1} = 190.9$ is the dimensionless measure of the strip's relative pretwist, $\lambda = (h_o/b_o)^2 = 0.0197$ with height $h_o = 0.295$ nm and width $b_o = 2b_{10} = 2.1$ nm, dimensions of the outer rectangular contour (Figs 6.1 and 6.3). The factor $k_{p0} = 2\pi/S_0 = 1.847\ \text{nm}^{-1}$ represents the initial twist parameter with helical pitch $S_0 = 3.4$ nm. The parameter E is Young's modulus in pN/nm^2, G is the shift modulus in pN/nm^2 of DNA molecule, and A_{csec} is its cross-sectional area in nm^2. Recall that the product of E and A_{csec} is equal to $E_{str} = 1{,}100$ pN as a so-called stretch modulus. $E/G = 2(1 + \nu_p)$ with ν_p as a conditional Poisson's ratio. Therefore, we can replace in Eq. (7.4) the ratio E/G by $2(1 + \nu_p)$, $GA_{csec} = E_{str}/2(1 + \nu_p)$ in Eq. (7.5), and use the corrected values of G and A_{csec}, including $A_{c\sec} = \pi b_{10}^2$ as for a round section or more comprehensive sandwich of pretwisted strip-coiled ribbon solution (see the following section) to satisfy the limit average stress value $\sigma_{av} = F/A_{csec} \leq \sigma_{lim}$ as an ultimate stress for DNA material. We have found in a previous study (Tseytlin, 2012) that Poisson's ratio for DNA in salt solutions may be assumed to be $\nu_p = 0.2\text{–}0.5$, and the stiffness form factor for the strip's cross section in that cases is $\nu_{if} = 0.14\text{–}0.18$. The limit stress for DNA is usually accepted as for plastics, namely, $\sigma_{lim} = 45$ MPa.

7.4.1 Force-Stretch Formula

Eq. (7.4) may be expressed (see Chapter 4) as a cubic equation, with respect to the angle θ_γ (in rad/nm):

$$A_3\theta_\gamma^3 + A_2\theta_\gamma^2 + A_1\theta_\gamma + A_0 = 0, \tag{7.6}$$

where $A_0 = -\,0.00715(ES_\gamma/G\lambda)\kappa_{p0}$;

$A_1 = 2[14.325\nu_{if}\chi_{kp} - 0.00715\,(ES_\gamma/G\lambda) + 14.325] + 0.02145\,\chi_{kp}\,ES_\gamma/G$;

$A_2 = 28.65\,\chi_{kp}[(2\nu_{if} + 0.375) + 0.0015\,ES_\gamma/G]\,\kappa_{p0}^{-1}$;

$A_3 = ½\,b_1^2[57.3\,(\nu_{if} + 0.375) + 0.0429\,ES_\gamma/G]\,\lambda^{-1}$.

The solution of this equation allows us to find values of θ_γ corresponding to stretching S_γ in the range from 0 to 1.7. The latter value corresponds to S-DNA conformation. It is also known that, for dsDNA without overstretching, the stretching force should be less than 65 pN.

Eq. (7.6) can be converted to the following canonical form, after dividing by A_3:

$$X^3 + rX^2 + pX + c = 0, \tag{7.7}$$

where $r = A_2/A_3$, $p = A_1/A_3$, and $c = A_0/A_3$.

Here,

$$r = \frac{28.65\chi_{kp}[(2\nu_{if} + 0.375) + 0.0015ES_\gamma/G]k_{p0}^{-1}}{(b_1^2/2)[57.3(\nu_{if} + 0.375) + 0.0429ES_\gamma/G]\lambda^{-1}};$$

$$p = \frac{2[14.325\nu_{if}\chi_{kp} - 0.00715(ES_\lambda/G\lambda) + 14.325] + 002145\chi_{kp}ES_\gamma/G}{(b_1^2/2)[57.3(\nu_{if} + 0.375) + 0.0429ES_\gamma/G]\lambda^{-1}};$$

$$c = \frac{-0.00715(ES_\gamma/G\lambda)k_{p0}}{(b_1^2/2)[57.3(\nu_{if} + 0.375) + 0.0429ES_\gamma/G]\lambda^{-1}}.$$

If we choose $\nu_{if} = 0.41$ as for perforated helical strip, $b_1 = 1.05$ nm, $\nu_p = 0.5$ corresponding to $E/G = 2(1 + \nu_p) = 3$, the expressions for r, p, and c will be as follows:

$$r = 105.823\frac{1.195 + 0.045S_\gamma}{44.9805 + 0.1287S_\gamma};$$

$$p = 81.1609\frac{1 + 0.00445S_\gamma}{44.9805 + 0.1287S_\gamma};$$

$$c = \frac{-0.0719S_\gamma}{44.9805 + 0.1287S_\gamma}.$$

If we choose $\nu_p = 0.2$ corresponding to $E/G = 2.4$, the expressions for r, p, and c are equal to

$$r = 105.819\frac{1.195 + 0.0036S_\gamma}{44.9885 + 0.10296S_\gamma};$$

$$p = 81.1607\frac{1 + 0.00356S_\gamma}{44.9885 + 0.10296S_\gamma};$$

$$c = \frac{-0.0575S_\gamma}{44.9885 + 0.10296S_\gamma}.$$

The solution of Eq. (7.7) with negative discriminant D_{sc} is as follows:

$$\theta_\lambda = 2\sqrt[3]{\rho}\cos(\varphi/3) - (r/3) \tag{7.8}$$

where $\rho = \sqrt{-w^3/27}$, $w = \frac{3p - r^2}{3}$, $\cos\varphi = q/(2\rho)$, and $q = \frac{2r^3}{27} - \frac{rp}{3} + c$.

$$D_{sc} = (w/3)^3 + (q/2)^2 < 0.$$

As a result, the stretching force F_q in picoNewtons equals

$$F_q(S_{\gamma 1}) = 57.3£_i \psi_2 j_{\theta\eta} L_c \theta_\gamma(S_\gamma) S_{\gamma 1}, \tag{7.9}$$

where $j_{\theta\eta}$ in Eq. (7.5) has a product $GA_{csec} = E_{str}/[2(1 + \nu_p)]$; $S_{\gamma 1} = \psi_1 S_\gamma$, $\psi_1 = 2$, $\psi_2 = 0.25$ are the scale factors which are necessary because Eq. (7.4) and (7.5) were originally developed for a strip sensor with the two pretwisted sections (Tseytlin, 2006), and $£_i$ is the scale factor for adjustment to the experimental data set.

Here, $GA_{csec} = E_{str}/3$ at $\nu_p = 0.5$ and $GA_{csec} = E_{str}/2.4$ at $\nu_p = 0.2$.

7.4.2 Calculation Results

The results of our calculations with Eq. (7.9) are shown in Figure 7.2a and Table 7.1.

We compared the plotted chart on Figure 7.2a F_q–$S_{\gamma 1}$ data with an average of experimental measurements F_b–$S_{\gamma 1}$ presented in Bustamante et al. (2000); Wenner et al. (2002); and Xiao (2013) (see also Tseytlin, 2013, 2014). In accordance with our estimations, the factor $£_i$ for this case is in the range 0.92…1.08 or an uncertainty within ±8%.

Some experimental results have a wide quasi-horizontal middle section (plateau) of the nonlinear B→S transition function. We partly achieve that with our

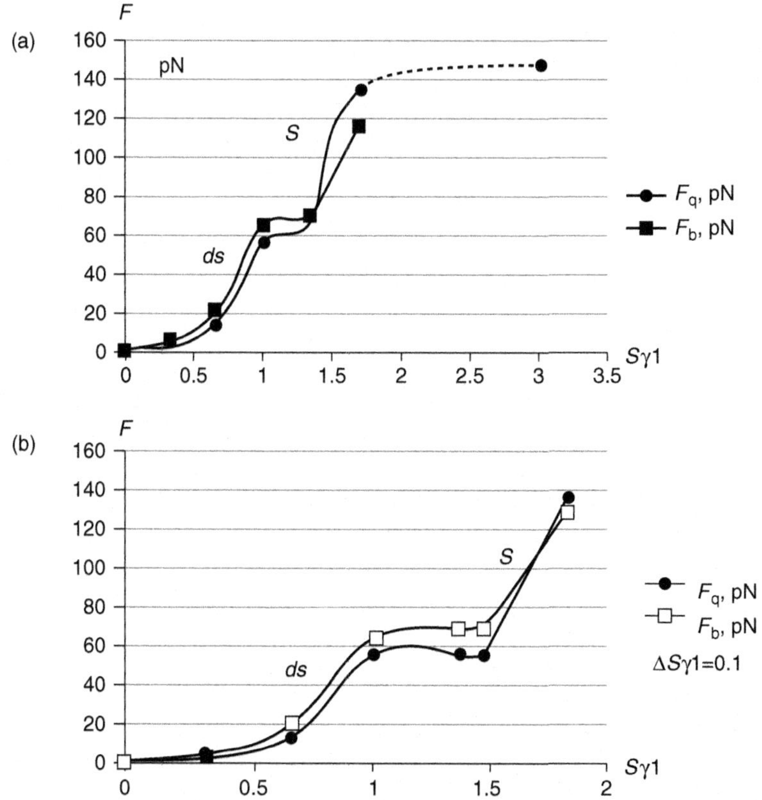

FIGURE 7.2 Chart (a) of the nonlinear elasticity in force F versus stretch $S_{\gamma 1}$ field. F_q denotes the stretch calculated plot using the EHM model, and F_b is the average experimental data from references Bustamante et al. (2000), Wenner et al. (2002), Xiao (2013), and Tseytlin (2013, 2014). Chart (b) has corrected data with $\Delta S_{\gamma 1} = 0.1$. This figure shows the regions dsDNA and S-DNA, labeled as *ds* and *S* respectively.

TABLE 7.1 Force-Stretch Data for DNA EHM Model with Helicoidal Pretwisted Nanostrip

$S_{\gamma 1}$	F_q, pN (7.9)	n_j (7.5a)	$£_j$ Assumed
0	0	–	–
0.34	2.82	−6.27	0.92
0.68	13.25	−4	1.08
1.02	56.75	−1	1.016
1.36	56.83	−2	1.08
1.5	120.07	0.21	1.00
1.7	137.6	0.5	0.92

$V_p = 0.2$

TABLE 7.2 Force-Stretch Corrected Data for DNA EHM Model With Helicoidal Pretwisted Nanostrip at $\Delta S_{\gamma 1} = 0.1$

$S_{\gamma 1}$	F_q, pN	F_b, pN
0	0	0
0.34	2.82	5
0.68	13.25	20
1.02	56.75	65
1.36	56.85	70
1.46	56.85	80
1.8	137.6	130
$v_p = 0.2$		

helicoidal model by transformation of Eq. (7.9) with the unit singular function U as follows:

$$F_q = F_q(S_{\gamma 1} => 1.4) + [F_q(S_{\gamma 1} > 1.4) - F_q(S_{\gamma 1} = 1.4)]U(S_{\gamma 1} --> (1.4 + \Delta S_{\gamma 1})), \quad (7.10)$$

where $U(S_{\gamma 1} \leq (1.4 + \Delta S_{\gamma 1})) = 0$, $U(S_{\gamma 1} > (1.4 + \Delta S_{\gamma 1})) = 1$, and $\Delta S_{\gamma 1}$ is an extension of the DNA nonlinear elastic function's middle section (plateau). Results of our calculation with Eq. (7.10) are presented in Table 7.2 and on Figure 7.2b.

Our model shows the similarity of nonlinear elasticity in the helicoidal mechanical nanosensor and the DNA molecule as a biomechanical object. This properly scaled model will be useful for the engineering community working with DNA applications in nanotechnology, as well.

7.5 BUCKLING OF DNA MOLECULE

Recent research on DNA molecule buckling by a compression force was provided by Fields et al. (2013), on the basis of FRET (Föster Resonance Energy Transfer) data analysis. A statistical mechanical model developed by Fields et al. (2013) for Euler's buckling showed only partial agreement with the experimental data. Our analysis shows that the model of Fields et al. (2013) does not reflect explicitly the twisted (helical) condition of the molecule.

If we put the expression for the persistence length $l_p = EJ/2(k_B T)$ into Fields' model, we find that the buckling length of the molecule, under compression force F, is equal to

$$L_{\text{buckle}} = \pi\sqrt{EJ/2F}. \quad (7.11)$$

The buckling length for our helicoidal model, in accordance with Table 4.4, equals

$$L_{\text{buckle(h)}} = \pi\sqrt{v_{\text{tp}}/4v_\eta F_x}, \quad (7.12)$$

where F_x is the applied force; $\nu_{\text{tp}} = \dfrac{2\nu_\eta}{\nu_\eta + \nu_\xi}$; $\nu_\eta = \dfrac{1 + \nu_\theta j_{\eta\theta}^2 / j_\eta}{j_\eta}$; $\nu_\xi = \dfrac{1 + \nu_\theta j_{\xi\theta}^2 / j_\xi}{j_\xi}$ from Table 4.2 and $\nu_\theta = [j_\theta - (j_{\varepsilon\theta}^2 / j_\varepsilon) - (j_{\xi\theta}^2 / j_\xi) - j_{\eta\theta}^2 / j_\eta]^{-1}$.

Table 4.1 shows that $j_\xi = EJ_\xi; j_\eta = EJ_\eta; j_{\eta\theta} = -k_{p0}E(J_{p\eta} - T_\eta)$ and $j_{\xi\theta} = k_{p0}E(j_{p\xi} - T_\xi)$.

Here, $j_{p\xi} = \int_{A_C} R^2 \eta \, dA_C$ and $J_{p\eta} = \int_{A_c} R^2 \xi \, dA_c$ are the polar-axial moments and A_C is the cross-sectional area.

After substitution into Eq. (7.12) of the expression for ν_{tp}, we finally have the following formula for the buckling length in our EHM model:

$$L_{\text{buckle(h)}} = \pi \sqrt{\frac{1}{\nu_\eta + \nu_\xi} / 2F}. \tag{7.13}$$

Analysis shows that the numerators of the expressions for ν_η and ν_ξ are the corresponding functions of the initial helix parameter k_{p0}. Therefore we can write them as follows:

$$\nu_\eta = \frac{1 + \varphi_\eta(k_{p0})}{j_\eta} \quad \text{and} \quad \nu_\xi = \frac{1 + \varphi_\xi(k_{p0})}{j_\xi}.$$

If the initial helix parameter is zero or $k_{p0} = 0$ for the straight beam, the functions $\varphi_\xi = 0$, $\varphi_\eta = 0$, and then $\nu_\eta = \dfrac{1}{j_\eta} = \dfrac{1}{EJ_\eta}$, $\nu_\xi = \dfrac{1}{j_\xi} = \dfrac{1}{EJ_\xi}$.

In our model with cross-section stretched 10–15 times in one direction (see Fig. 6.6b), one axial moment of inertia J_η is much larger than another one, J_ξ. Therefore, after substitution of the expressions of terms ν_η and ν_ξ into Eq. (7.13), we find an equation similar to Eq. (7.11), that is,

$$L_{\text{buckle(h)}} \approx \pi \sqrt{EJ_\xi / 2F}. \tag{7.14}$$

However, it is clear that a more comprehensive approach with appropriate value of helix parameter k_{p0} and functions $\varphi_\xi \neq 0$, $\varphi_\eta \neq 0$ in this case is necessary.

7.6 ELASTICITY OF ssDNA, ZZM, DPC, AND EHMR MODELS

In the case of ssDNA and RNA molecules, the choice of an adequate elastic model remains an open problem, unlike in dsDNA that has a well-established elastic description. On the other hand, the chemical structure of ssDNA strongly supports the picture with a finite bending modulus, but in different way than in WLC models. The Discreet Persistent Chain (DPC) model proposed by Storm and Nelson (2003) has advantages in calculating the response of ssDNA to stretching forces as great as 400 pN by combining the transfer matrix method with a variational procedure. However, the AFM experiments done in the limit of very high forces (up to 1000 pN) exhibit saturation of the end-to-end distance at a certain value. Any model that treats deformations of the ssDNA backbones in a linear manner is unable to capture this phenomenon.

7.6.1 Zigzag (ZZM) and DPC Models

Mod 8b. Tkachenko (2007) proposed a zigzag model (Type 8b, Table 1.2) for the low- and high-stretching force regimes of ssDNA elasticity, motivated by the actual microscopic architecture of this molecule nucleotide.

The "zigzag" model (ZZM) is described by Tkachenko with the following equations:

$\frac{z}{L} = x = \frac{b^*(F)}{b^*(0)} x(F)$ is the stretching curve in a DPC model with renormalization, z and L are the displacement distance and the original chain contour length, respectively; $l_p = b^*(0)J^*/k_B T \approx Jb/k_B T$ is the persistence length;

$$b^*(F) = 2b\cos\theta^* \approx 2b\left(1 - \left(\frac{bF}{2J} + 1\right)^{-2} \frac{\theta_{0p}^2}{8}\right),$$

where θ_{0p} is the same preferred angle for each bond pair.

Similar results are obtained by Storm and Nelson (2003) using a DPC model: $\left\langle \frac{z}{L} \right\rangle \to \frac{F}{k^{DPC}} + O(F^2)$ with the effective spring constant for the DPC model equal to

$$k^{DPC} = \frac{3}{2}\frac{k_B T}{l_p}\left(1 - \frac{b}{2l_p}\right).$$

Here, $b \approx 0.6$ nm for ssDNA. This DPC model fits experimental data only up to 400 pN.

The zigzag model fits the experimental results up to 800–1000 pN, but Tkachenko (2007) recommended relating $x = z/L$ to F using the known WLC formula

$\frac{Fl_p}{k_B T} = \frac{1}{4}\left(\frac{1}{(1-x)^2} - 1\right) + x$, which meets the unfeasible situation at $x = 1$ with an infinite value of the stretching force $F = \infty$.

7.6.2 Application of Coiled Ribbon

Mod 16c. The elasticity of the dsDNA molecule at strong stretching may be modeled with the addition of a coiled ribbon (EHMR model) after dsDNA melting that will give a wider quasi-horizontal middle section (plateau) of the nonlinear function for dsDNA transition seen in some experiments. The coiled

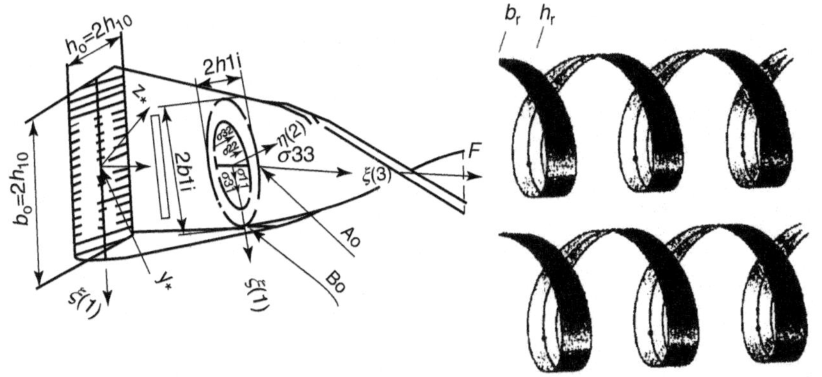

FIGURE 7.3 EHMR model with coiled ribbon strands of ssDNA, b_r, and h_r are the width and thickness of the ribbons, respectively.

ribbon nonlinear elasticity is close to a quasi-elliptical function. Therefore, a more comprehensive model may include a chain (Fig. 7.3) consisting of both structures: pretwisted strip, and coiled ribbons. This is important for stress estimation, as well.

The mathematical apparatus for the nonlinear coiled ribbon sensor, however, is not developed enough for our purposes to model ssDNA. But we will try to find some acceptable solution.

The rigidity of a coiled ribbon (Table 4.8) under extension is related to the untwist angle, as follows:

$$F/\theta = \frac{1}{2}6.3\frac{b_r h_r^3 E}{D_r L} = \frac{1}{2}6.3\frac{h_r^2 E_{str}}{D_r L}, \tag{7.15}$$

where F is the stretching force, b_r and h_r are the width and the height of the ribbon cross-section, respectively, E is Young's modulus, E_{str} is the stretch modulus, D_r is the diameter of the single strand molecule circumference of the cross section, and L is its end-to-end length.

If we use the elliptical representation similar to Eq. (4.81a), we will find an expression for the untwist angle θ as follows:

$$\theta = \sqrt{\left\{1 - \frac{\aleph_\theta}{1-x}\left[(1-x)^2 L^2 / 2L_r^2\right]\right\}(2L_r/D_r)^2}, \tag{7.16}$$

where $x = (L - z)/L$, $L_r = 2\pi R_0 n_{r0}/\cos\alpha_0$ is the full length of the ribbon, r_0 is the radius of ssDNA molecule cross-section outer contour, $n_{r0} = L/S_{av}$ is the number of the coiled strip's turns with the average pitch S_{av}, $\alpha_0 = S_{av}/2\pi r_0$ is the helix angle of coiled ribbon, and $\aleph_\theta = 125$ is the scale factor.

The properties of the ssDNA molecule are not known as definitely, as for dsDNA. Different publications show somewhat different data for ssDNA. In our case, we will use experimental data from Storm and Nelson (2003), Tkachenko (2007), and data for the ssDNA molecule from other sources. Here, those parameters are:

L = 3.9 μm, $E_{\text{str}} = 4.5 \times 10^3$ pN, r_0 = 1.07 nm (see Kim et al., 2007), D_{r} = 2.14 nm, S_{av} = 5.9 nm, h = 0.0375 nm (in our helicoidal model).

Therefore,

$$\alpha_0 = \frac{5.9}{2\pi 1.07} \times 57.3 = 50.31 \text{ (angular degree)}; \; n_{r0} = 3.9/0.0059 = 661;$$

$$L_r = 2\pi r_0 n_{r0} / \cos\alpha_0 = 2\pi 1.07 \times 661 / \cos 50.31 = 6.955\,\mu n.$$

Eq. (7.15) and (7.16) work in the range of stretching force F = 117–889 pN and x = 0.95–1.00.

Hence, at x = 1.00 we have

$$\theta = 2 \times 6.955/2.14 \times 10^{-3} = 6500$$

and

$$F = \frac{1}{2} 6.500 \frac{6.3(0.0375)^2 4.5 \times 10^3}{2.14 \times 3900} 57.3 = 889.56 \text{ pN}$$ that corresponds (Fig. 7.4) to Storm and Nelson (2003), experimental data, with the deflection less than 2.5%.

At x = 0.95

$$\theta = \theta = \sqrt{\left\{1 - \frac{125}{1-0.95}\left[(1-0.95)^2 3.9^2 / 2 \times 6.955_r^2\right]\right\}(2 \times 6.955/2.14 \times 10^{-3})^2} = 856.9,$$

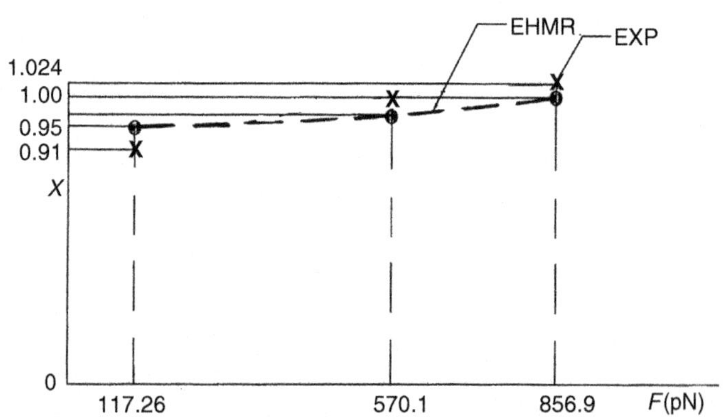

FIGURE 7.4 Stretching behavior of ssDNA represented by the EHMR model (circles) and experimental data – EXP (crosses). Data from Storm and Nelson (2003), with L = 3.9 mm and $\boldsymbol{x} = z/L$.

$$F = \frac{1}{2}856.9\frac{6.3(0.0375)^2 4.5\times 10^3}{2.14\times 3900}57.3 = 117.26 \text{ pN}.$$

This agrees with Storm and Nelson (2003) data within 4%.

At $x = 0.97$,

$$\theta = \theta = \sqrt{\left\{1 - \frac{125}{1-0.97}\left[(1-0.97)^2 3.9^2/2\times 6.955_r^2\right]\right\}(2\times 6.955/2.14\times 10^{-3})^2} = 4166$$

and

$$F = \frac{1}{2}84166\frac{6.3(0.0375)^2 4.5\times 10^3}{2.14\times 3900}57.3 = 570.1 \text{ pN}.$$

This agrees to Storm and Nelson (2003) experimental data within 3%.

Stress in coiled ribbon. Now we can estimate the stress value in the ribbon with radius 1.07 nm, $L = 3.9$ μm, $L_r = 6.955$ μm, stretched pitch $S_{av} = 5.9$ nm, number of turns $i = 661$, $\tan\alpha = 0.878$ at stretching it to $s = 3.9$ μm. We will use equations (4.74), (4.75) and (4.76) for that:

$$\sigma_{y1} = \frac{E}{1-\mu^2}(\varepsilon_{y1} + \mu\varepsilon_{x1}) = -\frac{E}{1-\mu^2}\frac{\Delta}{R_0^2}z_h,$$

where, for this model, $E = 3470$ pN/nm^2, $\mu = \nu_p = 0.2$ is the Poisson's ratio, $z_h = 0.375/2$ nm is the distance from the ribbon middle surface to a point of outer contour, $R_0 = 1.07$ nm, $D_r = 2.14$ nm,

$$\Delta = \left(\frac{\theta}{2\pi i} - \omega t \tan\alpha\right)R_0 \text{ and } \omega = s/(2\pi R_0 i) = 3900/(6.28\times 1.07\times 661) = 0.878.$$

We will estimate the corresponding angle θ from the elliptical Eq. (7.16)

$$\theta = \sqrt{\{1-[2^2 3.9^2/2\times 6.955^2]\}(2\times 6.955^2/2.14\times 10^{-3})^2} = 3950.$$

Therefore, $\Delta = \left(\frac{3950}{6.28\times 661} - 0.878\times 0.878\right)1.07 = 0.194$ nm

and

$$\sigma_{y1} = -\frac{E}{1-\mu^2}\frac{\Delta}{R^2}z = \frac{3470}{1-0.2^2}\frac{0.194}{1.07^2}0.0375/2 = 11.5 \text{ MPa}.$$

This stress data satisfies the properties of polymer materials (see Chapter 1, Sec. 1.7), but the calculated stress corresponds only to the first phase of the coiled ribbon deformation, without change to the cylindrical shell, and without extension of the middle surface (see Chapter 4, Sec. 4.5). These assumptions correspond to a small deformation of the coiled ribbons.

7.7 MECHANICAL STABILITY

7.7.1 Experimental Data

Clausen-Schaumann et al. (2000) investigated the mechanical stability of single dsDNA molecules as a function of stretching velocity, buffer composition, temperature, and DNA sequence, using AFM-based single molecule force spectroscopy. In addition, the kinetics of the force-induced melting of the double helix as reannealing kinetics was also investigated.

When a single dsDNA molecule is stretched beyond the previously investigated highly cooperative B-S transition at 65 pN, a new conformational transition occurs at 150 pN. At the end of this second transition, the force increases drastically upon further extension of the molecule. At subsequent relaxation of the molecule, the force drops continuously, following a smooth curve (Fig. 7.5), exhibiting none of the conformational transition of the extension curve. The relaxation curve of the molecule resembles strongly the force versus extension of a single-stranded ssDNA that is free of hysteresis, and follows a simple polymer elastic model, without conformational transitions.

7.7.2 EHM with Stretch and Relaxation Difference

Mod 16a (Type 6a, Table 1.3). This behavior of stretched and relaxed DNA molecule is similar to the behavior of a stretched and relaxed helicoidal microstrip sensor (of course, without melting and recombination, but with absorption

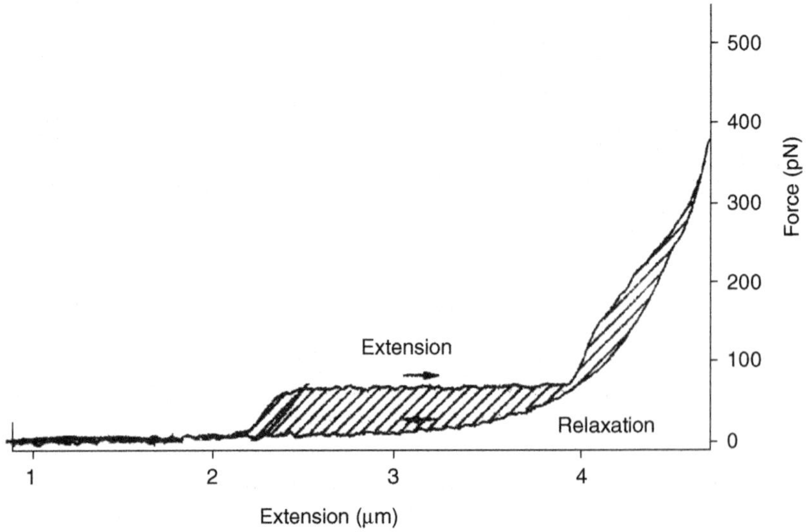

FIGURE 7.5 Force versus extension and relaxation cycles of a l-DNA segment in 100 mM NaCl, 10 mM Tris (pH 8), 1 mM EDTA at 20°C. *(Modified from Clausen-Schaumann et al., 2000, with permission.)*

damping ψ_Δ), discussed in Chapter 5, in which stretching at different speeds involves some kind of hysteresis, and relaxation at different speeds does not show any significant hysteresis (Fig. 7.6). This asymmetric phenomenon was studied by us in detail (see Chapter 5).

This study was carried out on the pretwisted microstrip form Br4Sn-3Zn with the following parameters:

$$h \times b \times L = 0.004 \times 0.12 \times 18 \text{ mm with pretwist factor } r_{p0} = \pi \, \text{mm}^{-1}.$$

After effect. The elastic after effect in strings corresponds to a stress variation described (see Tseytlin, 2006) by the empirical expression

$$\Delta\sigma_Q = N_c(\sigma_Q - \sigma_{Qo})(t^{0.23} - M_N).$$

This variation of the stresses in the strip of a helicoidal multivibrator (see Chapter 5, Fig. 5.6b,c,d) corresponds to a lower frequency at the zero mark on the dial, as

$$\Delta f_e = ½\Delta\sigma_Q(\sigma_Q - \sigma_{Qo})^{-1}(f^2 - f_0^2)f_0^{-1}. \qquad (7.17)$$

It is assumed that $N_c = 2{\cdot}10^{-3}$, $M_N = 0.6$ for artificially aged strings. Therefore, we obtain the variation that is equal after time $t = 1$ day to: $\Delta f_e = 2{\cdot}10^{-3} \times (1\text{–}0.6)\,[(4200/3500) + 1] \times 700 = 1.2$ Hz for $f = 4200$ Hz at the extreme position of the readout indicator, and $f_0 = 3500$ Hz at the zero mark. Similar values were observed in devices that allow limited rotation angles for the readout

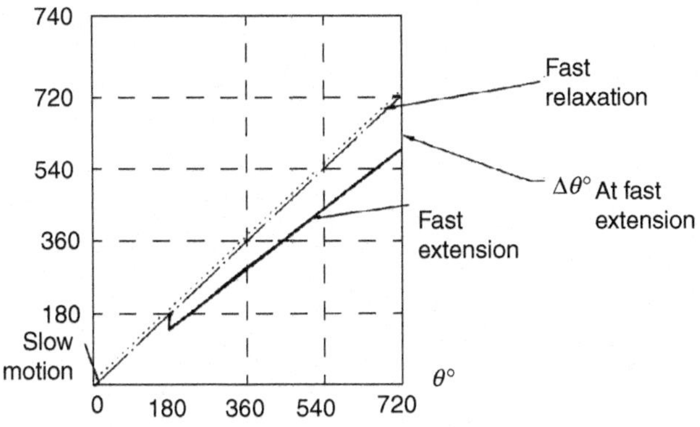

FIGURE 7.6 Fast stretch (step-function) extension (solid line) of helicoidal microstrip, fast relaxation (dotted line), and slow motion-extension and relaxation (dash-dot line); $\theta°$ represents a following untwist angle and $\Delta\theta°$ represents its jump change. $\Delta\theta°(180°) = -[5.8 \pm 0.9]°$ at a fast extension, $\Delta\theta°(180°) = [1.0 \pm 0.2]$ at a fast relaxation; $\Delta\theta°(720) = -[20.0 \pm 1.3]°$ at a fast extension, and $\Delta\theta°(720) = [1.5 \pm 0.5]°$ at a fast relaxation.

indicator, and the helicoid crosspiece in the range $\theta = \pm 60°$. In optical devices with no limiters (stops) on the rotation of the reading mirror attached to the crosspiece (see Fig. 5.6c), the angle of the mirror rotation from the zero mark may exceed $\theta = 300°$ with the measuring rod (spindle) retraction. After holding the measuring rod in the lifted (arrested) position, the frequency at the zero mark may change by 8–12 Hz, when the retractor is released and the rod comes to the contact with the datum surface. This frequency shift cannot be observed with the usual brief retractions. If the elastic aftereffect in a helicoidal multivibrator can be modeled as a nonrelaxing Kelvin body, the relationship between the stress tensor deviator dev $\hat{\sigma}$ and the deviator of micro-displacements tensor e_s can be described by the formula

$$\operatorname{dev}\hat{\sigma} = 2G \operatorname{dev} e_s + 2\eta_{VR} \operatorname{dev} \dot{e}_s, \tag{7.18}$$

where η_{VR} characterizes the "viscosity" (internal damping) of the rigid body deformation. The helicoidal multivibrator usually does not have a significant stress relaxation in time. This follows from the fact that the sensors based on the initially pretwisted strip can be stored for a long time in the tensioned (arrested) state, without loss of quality. Therefore, we can assume, in the case of the multivibrator with an arrow, that displacement derivative of e_s with the respect to time $\dot{e}_s = 0$ for the arrested state. Therefore dev $\hat{\sigma}$ = constant is irrespective of the holding time t. For an optical device, $\dot{e}_s \neq 0$ since the crosspiece of the double helicoid does not have the rotational stops. This means that the frequency shift at the zero mark on the scale is an exponential function of the holding time in multivibrators with mirrors at arrested state. Therefore,

$$\operatorname{dev}\hat{\sigma} = 0 \text{ and } \operatorname{dev} e_s = [\operatorname{dev} e_{st} = 0]e^{-Gt/\eta_{VR}}. \tag{7.19}$$

This type of elastic after effect is confirmed by the experimental data (Fig. 7.7a), where $\Delta f_e \approx \Delta f_{emax}(1 - e^{-t/20})$ corresponds to the frequency shift, t is the time in the arrested state, in minutes, and $\Delta f_{emax} \approx 10$ Hz. When a multivibrator is held arrested in the extreme tensioned position, the frequency adequate for

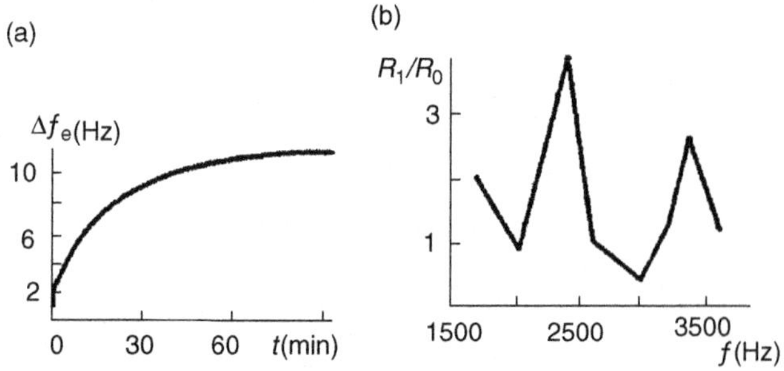

FIGURE 7.7 After effect shift (a) and damping frequency dependence (b) in the helicoidal string multivibrators.

return to the zero mark without resetting is lowered by $\Delta f_e \leq 10$ Hz. However, it increases by $\Delta f_e \leq 10$ Hz, if the device is stored with completely relaxed tension. The experiments showed that the shift in frequency for the multivibrator with a mirror is approximately equal to:

$$|\Delta f_{\text{opt}}| \leq 4(|\Delta\theta_{\text{opt}}| i_\theta^{-1} + (2/3)L|\varepsilon^{33}|)0.0013m_\sigma \tag{7.20}$$

and the multivibrator with an arrow has a shift in frequency that equals

$$|\Delta f_{\text{mic}}| \leq 2(|\Delta\theta_{\text{mic}}| i_\theta^{-1} + (4/3)L|\varepsilon^{33}|)0.0013m_\sigma, \tag{7.21}$$

where $|\varepsilon^{33}|$ is the module of the longitudinal strain in the helicoid at its retraction, $\Delta\theta$ is the angle of untwist during the measuring rod retraction, and $|\Delta\theta_{\text{mic}}| < |\Delta\theta_{\text{opt}}|$. It should also be noted that a damper with fluid of comparatively low viscosity (10^{-2} m^2/s) was used in the experimental devices with an arrow pointer. The damping fluid in the optical device has higher viscosity (the order of 10×10^{-2} m^2/s), a fact that might also influence the after effect. It would be helpful for the reduction of the aftereffect in the latter case to limit the crosspiece rotation angle inside the dial, and reduce the viscosity of the damping fluid.

Internal friction's dependence on frequency. It was experimentally found (Tseytlin, 1988) that internal friction may be lowered in a helicoidal multivibrator when it is excited with high-frequency vibrations in the range of 2.5–3.5 kHz. This effect is manifested in the uncertainty of readings from an appropriate spring-type sensor during multiple lifting (retraction) and dropping of the measurement rod, whose tip rests in its working position on a fixed datum (e.g., gage block) of the thick column in which the sensor is mounted. The smaller this uncertainty range of reading, the lower the internal friction in the precision mechanical system. Comparison of an uncertainty of readouts, at zero exciting frequency R_o and R_f at various frequencies f of oscillations for a helicoid, indicates that an uncertainty $R_f < R_o$ in certain frequency bands (Fig. 7.7b), and that $R_f/R_o < ½$ for the frequency bandwidth of $f = 2800$–3200 Hz.

Facts described in this section about helicoidal multivibrators may be helpful in experimental study of the DNA molecules and protein dynamics, as well.

7.8 NONLINEAR DYNAMICS AND TRAVELING WAVES

Mod 19a, b. Nonlinear dynamics in a double chain model of DNA (Table 1.4, Type 3a, Fig. 1.6c) and traveling wave solution (Table 1.4, Type 3b) for a double-chain model of DNA are resolved (Forinash et al., 1991; Peyrard et al., 2008; Ouyang and Zhang, 2014) by nonlinear dynamical equations:

$$\begin{aligned} u_{tt} - c_1^2 u_{xx} &= \lambda_1 u + \gamma_1 uv + \mu_1 u^3 + \beta_1 uv^3, \\ v_{tt} - c_2^2 v_{xx} &= \lambda_2 v + \gamma_2 u^2 + \mu_2 u^2 v + \beta_2 v^3 + c_0, \end{aligned} \tag{7.22}$$

where u denotes the difference of the longitudinal displacements of the bottom and top strands, and v represents the difference of the transverse displacements of the bases from their equilibrium positions, along the direction of the hydrogen bond that connects the two bases of the base pair. The constants c_i, λ_i, γ_i, μ_i, β_i and c_0 are written as follows, with $i = 1,2$:

$$c_i = \pm\sqrt{\frac{E}{\rho}},\quad \lambda_1 = \frac{-2\mu}{\rho A_\sigma h}(h-l_0),\quad \lambda_2 = \frac{-2\mu}{\rho A_\sigma},\quad \mu_i = \frac{-2\mu l_0}{\rho A_\sigma h^3},$$
$$\beta_i = \frac{4\mu l_0}{\rho A_\sigma h^3},\quad c_0 = \frac{\sqrt{2}\mu(h-l_0)}{\rho A_\sigma h}$$

where $\rho, A_\sigma, E, F, \mu, l_0$, and h are the mass density, the area of transverse cross-section, the Young's modulus, the tension density of the strand, the rigidity of elastic membrane, the distance of the two strands, and the height of the membrane in the equilibrium position, respectively.

Mod 19b. Using the physical quantities and assuming $b = h/\sqrt{2}$ and $E = F$, the following equation was derived (Ouyang and Zheng, 2014) from Eq. (7.22):

$$u_{tt} - c_1^2 u_{xx} = Au^3 + Bu^2 + Cu + D,$$

where $A=((-2\alpha/h^3)+(4a^2\alpha/h^3))$, $B=6\sqrt{2a}\alpha/h^3$, $C=((-2\alpha/l_0)+(6\alpha/h))$, and $D = 0$ with $\alpha = \mu l_0/\rho A_\sigma$.

Finally, assuming $u(x, t) = \varphi(\xi)$ with $\xi = (x/c_1 - \sqrt{2}t)$ and substituting $u(x,t) = \varphi((x/c_1) - \sqrt{2}t)$, the following two-dimensional system is obtained:

$$\frac{d\varphi}{d\xi} = y,\quad \frac{dy}{d\xi} = A\varphi^3 + B\varphi^2 + C\varphi, \tag{7.23}$$

analysis of which reveals six shapes of bell-shaped solitary and periodic traveling waves (see Ouyang and Zheng, 2014).

Mod 16a, b. may be used for the problem of the flexural waves propagation as well (see Chapter 5).

7.9 EHM IN CHAINS

Yamakawa provided a fundamental theoretical study of the helical wormlike chains (**Mod 4b,** Table 1.2) in polymer solutions, and compared the corresponding models (Yamakawa, 1997): KP (Kratky-Porod) with two parameters (TP) and HW (Helical Wormlike Chain-HWLC). The HW model focuses on the influence of the dilute solutions on the geometric parameters (radius, pitch, gyration radius, characteristic ratio, static stiffness parameter-Kuhn segment ($\lambda_k^{-1} = 1360\,\text{Å} = 2l_p$ at 25°C for DNA) of the helical wormlike chain, at different molecular weights, repeat of units (polymerization), and the chain interactions with those solutions (intrinsic viscosity, excluded-volume effects, friction, hydrodynamic volume, etc.). The excluded volume interaction of the protein with the polymer equals the sum of protein volume V_p and a term $V_p h_{0p}/R$, where h_{0p} is the rms end-to-end distance of the polymer, and R is the radius of the protein (Hermans, 1982). Theoretical analysis and evaluation of experimental data in the Yamakawa study were done with application of mathematical statistics methods: mean-square estimation, Fokker-Planck equation, multivariate distribution functions, topoisomer statistics, distribution of the writhe and linking number, Monte Carlo procedures. Yamakawa also studied dynamical properties, such as elementary process of chain motions, circular DNA ring closure probabilities, dynamic versus static chain stiffness with strong correlation, etc.

On the basis of this HWLC theory, Swigon (2009) showed that the linking number (*Lk*), which is the sum of the writhe *Wr* (bending) of one curve and the twist (*Tw*) of the second curve about the first, has an approximately Gaussian distribution:

$$P(Lk) \cong \exp(-G(Lk)/k_B T),\ G(Lk) = K(N)(Lk - N/h_p)^2,$$

where K is the apparent twisting coefficient, a function of linking number variance σ_{Lk}^2, and DNA length $N = L$, which is the plasmid size (in bp), and h_p is the helical pitch (10.5 bp/turn). Theoretical predictions of this distribution and the dependence of K on N by HWLC theory were found to be in excellent agreement with the experimental results. For HW chain $K(L) = \frac{4}{3} - \frac{2.711}{L^{1/2}} + \frac{7}{6L}$ at $L > 6$ and $K(L) = \frac{1}{L^{1/2}} \exp\left(-\frac{6.611}{L} + 0.9198 + -.-3516L\right)$ at $L \leq 6$ (see Yamakawa (1997), p. 271).

The DNA cyclization experiment is one of the most sensitive methods of measuring DNA structural and elastic properties in solution. The rates of cyclization and dimerization can be measured and their ratio, called the Jacobson-Stockmayer factor (or the *J* factor), can be plotted as a function of *N* to obtain the characteristic *J* curve. Yamakawa (1997) showed, for example, the following option of DNA cyclization: if the linear chain is deformed so that its contour is always confined in a plane, we must twist one end by $\Delta Lk = Lk - \tau_0/2\pi$ turns with respect to the other, in order to join them and obtain the closed DNA with linking number *Lk*. Variance of the twist linking number is equal to $<(\Delta Tw)^2> = (2\pi^2)^{-1}(1 + \nu_p)L$ and variance of writhe equals $<Wr^2> = <\sigma_{Lk}^2> = <(\Delta Lk)^2> - <(\Delta Tw)^2>$, where $\Delta Lk = Lk - \tau_0 L/2\pi$. Distribution of the topoisomers is equal to $f_N = \text{const}\ [-K(\Delta Lk)^2/RT]$ where R is the molar gas function. Levene and Crothers (1986) studied the dependence of topological distribution (shapes of molecular circles) and torsional rigidity of DNA on the linking number constituents (writhing number and twist) and the ratio of molecular length and its persistence length.

A DNA molecule may become knotted during a closure experiment. The probability of DNA knotting can be estimated using the HWLC model, and the results are sensitive to DNA electrostatic repulsion.

But the helical wormlike model does not show a transition of DNA molecules from B-DNA to S-DNA, and ssDNA forms. Swigon (2009) showed at page 307: "When large forces and/or twist is imposed on DNA, the molecule changes its secondary structure into alternative structures, or Pauli structure with backbone on the inside and bases on the outside – none of which are governed by HWLC theory." In addition, the known formula (see **Mod 1** in Chapter 1) for dependence of force F on extension x

$$\frac{Fl_p}{k_B T} = \frac{1}{4}\left(1 - \frac{x}{L}\right)^{-2} - \frac{1}{4} + \frac{x}{L},$$

that corresponds to HWLC and WLC models (see Swigon (2009) on the page mentioned above), can give unfeasible results at $x/L = 1$ with an infinite force $F = \infty$.

Yamakawa (1997) showed the potential energy of an HWLC chain as a function of the curvatures in Cartesian coordinates ξ, η, ς and bending α_b and torsional β_τ stiffness or force constants:

$$U = \frac{1}{2}\int_0^L [\alpha_b[\omega_\xi^2 + (\omega_\eta - k_0)^2] + \frac{1}{2}\beta_\tau(\omega_\varsigma - \tau_0)^2]ds,$$

where $\omega(s) = (\omega_\xi, \omega_\eta, \omega_\varsigma)$ is an angular velocity vector, $\beta_\tau = \alpha_b(1 + \nu_p)$, and k_0 and τ_0 are constants independent of contour point s in the segment $0 \le s \le L$.

Later, Liu et al. (2011) presented this equation in the following form:

$$U = \frac{k_b}{2}\int_0^L k^2\, ds + \frac{k_\tau}{2}\int_0^L (\tau - \tau_0)^2\, ds,$$

where $k_b = \alpha_b$, $k_\tau = \beta_\tau$, k is the curvature, τ is the torsion, and L is the polymer length, respectively.

These formulas, in some respects, are similar to the stress and strain tensors formulas of potential energy for helicoids with linear elastic materials in Chapter 4, but the latter have an effective option for this energy specification at different cross-section profiles.

Yamakawa (1997), p. 242 showed that the addition of ethidium bromide (*Et*) ν molecules at the base pair to a helical chain of DNA would unwind it by $\delta_{et} = \nu\phi_{Et} n_{bp}$ turns, and change the average linking numbers in the absence and presence of ethidium bromide in corresponding steps. Note that $\phi_{Et} = 26°$ and $n_{bp} = L / l_{bp}$ is a number of base pairs in the DNA fragment with l_{bp} as a distance between them.

The explicit functional helicoidal model (**Mod 16**, EFHM or briefly EHM) that has linear and nonlinear transition options from B-DNA to S-DNA, in the case of a long DNA molecule may be included in the extensible freely jointed chain (eFJC) or in the extensible worm like chain (eWLC). The first option is

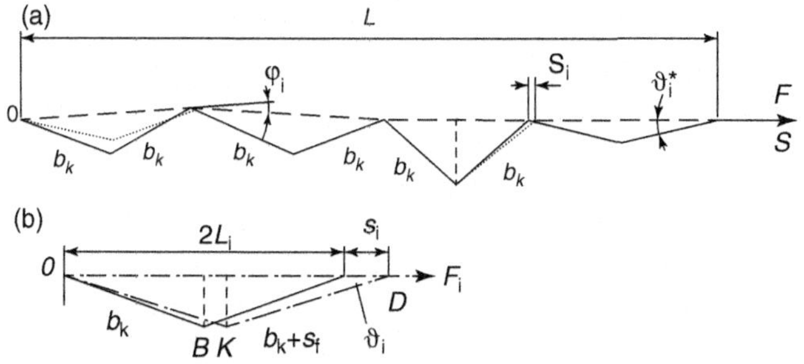

FIGURE 7.8 Schematic of EHM in an extensible freely jointed chain (eFJC) with Kuhn length b_k (a) and its elementary segment (b), with $s_i = S/2(N_{min} + n_i)$.

a simpler one because it uses only the stretching rigidity of the helicoidal strip. The second option requires application of the stretching and bending rigidities of the helicoidal strip. In the latter case we need to use the whole matrix of strains (4.7) and corresponding double integrals (4.8–4.11).

Let us consider the eFJC model example, which is shown in Figure 7.8a.

Here, each side of the elementary isosceles triangles represent a Kuhn segment $b_k = 2l_p$ before deformation. Each of those sides elongates by S_{fi} after stretching the chain by S with the force F. We assume, for the ease of calculations, that angle $\varphi_i = 0$. We see from triangle OKD in Figure 7.8b that, in accordance with the cosine theorem

$$OD^2 = 2(KD)^2[1-\cos(\pi - 2\vartheta_i)], \tag{7.24}$$

where $KD = b_k + S_f$ and $\cos(\pi - 2\vartheta_i) = -\cos 2\vartheta_i = \sin^2\vartheta_i - \cos^2\vartheta_i$, but $\sin^2\vartheta_i \approx \vartheta_i^2$, $\cos\vartheta_i = 1 - \frac{1}{2}\vartheta_i^2$ and $\sin^2\vartheta_i = 2(1-\cos\vartheta_i)$ because $\vartheta_i \ll 1$.

Therefore, $1-\cos(\pi - 2\vartheta_i) = 2(1-\cos\vartheta_i) - \cos^2\vartheta_i$. In addition, we assume that $S_i = S/(N \pm n_i)$ and $OD^2 = \left(\frac{L+S}{N_{min} \pm n_i}\right)^2$,

where $N_{min} = L/2b_k$ and $n_i < 0.5$ reflects a probable shift of the apex of an individual triangle.

We find the value of $\cos\vartheta_i$ by using the sine theorem that states for triangle OKD that

$$(b_k + S_f)/\sin\vartheta_i = OD/\sin(\pi - 2\vartheta_i),$$

where

$$\sin(\pi - 2\vartheta_i) = \sin 2\vartheta_i = 2\sin\vartheta_i \cos\vartheta_i .$$

Hence, $(b_k + S_f) = OD/2\cos\vartheta_i$ and $\cos\vartheta_i = OD/2(b_k + S_f)$.

Now, we can substitute all values found into Eq. (7.24) that gives us the following quadratic equation:

$$\left(\frac{L+S}{N_{\min} \pm n_i}\right)^2 = 2(b_k + S_f)^2(2\cos\vartheta_i + \cos^2\vartheta_i - 1)$$
$$= 2(b_k + S_f)^2\left(\frac{L+S}{(N_{\min} \pm n_i)(b_K + S_f)} + \frac{(L+S)^2}{4(N_{\min} \pm n_i)^2(b_k + S_f)^2} - 1\right)$$

which after multiplications and certain cancelations, transforms to a canonical form

$$(b_k + S_f)^2 - \frac{(L+S)(b_k + S_f)}{(N_{\min} \pm n_i)} + \frac{(L+S)^2}{4(N_{\min} \pm n_i)^2} = 0. \quad (7.25)$$

The solution of Eq. (7.25) is equal to

$$(b_k + S_f) = \frac{(L+S)}{2(N_{\min} \pm n_i)} \pm \sqrt{\frac{(L+S)^2}{4(N_{\min} \pm n_i)^2} - \frac{(L+S)^2}{4(N_{\min} \pm n_i)^2}} = \frac{(L+S)}{2(N_{\min} \pm n_i)}$$

and

$$S_f = \frac{(L+S)}{2(N \pm n_i)} - b_k.$$

Therefore, the function of stretching force F versus stretching displacement S may be written as

$$F = \sum_{i=1}^{\operatorname{int} N_{\min}} < \frac{L}{2(N_{\min} \pm n_i)b_k} > S_f i_\theta j_\theta(b_k)$$
$$= \sum_{i=1}^{\operatorname{int} N_{\min}} < \frac{L}{2(N_{\min} \pm n_i)b_k} > \left(\frac{(L+S)}{2(N_{\min} \pm n_i)} - b_k\right) i_\theta j_\theta(b_k)/L, \quad (7.26)$$

where $< \frac{L}{2(N_{\min} \pm n_i)b_k} >$ corresponds to the average of $\cos\vartheta_i^*$.

Hence, the stretching stiffness of the chain under study equals

$$j_{sL} = F/S = \sum_{i=1}^{\operatorname{int} N_{\min}} < \frac{L}{2(N_{\min} \pm n_i)b_k} > \left(\frac{(L+S)}{2(N_{\min} \pm n_i)} - b_k\right) i_\theta j_\theta(b_k)/LS. \quad (7.27)$$

We can estimate on this basis (see Sec. 6.8.3) the variance of thermomechanical length fluctuation of the chain as follows:

$$\operatorname{Var}(L) = \aleph k_B T / j_{sL},$$

where $\aleph$ is a dimensional factor, and j_{sL} is the chain stretch stiffness.

We see that the explicit helicoidal model – EHM – is more comprehensive in comparison with other known mechanical models of the DNA, not only in the mathematical representation of main static and dynamical features of its elasticity and transitions, but also in the adequate presentation of this molecule's helical structure and geometric parameters.

This provides effective means for the better understanding of DNA mechanical features and functions.

REFERENCES

Bustamante, C., Smith, S.B., Liphardt, J., Smith, D., 2000. Single-molecule study of DNA mechanics. Curr. Opin. Struct. Biol. 10, 279.

Cizeau, P., Viovy, J.-L., 1997. Modeling extreme extension of DNA. Biopolymers 42, 383–385.

Clausen-Schaumann, H., Rief, M., Tolksdorf, C., Gaub, H.E., 2000. Mechanical stability of single DNA molecules. Biophys. J. 78, 1997–2007.

Cluzel, P., Lebrun, A., Heller, C., Lavery, R, Viovy, J.-L., Chateney, D., Caron, F., 1996. DNA: an extensible molecule. Science 271 (5250), 792.

Fields, A.P., Meyer, E.A., Cohen, A.E., 2013. Euler buckling and nonlinear kinking of double-stranded DNA. Nucleic Acids Res. 41 (21), 9881–9890.

Forinash, K., Bishop, A.R., Lomdahl, P.S., 1991. Nonlinear dynamics in double-chain model of DNA. Phys. Rev. B 43 (13), 10743.

Hermans, J., 1982. Excluded-volume theory of polymer-protein interactions based on polymer chain statistics. J. Chem. Phys. 77, 2193.

Kim, Y.-R., Li, C.M., Wang, Q., Chen, P., 2007. Detecting translocation of individual single stranded DNA homopolymere through a fabricated nanopore chip. Frontiers Biosci. 12, 2978–2983.

Levene, S.D., Crothers, D.M., 1986. Topological distributions and the torsional rigidity of DNA. A Monte Carlo study of DNA circles. J. Mol. Biol. 189, 73–83.

Liu, Y., Perez, T., Li, W., Gunton, J.D., Green, A., 2011. Statistical mechanics of helical wormlike chain model. J. Chem. Phys. 134, 065107, 6 p.

Ouyang, Z.-Y., Zheng, S., 2014. Traveling wave solutions on nonlinear dynamical equations in a double-chain model of DNA. Abs. Appl. Anal. 2014(2014), Article ID 317543.

Peyrard, M., Cuesta-Lopez, S., James, G., 2008. Modeling DNA at the mesoscale: challenge for nonlinear science? Nonlinearity 21, T91–T100.

Smith, S.B., Cui, Y., Bustamante, C., 1996. Overstretching B-DNA: the elastic response of individual double-stranded and single-stranded DNA molecules. Science 271, 795.

Storm, C., Nelson, P.C., 2003. Theory of high-force DNA stretching and overstretching. Phys. Rev. E 67, 051906.

Swigon, D., 2009. The mathematics of DNA structure, mechanics and dynamics. In: Benham, C.J., Harvey, S., Olson, W.K. (Eds.), Mathematics of DNA structure, function and interactions. Springer, New York, USA.

Tkachenko, A.V., 2007. Elasticity of strongly stretched ssDNA. Phys. A Stat. Mech. Appl. 384 (1), 133–136.

Tseytlin, Y.M., 1988. Internal friction in a helicoidal multivibrator. Soviet Mach. Sci. 3, 105–109, Allerton Press, New York.

Tseytlin, Y.M., 2006. Structural Synthesis in Precision Elasticity. Springer, New York, NY.

Tseytlin, Y.M., 2012. Flexible helicoids, atomic force microscopy (AFM) cantilevers in high mode vibration, and concave notch hinges in precision measurements and research. Micromachines 3, 480.

Tseytlin, Y.M., 2013. Functional helicoidal model of DNA molecule with elastic nonlinearity. AIP Adv. 3, 062110.

Tseytlin, Y.M., 2014. DNA molecule elastic nonlinearity: a functional helicoidal model. ZAMM 94 (6), 505–508.

Wenner, J.R., Wiliams, M.C., Rouzina, I., Bloomfield, V.A., 2002. Salt dependence of the elasticity and overstretching transition of single DNA molecules. Biophys. J. 82, 3160.

Xiao, H., 2013. DNA elastic nonlinearities as multiple smooth combination of soft and hard linear springs. ZAMM 93, 50.

Yamakawa, H., 1997. Helical Wormlike Chains in Polymer Solutions. Springer, Berlin.

Index

D

E

U

V

W

X

Y

Z

Printed in the United States
By Bookmasters